Imre J. Rudas, János Fodor, and Janusz Kacprzyk (Eds.)

Computational Intelligence in Engineering

T0135169

Studies in Computational Intelligence, Volume 313

Editor-in-Chief

Prof. Janusz Kacprzyk
Systems Research Institute
Polish Academy of Sciences
ul. Newelska 6
01-447 Warsaw
Poland
E-mail: kacprzyk@ibspan.waw.pl

Imre J. Rudas, János Fodor, and Janusz Kacprzyk (Eds.)

Computational Intelligence in Engineering

 Springer

Prof. Imre J. Rudas
Óbuda University
Bécsi út 96/B
H-1034 Budapest
Hungary
E-mail: rudas@uni-obuda.hu

Prof. János Fodor
Óbuda University
Bécsi út 96/B
H-1034 Budapest
Hungary
E-mail: fodor@uni-obuda.hu

Prof. Janusz Kacprzyk
Systems Research Institute
Polish Academy of Sciences
Ul. Newelska 6
01-447 Warsaw
Poland
E-mail: kacprzyk@ibspan.waw.pl

ISBN 978-3-642-42301-7   ISBN 978-3-642-15220-7 (eBook)

DOI 10.1007/978-3-642-15220-7

Studies in Computational Intelligence   ISSN 1860-949X

© 2010 Springer-Verlag Berlin Heidelberg

Typeset & Cover Design: Scientific Publishing Services Pvt. Ltd., Chennai, India.

Printed on acid-free paper

9 8 7 6 5 4 3 2 1

springer.com

Preface

The *International Symposium of Hungarian Researchers on Computational Intelligence and Informatics* celebrated its 10th edition in 2009. This volume contains a careful selection of papers that are based on and are extensions of corresponding lectures presented at the jubilee conference.

This annual Symposium was launched by Budapest Tech (previously Budapest Polytechnic) and by the Hungarian Fuzzy Association in 2000, with the aim to bring together Hungarian speaking researchers working on computational intelligence and related topics from all over the world, but with special emphasis on the Central European Region.

The Symposium of the 10th jubilee anniversary contained 70 reviewed papers. The growing interests, the enthusiasm of the participants have proved that the Symposium has become an internationally recognized scientific event providing a good platform for the annual meeting of Hungarian researchers.

The main subject area called Computational Intelligence includes diverse topics. Therefore, we offer snapshots rather than a full coverage of a small particular subject to the interested reader. This principle is also supported by the common national root of the authors.

The book begins with *Information Systems and Communication*. This part contains papers on graphs of grammars, software and hardware solution for Mojette transformation, statistical intrusion detection, congestion forecast, and 3D-based internet communication and control.

Robotics and Control is treated in the second part. It includes global, camera-based localization and prediction of future positions in mobile robot navigation, new principles and adequate robust control methods for artificial pancreas, modern control solutions with applications in mechatronic systems, adaptive tackling of the swinging problem for a 2 DOF crane – payload system, and finally about robots as in-betweeners.

The third part is devoted to *Computational Techniques and Algorithms*. It treats subjects such as comparative investigation of various evolutionary and memetic algorithms, a novel approach to solve multiple traveling salesmen problem by genetic algorithm, some examples of computing the possibilistic correlation coefficient from joint possibility distributions, and also a novel bitmap-based algorithm for frequent item sets mining.

Fuzzy Systems and Neural Networks belong to the core of CI, and form the fourth part of the volume. Topics belonging to this part are neural networks adaptation with NEAT-like approach, incremental rule base creation with fuzzy rule interpolation-based Q-learning, protective fuzzy control of hexapod walking robot driver in case of walking and dropping, a survey on five fuzzy inference based student evaluation methods, fuzzy hand posture models in man-machine communication, computational qualitative economics computational intelligence assisted building, writing and running virtual economic theories.

The last (fifth) part of the book deals with *Modeling and Optimization*. It contains topics such as spectral projected gradient optimization for binary tomography, cost model for near real-time recommender systems, modeling and simulation of incremental encoder in electrical drives, real-time modeling of an electro-hydraulic servo system, mathematical model of a small turbojet engine MPM-20, performance prediction of web-based software systems, and finally optimization in fuzzy flip-flop neural networks.

The editors are grateful to the authors for their excellent work. Thanks are also due to Ms. Anikó Szakál for her editorial assistance and sincere effort in bringing out the volume nicely in time.

Imre J. Rudas, János Fodor and Janusz Kacprzyk
Budapest, Hungary and Warsaw, Poland

Contents

Graphs of Grammars –
Derivations as Parallel Processes

Benedek Nagy

Faculty of Informatics, University of Debrecen
Egyetem tér 1, H-4032 Debrecen, Hungary
nbenedek@inf.unideb.hu

Abstract. In this paper the concepts of the graph of generative grammars and Lindenmayer systems are investigated. These special and-or graphs represent all information about the grammar. For regular grammars the new concept is equivalent to the finite automata. In context-free case it is an extension of the dependency graph. We also analyze how it is related to programmed grammars (using context-free rules). Moreover the graph of the grammar is more general; it can be defined for generative grammars to any recursive enumerable languages containing all information of the grammar. By the help of tokens the new concept can effectively be used to follow derivations. Moreover parallel derivations can be simulated in non-linear grammars. While in regular and linear case only sequential derivations are going, in context-free case a kind of maximal parallelism can be involved. In context-sensitive case communication/synchronization is also needed in these processes. The Lindenmayer systems are well-known parallel rewriting systems; our concept also can be defined and used for various Lindenmayer systems. Especially the main difference is not in the graph form, but in the simulation process.

Keywords: formal languages, rewriting systems, graphs, grammars, derivations, automata, programmed grammars, dependency graphs, L systems, parallel derivations, parallel processes, Petri nets.

1 Introduction

The Chomsky type (generative) grammars and the generated language families are one of the most basic and most important fields of theoretical computer science [3, 4]. In these systems the derivation relation is defined in sequential manner, however in several cases some parallelism can be involved to speed up the process without changing the generated word/language [6]. In other side the Lindenmayer systems (L systems) are rewriting systems (grammars) having parallel derivations (in correlated way) [12, 14]. The graph theory is a well-known and widely used part of discrete mathematics. There are several connections between these fields, such as derivation trees and dependency graphs for context-free grammars. Graphs can control the derivation process at programmed grammars [2]. In this paper, followed the results of

I.J. Rudas et al. (Eds.): Computational Intelligence in Engineering, SCI 313, pp. 1–13.
springerlink.com   © Springer-Verlag Berlin Heidelberg 2010

[10], we are using a new relation, namely the graph of a grammar. These graphs give graphical representations of the grammars. This concept is very helpful to understand and model how the production rules of grammars work in derivation processes. We show that the graph of the grammar can be considered as a generalization of the finite automata and of the dependency graphs in the same time. Moreover, by using context edges, it is defined for all context-sensitive and for all recursive enumerable languages as well. The derivation processes can also be followed on these graphs by the help of tokens. In regular and linear grammars there is a unique place where the derivation is continued; therefore the derivations are sequential and easy to extract the derived word. In other systems we show that by using alphabetically ordered tokens our graphs can effectively be used to follow derivations and read their results.

2 Definitions

First we recall some definition about the generative grammars and formal languages ([3]). We are fixing our notations as well.

A grammar is a construct $G = (N,T,S,H)$, where N, T are the non-terminal and terminal alphabets. $S \in N$ is a special symbol, called initial letter (or start symbol). H is a finite set of pairs, where a pair uses to be written in the form $v \rightarrow w$ with $v \in (N \cup T)^* N (N \cup T)^*$ and $w \in (N \cup T)^*$. H is the set of derivation (or production) rules. The sign λ refers for the empty word. Let G be a grammar and $v,w \in (N \cup T)^*$. Then $v \Rightarrow w$ is a direct derivation if and only if there exist $v_1, v_2, v_0, w_0 \in (N \cup T)^*$ such that $v = v_1 v_0 v_2$, $w = v_1 w_0 v_2$ and $v_0 \rightarrow w_0 \in H$. The derivation $v \Rightarrow^* u$ holds if and only if either $v = u$ or a finite sequence connect them as $v = v_0, v_1,...,v_m = u$ in which $v_i \Rightarrow v_{i+1}$ is a direct derivation for each $0 \le i < m$. A sequence of letters $v \in (N \cup T)^*$ is a sentential form if $S \Rightarrow^* v$. The language generated by a grammar G is the set of (terminal) words can be derived from the initial letter: $L(G) = \{w | S \Rightarrow^* w, \ w \in T^*\}$. Two grammars G_1, G_2 are called equivalent if $L(G_1) \setminus \{\lambda\} = L(G_2) \setminus \{\lambda\}$. Therefore we do not care about the fact if $\lambda \in L$, or not.

Depending on the possible structures of the derivation rules we have the following classes of grammars and languages (for more details we refer for [3]):

- Phrase-structure (type 0) grammars – recursive enumerable languages: arbitrary generative grammars;
- context-sensitive (type 1) grammars and languages: $v_1 A v_2 \rightarrow v_1 v v_2$ ($A \in N$, $v_1, v_2, v \in (N \cup T)^*$, $v \ne \lambda$);
- context-free (type 2) grammars and languages: $A \rightarrow v$ ($A \in N$, $v \in (N \cup T)^*$);
- linear grammars and languages: $A \rightarrow v$ ($A \in N$, $v \in T^* N T^* \cup T^*$); and
- regular (type 3) grammars and languages: $A \rightarrow v$ ($A \in N$, $v \in T^* N \cup T^*$).

There are known subclasses of regular languages, such as the finite languages and the union-free languages [7]. The finite languages can be generated by rules of the form $S \rightarrow w$, $w \in T^*$. The union-free languages can be described by regular expression without union (only concatenation and Kleene-star can be used).

For our convenience we present some widely used normal forms of various types of grammars.

Fact 1. Each regular grammar has an equivalent one containing rules of forms $A{\rightarrow}aB$ and $A{\rightarrow}a$. Each linear language can be generated using only rules of type $A{\rightarrow}aB$, $A{\rightarrow}Ba$ and $A{\rightarrow}a$. Every context-free language can be generated by rules of the forms $A{\rightarrow}BC$, and $A{\rightarrow}a$ (Chomsky normal form). Every context-sensitive language can be generated by rules of the forms $A{\rightarrow}BC$, $AB{\rightarrow}AC$ and $A{\rightarrow}a$ (Penttonen's one-sided normal form, [11]). Each (phrase-structure) grammar can be generated by rules of the following types ([5]): $A{\rightarrow}BC$, $AB{\rightarrow}AC$, $A{\rightarrow}a$, $A{\rightarrow}\lambda$, where $A,B,C{\in}N$ and $a{\in}T$.

Note that in this paper we generate recursive enumerable languages using only context-sensitive (or context-free) rules allowing deletion-rules (in which only the substituted non-terminal will be deleted).

L-systems are investigated to model parallel computations [14]. The simplest class of L-systems are termed D0L-systems. Now we recall its formal definition:

A construct (T,w,P) is a 0L system, where T is the alphabet, $w{\in}T^*$ is the axiom and $P{\subset}T{\times}T^*$ is the finite set of (production) rules. The rules are usually written in the form $a{\rightarrow}v$. A 0L system is deterministic, i.e., D0L, if and only if for every letter a of T there is a unique v such that $a{\rightarrow}v \in P$ (where a may equal to v).

In Chomsky grammars rules are applied sequentially, whereas in L-systems they are applied in parallel and simultaneously replace all letters in a given word at the same time. This difference reflects the biological motivation of L-systems [12].

Now we recall the concept of dependency graph of non-terminals of context-free grammars based on [4]. The nodes are the non-terminal symbols. There is an edge from the node labelled by the non-terminal symbol A to the node labelled by B if there is a rule in the grammar $A{\rightarrow}v_1Bv_2$ for some $v_1,v_2{\in}(N{\cup}T)^*$. We will extend this definition in the following way.

We define the graph of a grammar for context-free, context-sensitive and phrase-structured (special formed) grammars. The nodes will be labelled by the letters of the alphabet (the non-terminal and terminal symbols) and by the empty word. The edges will represent the derivation rules. For each rule in H let directed edges be from the substituted non-terminal to all letters introduced in the rule (we use multiplicities as well); these edges are in an 'and'-relation which is represented by a directed arc gathering them in the correct order (see Figure 1, for instance, representing a context-free grammar). Moreover at non-context-free rules the context node(s) are connected to the bundle of the edges; these context-edges are shown by broken lines. Formally:

Definition 1. Let $G = (N,T,S,H)$ be a grammar. Let the graph $\Gamma(V,E,Ec)$ be defined in the following way: $V=T{\cup}N{\cup}\{\lambda\}$. There are two types of edges, E contains the directed bundles of 'derivation'-edges as follows. For each rule $uAv{\rightarrow}ua_1...a_nv$ $(u,v{\in}(N{\cup}T)^*, A{\in}N, a_i{\in}N{\cup}T$ for $i{\in}\{1,...,n\})$ the bundle of (directed) edges is in the graph: $((A,a_1),...,(A,a_n)){\in}E$ in the appropriate order. In case of $n = 0$ (deletion-rule) the bundle contains the only edge (A,λ). The set Ec contains the 'context'-edges. They are between nodes and bundles of edges in E. For each rule with $uv{\neq}\lambda$, there are context edges connecting each element of uv (with multiplicities) to the bundle of edges representing this rule.

Similar graph can be constructed for an 0L or a D0L system (T,w,P).

Note that the node λ will play only if there is a rule in which the right-hand-side is strictly shorter than the left-hand-side.

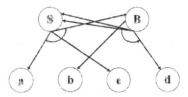

Fig. 1. An example for the graph of a context-free grammar

The concept of the graph of a grammar can be used as a visualization of derivations. Initially put a token to the node S of the graph of a generative grammar. In case of L-systems there are exactly k tokens in the initial token distribution on a letter a of the axiom w, if it occurs k times in w. The set of the existing tokens refers for the letters of the current sentential form at each time. The essential difference of generative grammars and L systems is the mode of the word generation.

In generative grammars, when a derivation rule is applied, a token is deleted from the node labelled by the substituted non-terminal and the number of tokens is increasing at every end-node of the bundle of the applied rule by the multiplicities of the arrows (these new tokens are the children of the deleted one). A context-sensitive rule can be applied if there are (appropriate) tokens at the nodes which are connected to the bundle by context edges. Appropriate means here that those tokens must represent those letters of the sentential form which are neighbours of the replaced non-terminal as it is given by the production rule. A derivation is successfully finished when all the tokens are on the terminals (and maybe on the empty word).

Opposite to this, in D0L systems all the tokens are deleted and evolved in each step by the only bundle of edges starting from the actual node. In these systems, since there are no non-terminals, after every step a (new) word of the generated language is obtained.

Now we recall the concept of programmed grammars (without appearance checking) from [2]. Let a directed graph be given. At each node there is a context-free rule. The derivation starts at any rule in which the start symbol (S) is in the left-hand-side. The derivation can be continued only with a rule which is connected to the node containing the previous applied rule. So, the directed edges of the graph drive the derivation. The process is finished with a generated word if there are not any non-terminal symbols in the sentential form.

In this paper we will use Greek letters referring for the nodes of the graphs.

3   Results about Regular Grammars

In this section we show how the three different graph approaches represent the grammars, translation algorithms are given among them. Some grammars with special structured graphs are also detailed.

In the graph of a regular grammar there is at most 1 edge going to a non-terminal in each bundle of edges (and there are no context-edges in the graph).

Now let us see the special case, when the grammar is in normal from, i.e., each rule is one of the following type: $A{\rightarrow}a$, $A{\rightarrow}aB$. The graph of a regular grammar in normal form holds all information without signing the direction of the bundles of edges, since at each bundle having more than one edge the edge going to a terminal symbol precedes the edge going to a non-terminal symbol. So in regular case we can have a very simple graph representation for every language. Moreover, using the normal form, the relation between the graph of the grammar and the automaton accepting the same language is straightforward.

So, there are three variations of graphs which represent regular languages. In the first ones the nodes represent the rules of the regular grammar and edges between them represent the possible ways of the derivation (i.e., which rule can be the next one). It is the programmed grammar graph form. There is an arrow from the node having rule $A \rightarrow aB$ to all rules having rules starting with B. (Other rules cannot continue the derivation.)

The second type of graphs are and-or graphs having bundles of edges. In them the non-terminals and terminals are the nodes, they are connected by 'and'-edges if a rule introduce more letters. It is our new form, called the graph of a grammar, moreover this graph includes the dependency graph in the following way. If from node γ there is an arrow to node δ, then the letter labelling the node δ is dependent by the non-terminal labelling γ. It means that the letter of δ can be derived from the non-terminal of γ.

The third graphs have only non-terminal nodes (and a node representing λ) and has only 'or'-edges. The terminals are written on the edges. This graph is the finite automaton accepting the language with initial state S and final state λ.

The graph of a grammar, the nondeterministic finite automaton of the language and the programmed grammar form can be converted to each-other in constructive ways. Now we detail these processes.

We start from the finite automata. Let the nodes of the graph of the grammar be the states of automata and the terminal labels of the transitions. If there is a transition from state A to a final state by terminal a, then the graph of the grammar has the edge (A, a) (it is a bundle with 1 edge). For all transitions: A to B by a the graph has the bundle $((A,a),(A,B))$. It is obvious that the graph representing the same grammar as the automaton. (We refer the non-terminals by A,B and the terminal symbols by a.)

Now we translate the graph of the grammar to the programmed grammar form. The rules at the nodes of the programmed grammar are representing the bundles of the graph of the grammar. So, for each bundle having only one edge in the form (A,a) a node is created with rule $A \rightarrow a$. For bundles having 2 edges as $((A,a),(A,B))$ the rule is created $A \rightarrow aB$. Now, let an edge be from each node having A in the right-hand-side of its rule to each node having A at the left-hand-side (so to each node representing the bundles of the edges from A in the graph of the grammar).

Finally, we translate the programmed grammar form to finite automaton. Let the non-terminals be the states and S be the initial state of the automata. Moreover we put an additional state to the automaton as final state. If there is a rule in the programmed grammar $A \rightarrow a$, then let a transition be from the state of A to the final state by a. For all rules of the form $A \rightarrow aB$ let a transition be from the state of A to the state of B by the terminal symbol a. (In this way the number of transitions from each node will be the same as the number of connections from the nodes of the programmed grammar graph which has A in the left-hand-side of the rules. Moreover the transitions are going to those non-terminal labelled states (or the final state) which occur on the right-hand-side of the rules by the appearing terminal symbols.)

Note that using only regular rules a programmed grammar cannot go beyond regular languages even if the edges are varied.

Figure 2 shows examples for all the three graphs of the language $a*b(bc)*$ generated by grammar ($\{S,B,C\},\{a,b,c\},S,\{S{\to}aS,S{\to}b,S{\to}bB,B{\to}bC,C{\to}c,C{\to}cB\}$).

Now we define the concept of (directed) cycle for these graph-representations. In case of finite automata and programmed grammar forms the usual graph-theoretical definition works. In graph of a grammar there is a cycle if a descendant of a token will appear at the same node. In Figure 2 a direct cycle is shown at S and there is a 2-step cycle including nodes B and C.

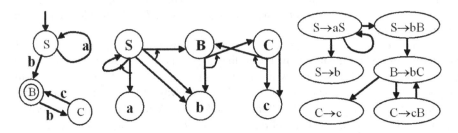

Fig. 2. An example for various concepts of graphs of a regular grammar (finite automaton – to the left, graph of grammar – in the middle, and programmed grammar – to the right)

Proposition 1. The minimal number of cycles is the same in finite automaton, in graph of grammar and in programmed grammar forms of the same grammar.

The grammars that can be represented by cycle-free graphs generate finite languages. All finite languages can be generated by such grammars. Moreover the generated language is finite even if any kinds of rules are allowed in a cycle-free graph of grammar.

With only regular rules the language is union-free if and only if a graph without alternating paths represents a grammar of the language. It means that there are not two cycle-free paths from a node γ to a node δ in automata and in programmed grammar forms. At automata and at the graph of the grammar form it means that there is at most one cycle free path from a non-terminal labelled node to any other non-terminal labelled node.

4   Results about Grammars with Context-Free Rules

The graph of the grammar is an extension of the well-known dependency graphs of context-free grammars. The extension goes in two steps. First the dependency relation is extended to terminals (and if deletion rules are used, to the empty word) as well. In second step, the edges coming from the same rule are gathered to a bundle, representing that they are dependent (they generate at the same time in a derivation). Sometimes several copies of edges of the original dependency graphs are needed to pack them in bundles.

Note that in context-free case the graph of the grammar holds all information of the grammar. The order of the edges in a bundle is important and it can be seen on the figures as well (see Figure 1).

4.1 Linear Languages

Now, let us analyse a subclass of context-free languages, namely the linear ones. This case is very similar to the previously described regular case. In linear grammars, such as in regular ones there is only one non-terminal in the sentential form in each step while the derivation procedure is going. By this reason the derivation can be continued only by a rule having the appearing non-terminal in its left hand side. Therefore the derivation process has a sequential order in the graph of the grammar by following it in non-terminal nodes.

Using only rules of the presented normal form all the three graphical representation of the grammars can easily be given. The programmed grammar and the graph of the grammar forms hold all information. The finite automata can be extended to carry all information [8], these automata have two heads, and the stepping head depends on the form of the used rule ($A{\rightarrow}aB$ and $A{\rightarrow}Ba$). Moreover, since there is at most 1 non-terminal in the sentential form, the following fact can easily be proven.

Proposition 2. Programmed grammars having only linear rules (even if the edges of the graph are varied) cannot generate more than linear languages.

4.2 Graphs of the Grammars and Programmed Grammars

Generally at context-free grammars the derivation cannot be followed by a sequence of non-terminal labelled nodes in the graph as it works in linear grammars. However, at the so-called programmed grammars (without appearance checking) a given graph controls the derivation. So the programmed grammars can be interpreted as an extension of finite automata in a certain sense. These graphs are also related to our graphs as we detail below. A programmed grammar can be seen on Figure 3 left. To get the new concept of graphs of these (programmed) grammars we need to have some nodes with the same labels. Following the paths of the possible derivations one can obtain the graph shown on the right.

Since the order of applications of rules are not arbitrary in a programmed grammar (one can have a complete directed graph – with n nodes and n^2 edges – to represent a context-free grammar), the graph of the grammar contains a non-terminal as many times as many roles it has in the generation.

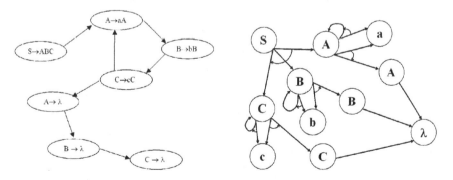

Fig. 3. A programmed grammar generating a non context-free language and its graph-of-grammar presentation

We need to gather all possibilities of the next applied rules of the given non-terminal (left-hand-side). If this set is not the same for the occurrences of a non-terminal at two nodes in the programmed grammar, then we need two distinct nodes for the non-terminal representing these two rules in our new graph.

This new graph does not have all information about the programmed grammar. At programmed grammars the rules and the roles of the non-terminals are synchronized.

4.3   Regular Languages and Programmed Grammars

Now, assume the opposite way: given the graph of the programmed grammar, what can be the generated language.

We are dealing programmed grammars with some special structured graphs.

The simplest ones are the trees and the directed acyclic graphs.

First let us consider that directed acyclic graphs, i.e., cycle-free graphs direct the derivations.

Theorem 1. The class of languages defined by programmed grammars using directed acyclic graphs are exactly the finite languages.

Proof. First, we show that a programmed grammar with a cycle-free graph can generate only a finite language. In a cycle-free graph there are only finitely many paths (starting and finishing at arbitrary nodes), therefore only finitely many generation sequence exist. These finite sequences can generate finitely many words. Now we prove the opposite direction. Let $L=\{w_1,\dots,w_n\}$ be a finite language. We construct a special programmed grammar with cycle-free graph to generate L. Let the structure be a tree with n nodes. The vertices have the rules $S \to w_i$ (for $1 \le i \le n$). Clearly, it is cycle-free and generates L.   □

The trees and the line-graphs (when each node has degree at most 2, but the two end-nodes having only degree 1) are special cases of the directed acyclic graphs.

Now let us see graphs with cycles (the used term 'ring' is coming form network theory). First we are analysing graphs having exactly 1 ring and some additional structure (these are the so-called ring-backbone graphs).

Lemma 1. Given a programmed grammar with a graph containing exactly 1 (directed) cycle. If the number of the non-terminals (as a multiset) are changing deriving through the cycle, then the generated language is finite.

Proof. There are finitely many paths into and from the cycle. Each pair of these paths coincides at most 1 multiset of non-terminals. When the multiset of the non-terminals of the sentential form is "growing" (i.e., at least one element is growing) in a derivation cycle, then after some cycles there will not be any terminal derivations. Therefore words can be generated with paths containing only a limited number of the cycle. It implies that only finitely many words can be generated.   □

Consider the language $a^n b^m c^k$, where n,m,k are independent variables. This regular language (defined by the regular expression $a*b*c*$) cannot be obtained by a programmed grammar having only 1 cycle in the graph.

In Figure 3 a programmed grammar shown which generates the language $a^n b^n c^n$, $n>0$. It is known that this language is not regular and not even context-free.

What does it mean that the programmed grammar have (or have not) cycle in terms of their graphs of grammars? We have the following result.

Theorem 2. The graph of a grammar form of the programmed grammar must have cycle if and only if the language cannot be generated by cycle-free programmed grammar.

Proof. We prove the theorem by directions. Assume that the language cannot be generated without cycle by programmed grammar (i.e., it is not finite). The programmed grammar has cycle and therefore there is a non-terminal symbol A such that its number does not change in a derivation-cycle, however there is a rule in the cycle with A in the left-hand-side. This fact means that one can go further in the derivation getting A in the same role, therefore it must be a cycle in the graph-of-grammar. (This cycle can be direct as Figure 3 shows at an occurrence of A, B and C, or it can be indirect as generated by for instance rules $A \rightarrow aB, B \rightarrow bA$.)

Other way, if the graph of the grammar has a cycle, then a rule can be applied in arbitrary many times. It means that must be a cycle in the programmed grammar. □

About the number of cycles in the programmed grammar we have the following more general theorem.

Theorem 3. The languages obtained by programmed grammar (without appearance checking) with a graph having (at most) n cycles (for any fixed natural number n) is incomparable with the regular languages.

Proof. The theorem is a consequence of the following facts. Only regular languages defined by regular expressions using at most n Kleene-stars can be described by these graphs. Some non-regular (and non-context-free) languages can be generated as well.

□

Note that in [1] there are several other kinds of graphs are analysed in this point of view, such as planar, bipartite, Hamiltonian, Eulerian and complete graphs, etc.

5 Graphs of Grammars in Context-Sensitive Case

The graph representation of the grammars exists for context sensitive case. The used contexts are represented in special way, the set Ec contains the context-edges, i.e., the letters which should be neighbours of the replaced one at the application of a given rule.

In these graphs the derivation cannot go independently at the nodes present in the sentential form, the context-condition plays an important role therefore the context-edges are needed. The generation usually cannot be maximal parallel, since some non-terminals/tokens need to wait for other ones [9]. They represent the fact, that some rules can be applied only if the contexts are present. The graph represents only that there is/are context condition(s), but it does not refer to their order. Using the presented special normal form the context must be 1-letter left-context, therefore the presentation is simple and the graph of the grammar has all information about the grammar. Figure 4 presents an example.

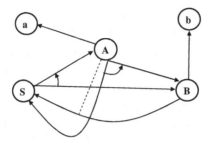

Fig. 4. Example for the graph of a context-sensitive grammar

6 Graphs of Grammars for Phrase-Structure Grammars

One can use the concept of the graph of a grammar for phrase-structured grammars as well. For phrase structured grammars one should use rules in context-sensitive form and allow deletion rules as well. To have graphs with relatively simple structure we recommend to use the special normal forms (see Fact 1).

In Figure 5 an example is shown. The graph represents the grammar $(\{A,B,S\}, \{a,b\}, S, \{S{\rightarrow}AB, S{\rightarrow}BB, AB{\rightarrow}AS, A{\rightarrow}AB, S{\rightarrow}a, A{\rightarrow}\lambda, BB{\rightarrow}BS, B{\rightarrow}b\})$. A derivation in this grammar is shown: it starts with a token on S. $S{\rightarrow}AB$: children tokens will be on A and on B. Now the rule $A{\rightarrow}AB$ can be used. Now $BB{\rightarrow}BS$ can be used since the two tokens on B are neighbours (context-edge from B, and this token represents the left neighbour of the other token in B in the sentential form): a token move from B to S. Now $AB{\rightarrow}AS$ can be used moving the other token from B to S. Then by $A{\rightarrow}\lambda$ the token on A is deleted and a token appears on λ. Now $S{\rightarrow}a$ is used for both tokens on S: the tokens of S are deleted and new tokens appear on a. There is no token on non-terminals, the derivation is finished, the derived word (can be formed by the tokens of the end-state): aa.

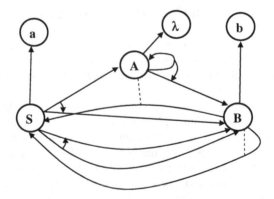

Fig. 5. An example for the graph of a phrase-structured grammar

7 Graph Representations of L-Systems

In this section we show how the graphs of L-systems can be used. First we present a well-known example: let $(\{a,b\}, a, \{ a{\rightarrow}b, b{\rightarrow}ab\})$ is a D0L system. It generates the Fibonacci words: $a, b, ab, bab, abbab, bababbab, \ldots$ (their lengths form the sequence of Fibonacci numbers). This language is not context-free. In Figure 6 its graph is shown. Initially there is a token of each letters of the axiom, and in each derivation step all of them evolve.

Fig. 6. Graph of an L-system

In the same way all 0L systems can be represented. The only difference is that it is allowed to start not only 1 bundle of edges from a node.

8 Derivations with Graphs of Grammars

As we already presented the simulation process of derivations may go in analogous way as Petri-nets work. Petri-nets are usually used to model parallel processes [13]. In our graphs the nodes are the places where tokens can be, and bundles of edges play the roles of transitions. While in Chomsky-type grammars the derivation is sequential by definition, in some cases some parallelism can be involved. In regular and linear grammars there is at most 1 token on non-terminals, and therefore the derivation process must be sequential. The derived word can easily been reconstructed by remembering to order of tokens reaching terminal nodes. In non linear grammars there can be several tokens on non-terminals at the same time allowing parallelism in the derivation process. In context-free grammars the tokens are independent of each other, therefore any of the tokens standing on non-terminals may evolve at the same time by appropriate rules. It is also possible to speed up the derivation process by rewriting all non-terminals parallely at the same time at each step. In non linear grammars the reconstruction of the derived word (or the actual sentential form) need a coding system. Each token will be indexed. At generative grammars initially the token on S has index 1, while in L systems the initial tokens are indexed in appropriate way from 1 to n, where n is the length of the axiom w. Then at each derivation step the new token(s) get the index of the old token (that eliminated and evolved to this/these new token(s)) concatenated with the number of this edge in its bundle. For instance, using the rule $A{\rightarrow}SB$ to the token with index 12112, this token disappears and new tokens on S and B appear with indices 121121 and 121122, respectively. In this way the alphabetical order of the tokens of the system gives the appropriate order of the letters and so, the actual sentential form or derived word. The power of the language generation of L-systems comes from the maximal parallel use of production rules, i.e., in each step all letters must be rewritten. In this way it is easy to generate non context-free languages, as our example showed. The real difference of the maximal parallel derivations of context-free and D0L systems is the following: in D0L systems the same letters must be rewritten in the same way at each time, while in context-free grammars various rules can be applied at the same time to rewrite the same non-terminals in various places of the actual sentential form. Moreover the terminals are not rewritten in generative grammars. The indexing method presented here is also useful in derivations of non-context-free grammars. Not only the derived word, but the actual context condition can also be checked in this way. For instance, the derivation rule $SA \rightarrow SB$ can be used for a token on A with index 12221 if there is a token on S with index (alphabetically) less than 12221 and there is no token on the system with index (alphabetically) between these two indices. In case of deletion rules, i.e., having node λ in the graph the tokens appeared on this node are not counted any more as letters, so they could be deleted from the system. In this way some tokens having not neighbour indices may became neighbours later on (eliminating all the tokens having indices between them by the help of node λ). Therefore the derivations of both generative and L systems can also be represented on our graphs in one-to-one way. Moreover possible parallel derivations can also be shown with synchronization/context-check.

9 Concluding Remarks

We introduced the graphs of the generative grammars which generalize the concepts of finite automata (for regular case) and dependency graphs (for context-free case). The nodes are the terminal and non-terminal symbols. The bundles of the edges of the graph demonstrate the rules of grammar. For regular grammars in normal form the directions of the bundles of arrows are uniquely determined, therefore our notation is slightly redundant in this case. In linear (and regular) grammars there is at most one arrow for non-terminals in each bundle, and therefore in simulation of derivations there is at most one token on non-terminals at each timepoint, therefore the derivation is entirely sequential. Opposite to this fact in context-free case there can be several tokens on non-terminals, the derivation may go in a maximal parallel way. The new concept is more general, one can use it in the case of context-sensitive and eliminating rules as well. In case of context-sensitive grammars the so-called context-edges are needed, they connect nodes and bundles of edges. At non-context-sensitive grammars the node labelled by λ must be used. Using normal forms the graph has all information about the represented (phrase structured) grammar, and its structure is simple. Moreover the same type of graphical representation was introduced for parallel grammars. By the help of tokens in our graph representation one can have a better understanding of the possible derivation processes both in generative grammars and L-systems. We believe that our concept can be extended to further variants of formal grammars and rewriting systems.

Acknowledgments. This paper is the extended version of paper [10]. The author thanks the comments and remarks of his colleagues. Discussion with Jozef Kelemen is gratefully acknowledged.

References

[1] Barbaiani, M., Bibire, C., Dassow, J., Delaney, A., Fazekas, S., Ionescu, M., Liu, G., Lodhi, A., Nagy, B.: Languages Generated by Programmed Grammars with Graphs from Various Classes. J. Appl. Math. Comput. 22, 21–38 (2006)

[2] Dassow, J., Paun, G.: Regulated Rewriting in Formal Language Theory. In: EATCS Mono-graphs on Theoretical Computer Science, vol. 18. Springer, Heidelberg (1989)

[3] Hopcroft, J.E., Ullmann, J.D.: Introduction to Automata Theory, Languages, and Computation. Addison-Wesley, Reading (1979)

[4] Linz, P.: An Introduction to Formal Languages and Automata. Jones & Bartlett Publisher, Sudbury (2001)

[5] Mateescu, A.: On Context-Sensitive Grammars. In: Martin-Vide, C., Mitrana, V., Paun, G. (eds.) Formal Languages and Applications, Studies in Fuziness and Soft Computing, vol. 148, pp. 139–161. Springer, Heidelberg (2004)

[6] Nagy, B.: Derivations in Chomsky-Type Grammars in Mirror of Parallelism. In: Proc. of IS-TCS 2004, Theoretical Computer Science - Information Society, Ljubljana, Slovenia, pp. 181–184 (2004)

[7] Nagy, B.: Union-Free Languages and 1-Cycle-Free-Path-Automata. Publ. Math. Debrecen. 68, 183–197 (2006)

[8] Nagy, B.: On 5'→3' Sensing Watson-Crick Finite Automata. In: Garzon, M.H., Yan, H. (eds.) DNA 2007. LNCS, vol. 4848, pp. 256–262. Springer, Heidelberg (2008)

[9] Nagy, B.: Derivation Trees for Context-Sensitive Grammars. In: Abstract volume of AFLAS 2008, International Workshop on Automata, Formal Languages and Algebraic Systems, Kyoto, Japan, Kyoto Seminar House (2008); Submitted to Ito, M., Kobayashi, Y., Shoji, K. (eds.): Automata, Formal Languages and Algebraic Systems. World Scientific, Singapore (2010)

[10] Nagy, B.: Graphs of Generative Grammars. In: Proc. of 10th International Symposium of Hungarian Researchers on Computational Intelligence and Informatics, Budapest, Hungary, pp. 339–350 (2009)

[11] Penttonen, M.: One-sided and Two-sided Context in Formal Grammars. Information and Control 25, 371–392 (1974)

[12] Prusinkiewicz, P., Lindenmayer, A.: The Algorithmic Beauty of Plants. Springer, New York (1990)

[13] Reisig, W., Rozenberg, G.: Lectures on Petri Nets I. In: APN 1998. LNCS, vol. 1491, Springer, Heidelberg (1998)

[14] Rozenberg, G., Salomaa, A.: The Mathematical Theory of L Systems. Academic Press, New York (1980)

MTTool Software Tool and Low Complexity Hardware for Mojette Transformation

Péter Szoboszlai[1], József Vásárhelyi[2], Jan Turán[3], and Péter Serfözö[4]

[1] Magyar Telekom Plc., 1541 Budapest, Hungary
[2] Department of Automation, University of Miskolc, H-3515 Miskolc, Hungary
[3] Department of Electronics and Multimedia Communications, University of Kosice,
  041 20 Košice, Slovak Republic
[4] Ericsson Hungary Ltd., 1139 Budapest, Hungary

Abstract. The Mojette (MOT) transform is a relatively new method for image transformation. With the use of MOT on images, multimedia files, distributed databases before information transmission became possible to reconstruct the data even if some part of information were damaged. The Mojette transform similarly to the Radon transformation calculate a group of projection on an image block. As result of this transform the processed image result in a file with redundant information about the image. There are several applications such as watermarking or applications on distributed databases. There were several implementations using personal computer, but these application were implemented only for locally stored images. The Mojette Transformation Tool which is developed in .Net environment with Csharp (C#) it is used for performance analyses on different computer platforms. While the hardware implementation called MoTIMoT developed as co-processing elements of an embedded processor using FPGAs tries to find a minimal implementation for low power applications such as mobile communications.

1 Introduction

The Mojette Transform (MOT) originates from France where J-P. Guédon referred to an old French class of white beans, which were used to teach children computing basics of arithmetic with simple addition and subtraction. He named it after the analogy of beans and bins. Bins contain the sum of pixel values of the respective projection line [2]. There are several different variations of MOT applications nowadays which are used in different areas, such as tomography [3], internet distributed data bases [4], encoding, multimedia error correction [5], or The Mojette Transform Tool (MTTool), which was created for testing purposes. Moreover, it can be used for demonstrations and training purposes as well.

  Although the MTTool development has not been finished yet, we have already gained much experience with it, and we can see how it may become more helpful for further projects both in software and hardware development. So the main purpose to build such an environment is that with its help we could try to compare MOT

I.J. Rudas et al. (Eds.): Computational Intelligence in Engineering, SCI 313, pp. 15–26.
springerlink.com   © Springer-Verlag Berlin Heidelberg 2010

software version with the hardware one. Due to this, we have many similarities between the two and in some cases we had to choose in MTTool a way which was not optimal, for the MOT and for its inverse transform.

In this paper first the Mojette transform theory is presented and the research work made by Guédon and others [1, 4, 6, 7]. After the theorethical presentation it is presented the MTTools nad the low power implementation of the MOTIMOT coprocessor, and there is a presented the MTTool performance analyses.

2 Mojette and Inverse Mojette Transform

Mojette Transform: The main idea behind the Mojette transformation (similarly to the Radon transformation) is to calculate a group of projections on an image block [6]. The Mojette transform (see [7], [8] and [9]) projects the original digital 2D image:

$$F = \left\{ F(i, j); i = 1, ..., N; j = 1, ..., M \right\} \tag{1}$$

to a set of K discrete 1D projections with:

$$M = \left\{ M_k(1); k = 1, ..., K; 1 = 1, ..., 1_K \right\}. \tag{2}$$

MOT is an exact discrete Radon transform defined for a set
$S = \{(p_k, q_k), k = 1, ..., K\}$ specific projections angles:

$$M_K(l) = proj(p_k, q_k, b_l) = \sum_{(i,j) \in L} F(i, j) \delta(b_l - iq_k - jp_k), \tag{3}$$

where *proj* (p_k, q_k, b_l) defines the projection lines p_k, q_k, $\delta(x)$ is the Dirac delta with the form:

$$\delta(x) = \begin{cases} 1, if \_ x = 0 \\ 0, if \_ x = 1 \end{cases} \tag{4}$$

and

$$L = \left\{ (i, j); b_l - iq_k - jp_k = 0 \right\} \tag{5}$$

is a digital bin in the direction θ_k and on set b_l.

So the projection operator sums up all pixels values whose centers are intersected by the discrete projection line *l*. The restriction of angle θ_k leads both to a different sampling and a different number of bins in each projection (p_k, q_k). For a projection defined by θ_i, the number of bins n_i can be calculated by:

$$n_i = (N-1)|p_i| + (M-1)|q_i| + 1 \tag{6}$$

The direct MOT is depicted in Figure 1 for a 4x4 pixel image. The set of three directions $S=\{(-1,2),(1,1),(0,-1)\}$ results in 20 bins.

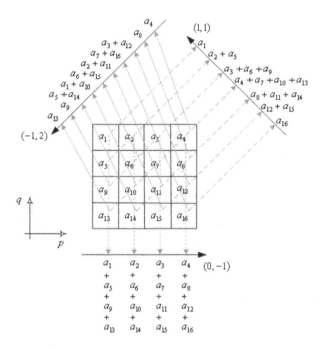

Fig. 1. The set of three projections computed from a 4x4 image

The MOT can be performed by direct addition of the image pixel values in grey scale images, and for bitmaps we can add the different bitmap color table values.

Inverse Mojette Transform: The basic principle of the inverse Mojette transform is the following. We start the image reconstruction with bins corresponding to a single pixel summation. This reconstructed pixel value is then subtracted from the other projections and the process is iterated for the N^2-1 pixels: the image is then completely decoded. In the case of a 4x4 pixel image reconstruction, if the directions of the MOT sets are $S=\{(-1,2),(1,1),(0,-1)\}$, then the minimum number of subtractions needed is 10, from the 20 bins. So should it happen to lose some of the bins we could still reconstruct the image due to the redundancy of the MOT.

3   Mojette Transform in C#

In MTTool the implementation of the MOT was applied in three different ways. This is due to the fact that this application is still under development and the three different ways were constructed not at the same time, but in the previous years.

The First Version: In the initial release one of the hardest decisions was to declare some rules, which had to be both flexible and at the same time not very complex. We had to declare the image sizes we had to work later with, and to look for a useful relationship between the picture size and the vectors we use in the MOT, Inverse Mojette Transform (IMOT). After calculating several file sizes, it was clear we had to select it

Table 1. MOT implementation and its main differences

Nr.	Image Format	Projections	MOT and Inverse MOT
1	PGM	p={1,-1,3,-3}; q={quarter of the image size}	addition and subtraction
2	BMP	p={2,-2}; q={1} and p={3,-3,2}; q={1}	addition and subtraction
3	BMP	p={2,-2}; q={1} and p={3,-3,2}; q={1}	Matrix

to be easy to remember and have something common with all the other images as well. We decided to take the picture size $2^n x2^n$, where n is equal to 8 and 9, but can be changed easily later on. So the picture size is 256 x 256 and 512 x 512. In the Picture Preview we can open and display any kind of PGM or BMP file irrespective of the picture size, but some of the images are increased or decreased to fit on the screen.

After checking the restrictions, the first step in the MOT is to make a vector from the pixels of the image. When following a simple rule $(1, 2^n x2^n)$, it is easy to define the size of this vector. If $n=8$, this result in the vector $(1, 65536)$, in which every line contains a pixel value from the picture. Because the PGM picture is a 256 grayscale image, a PGM file contains pixel values only from 0 to 255. In case of a BMP image, we could make it three times because of the different bitmap color table values.

Table 2. Image display

Original size	Displayed size	Ratio
1600 x 1200	400 x 300	0,25
1599 x 1199	799 x 599	0,5
1024 x 768	512 x 384	0,5
Height < 1024	Height +180	Other

In the second step we make the Mojette Transformation. The vector p is predefined for the four projection directions and the q vector has the same value in each case (quarter size of the $2^n x 2^n$ image). We generate four files for the four different projections, which are the following:

- originalfilename.pgm.moj1 $(1, q)$
- originalfilename.pgm.moj2 $(-1, q)$
- originalfilename.pgm.moj3 $(3, q)$
- originalfilename.pgm.moj4 $(-3, q)$.

From the existing MOT files (moj1, … moj4), we get the original PGM picture with the IMOT. In this case all of the four Mojette Transformed files are needed to rebuild the original image without any errors at all. If any of the Mojette Transform files is defect or incomplete, the Inverse Mojette Transform will not give back the original image. Each of the four files contains a vector described above. The next step of the IMOT is to read the first and last vectors of the third and fourth MOT files and put them in their place. So we have in all four corners of the picture the valid pixel values filled up. See steps 1, 2, 3 and 4 on the following figure:

After recreating the pixel values, we only need to add the new header for the file and the restoration of the original image is already performed.

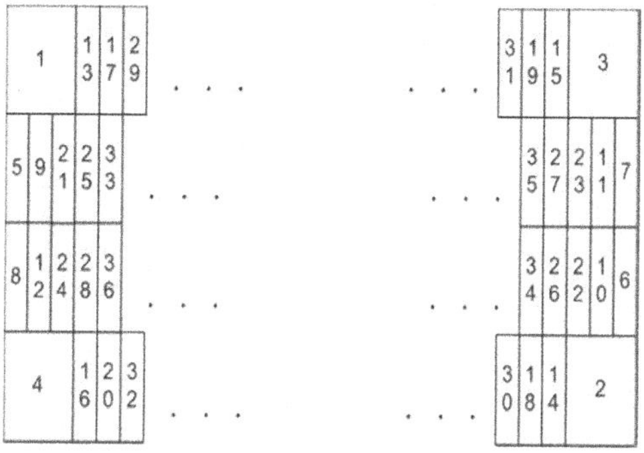

Fig. 2. First 30 steps of the IMOT

The Second and Third Version: These solutions differ from the previous one in such a way that these are applied on BMP images and in these cases we perform the MOT and IMOT on the three different bitmap color tables. We use the same algorithm for the three different color maps and collecting the bins into 3 separate files which differ in their extensions and of course in their content. On the bitmap images we use the directions $S_1=\{(2,1),(-2,1)\}$ and $S_2=\{(3,1),(-3,1),(2,1)\}$ for the block sizes 4 and 8. Although the MOT is also prepared for the block size 16 and 32, the implementation of the IMOT isn't done yet. In the second version, we use simple addition and subtraction – different from the one mentioned in the first version –, since here we have block sizes 4 and 8 and there we perform the MOT and IMOT on the whole image at once and not step by step. In the third version, instead of addition and subtraction, we use matrices for the MOT and IMOT on the above mentioned block sizes. The MOT with matrices is implemented in the following way, where b_i is the bin resulted from the following equation:

$$
\begin{bmatrix} b_1 \\ b_2 \\ b_3 \\ b_4 \\ b_5 \\ b_6 \\ b_7 \\ b_8 \\ b_9 \\ b_{10} \\ b_{15} \\ b_{16} \\ b_{17} \\ b_{18} \\ b_{19} \\ b_{20} \end{bmatrix}
=
\begin{bmatrix}
1 & 0 & 0 & 0 & 0 & 0 & 0 & 0 & 0 & 0 & 0 & 0 & 0 & 0 & 0 & 0 \\
0 & 1 & 0 & 0 & 0 & 0 & 0 & 0 & 0 & 0 & 0 & 0 & 0 & 0 & 0 & 0 \\
0 & 0 & 1 & 0 & 1 & 0 & 0 & 0 & 0 & 0 & 0 & 0 & 0 & 0 & 0 & 0 \\
0 & 0 & 0 & 1 & 0 & 1 & 0 & 0 & 0 & 0 & 0 & 0 & 0 & 0 & 0 & 0 \\
0 & 0 & 0 & 0 & 0 & 0 & 1 & 0 & 1 & 0 & 0 & 0 & 0 & 0 & 0 & 0 \\
0 & 0 & 0 & 0 & 0 & 0 & 0 & 1 & 0 & 1 & 0 & 0 & 0 & 0 & 0 & 0 \\
0 & 0 & 0 & 0 & 0 & 0 & 0 & 0 & 0 & 1 & 0 & 1 & 0 & 0 & 0 & 0 \\
0 & 0 & 0 & 0 & 0 & 0 & 0 & 0 & 0 & 0 & 1 & 0 & 1 & 0 & 0 & 0 \\
0 & 0 & 0 & 0 & 0 & 0 & 0 & 0 & 0 & 0 & 0 & 0 & 0 & 1 & 0 & 0 \\
0 & 0 & 0 & 0 & 0 & 0 & 0 & 0 & 0 & 0 & 0 & 0 & 0 & 0 & 0 & 1 \\
0 & 0 & 0 & 0 & 0 & 1 & 0 & 0 & 0 & 0 & 1 & 0 & 0 & 0 & 0 & 0 \\
0 & 0 & 0 & 0 & 1 & 0 & 0 & 0 & 0 & 0 & 1 & 0 & 0 & 0 & 0 & 0 \\
0 & 0 & 0 & 0 & 0 & 0 & 0 & 0 & 0 & 1 & 0 & 0 & 0 & 0 & 0 & 1 \\
0 & 0 & 0 & 0 & 0 & 0 & 0 & 0 & 1 & 0 & 0 & 0 & 0 & 0 & 1 & 0 \\
0 & 0 & 0 & 0 & 0 & 0 & 0 & 0 & 0 & 0 & 0 & 0 & 0 & 1 & 0 & 0 \\
0 & 0 & 0 & 0 & 0 & 0 & 0 & 0 & 0 & 0 & 0 & 0 & 1 & 0 & 0 & 0
\end{bmatrix}
*
\begin{bmatrix} a_1 \\ a_2 \\ a_3 \\ a_4 \\ a_5 \\ a_6 \\ a_7 \\ a_8 \\ a_9 \\ a_{10} \\ a_{11} \\ a_{12} \\ a_{13} \\ a_{14} \\ a_{15} \\ a_{16} \end{bmatrix}
=
\begin{bmatrix} 10 \\ 123 \\ 37 \\ 137 \\ 254 \\ 319 \\ 433 \\ 68 \\ 6 \\ 234 \\ 125 \\ 267 \\ 312 \\ 8 \\ 45 \\ 178 \end{bmatrix}
\tag{6}
$$

The inverse matrix for the previous example (for the 4x4 matrix size) is implemented as it is shown in the next equation, where a_i stands for the original values of the matrix:

$$
\begin{bmatrix}
a_1 \\ a_2 \\ a_3 \\ a_4 \\ a_5 \\ a_6 \\ a_7 \\ a_8 \\ a_9 \\ a_{10} \\ a_{11} \\ a_{12} \\ a_{13} \\ a_{14} \\ a_{15} \\ a_{16}
\end{bmatrix}
=
\begin{bmatrix}
1 & 0 & 0 & 0 & 0 & 0 & 0 & 0 & 0 & 0 & 0 & 0 & 0 & 0 & 0 & 0 \\
0 & 1 & 0 & 0 & 0 & 0 & 0 & 0 & 0 & 0 & 0 & 0 & 0 & 0 & 0 & 0 \\
0 & 0 & 1 & 0 & 0 & 0 & 1 & 0 & 0 & 0 & 0 & -1 & 0 & 0 & 0 & -1 \\
0 & 0 & 0 & 1 & 0 & 0 & 0 & 1 & 0 & 0 & -1 & 0 & 0 & 0 & -1 & 0 \\
0 & 0 & 0 & 0 & 0 & 0 & -1 & 0 & 0 & 0 & 0 & 1 & 0 & 0 & 0 & 1 \\
0 & 0 & 0 & 0 & 0 & 0 & 0 & -1 & 0 & 0 & 1 & 0 & 0 & 0 & 1 & 0 \\
0 & 0 & 0 & 0 & 1 & 0 & 0 & 0 & 1 & 0 & 0 & 0 & 0 & -1 & 0 & 0 \\
0 & 0 & 0 & 0 & 0 & 1 & 0 & 0 & 0 & 1 & 0 & 0 & -1 & 0 & 0 & 0 \\
0 & 0 & 0 & 0 & 0 & 0 & 0 & 0 & -1 & 0 & 0 & 0 & 0 & 0 & 0 & 0 \\
0 & 0 & 0 & 0 & 0 & 0 & 0 & 0 & 0 & -1 & 0 & 0 & 1 & 0 & 0 & 0 \\
0 & 0 & 0 & 0 & 0 & 0 & 1 & 0 & 0 & 0 & 0 & 0 & 0 & 0 & 0 & -1 \\
0 & 0 & 0 & 0 & 0 & 0 & 0 & 1 & 0 & 0 & 0 & 0 & 0 & 0 & -1 & 0 \\
0 & 0 & 0 & 0 & 0 & 0 & 0 & 0 & 0 & 0 & 0 & 0 & 0 & 0 & 0 & 1 \\
0 & 0 & 0 & 0 & 0 & 0 & 0 & 0 & 0 & 0 & 0 & 0 & 0 & 0 & 1 & 0 \\
0 & 0 & 0 & 0 & 0 & 0 & 0 & 0 & 1 & 0 & 0 & 0 & 0 & 0 & 0 & 0 \\
0 & 0 & 0 & 0 & 0 & 0 & 0 & 0 & 0 & 1 & 0 & 0 & 0 & 0 & 0 & 0
\end{bmatrix}
*
\begin{bmatrix}
b_1 \\ b_2 \\ b_3 \\ b_4 \\ b_5 \\ b_6 \\ b_7 \\ b_8 \\ b_9 \\ b_{10} \\ b_{15} \\ b_{16} \\ b_{17} \\ b_{18} \\ b_{19} \\ b_{20}
\end{bmatrix}
=
\begin{bmatrix}
10 \\ 123 \\ 25 \\ 35 \\ 12 \\ 102 \\ 252 \\ 241 \\ 2 \\ 78 \\ 255 \\ 23 \\ 178 \\ 45 \\ 6 \\ 234
\end{bmatrix}
\tag{7}
$$

4   Experiments and Results with MTTool

MTTool: We can decrease the size of any vectors which are created from the projections of MOT with the built in ZIP and Huffman coding opportunities. The Huffman lossless encoding and decoding algorithm was chosen due to its binary block encoding attribute and not because of its compression capability. Good data compression can be achieved with Zip and Unzip, which are also implemented. The possibility of time measuring with simple tools, such as labels or easily generated text files which include the test results, can give us a good insight into the MOT and IMOT. From these results we can estimate and predict the consumed time on hardware implementation and its cost as well.

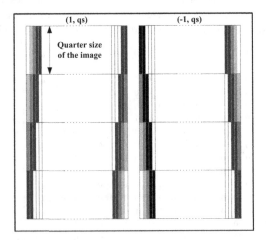

Fig. 3. MOT computing sequence for directions (p, q) and (p,-q) where p = quarter size of the image (qs) and q = 1

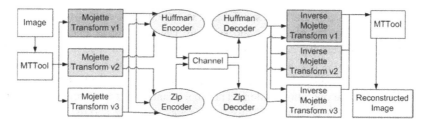

Fig. 4. Logical system architecture of the MTTool

The time measurement was applied on three different images with three different image sizes and with three different periods. The images were black and white PGM files with pixel values of 0 and 255 and the LENA.PGM. The first test ran only once, after which the second test ran for 6 times in a row, and the last test ran 24 times. Each test was performed with sizes of 16x16, 32x32 and 512x512. The results of the two smallest image sizes are nearly identical, and the results were nearly always under 20 milliseconds for MOT and IMOT, but we could see the following difference regarding the 512x512 image size:

Table 3. Test result of the MOT and IMOT with the first version

IMAGE:	Black (512x512)		White (512x512)		Lena (512x512)	
	[min:s:ms]	MOT and IMOT in [ms]	[min:s:ms]	MOT and IMOT in [ms]	[min:s:ms]	MOT and IMOT in [ms]
MT start	57:14:277		3:45:510		21:36:79	
MT end IMT start	57:15:439	1162	3:47:403	1893	21:37:762	1683
IMT end	57:15:910	471	3:47:964	561	21:38:303	541
MT start	57:22:259		4:0:822		21:49:749	
MT end IMT start	57:23:411	1152	4:2:555	1733	21:51:391	1642
IMT end	57:23:891	480	4:3:105	550	21:51:932	541

From this table we can see that the difference between black and white images is more than 50 percent, when it comes to the MOT, and only 20 percent when we apply the IMOT on the Mojette files.

5 Mojette Co-processor: MOTIMOT

The "MOTIMOT" co-processor denotes the hardware implementation of the direct and inverse Mojette transform as co-processing elements of an embedded processor.

To construct a new hardware for a special task is difficult and expensive both in time and costs. Estimating the hardware and software needs to implement a MOTIMOT processing system there [9] is necessary to map the tasks what such a system should implement. These tasks are not limited to the direct and inverse Mojette transformation, but also posses embedded computer functions with or without (depends on the application) real time operating system. The implementation target are several boards containing Xilinx FPGAs such as Virtex II PRO and Virtex IV and use powerPC 405 RISC processor (PPC). This general purpose processor can manage the MOTIMOT co-processor hardware, its driver and the incoming and outgoing data. The MOT and IMOT blocks are connected to the PPC via the processor local bus (PLB). Probably calculation of MOT or IMOT of an image is not necessary in the same time, therefore only one of the configuration files is loaded into the device.

The images are received in real-time thru an Ethernet connection or from a digital camera connected to the platform via an USB port, after processing the image or the frames these are returned to the sender or sent to a client (PC, PDA, etc.).

5.1 Low Complexity System Design for Stepping Window Method

Figure 5 shows the symbolic logic circuitry of MOT on a 4x4 image. The input registers ($IR_{x,y}$, $x = 1, .., 4$; $y = 1, .., 4$) contain the values of pixels of the image while the output registers ($OR_{i,j}$, $i = 1, .., 3$; $j = 1, .., max\_bin\_number_i$) contain the values of bins. Three projection lines (1,1; -1,1; 1,2) gives an adequate set of bins and the set of projections meets the reconstruction criteria. This size do not meet with real image sizes of course, even so the 4x4 pixel size window is usable in the stepping window method. The register size depends on the width of input data (byte, word, double word) but notice that the larger the data width the more hardware resources are required [9]. The symbolized logic circuit of the reconstruction (IMOT) is depicted in Figure 6.

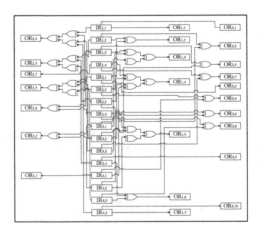

Fig. 5. Logical model of the MOT on a 4x4 image

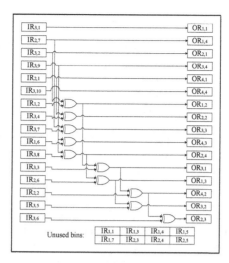

Fig. 6. Symbolized logic circuitry of a 4x4 size image reconstruction

The pixels are calculated in two ways. At first every pixel value calculable from *bin* values only and secondly a pixel value is calculable from *bin* values and the already calculated pixel values. The second version gives more simple (low-complexity) hardware. In Figure 6 the input registers (IR) contain the *bin*-values, the output registers (OR) contain the original pixel-values, while the IMOT operator is the XOR as in the MOT was. The number of registers is the same in both cases (MOT and IMOT). As a matter of fact the number of input registers can be smaller than the number of bins because of the redundancy.

The single pixel-*bin* correspondences (most of projection lines contain some) give the original pixel value without any calculation this is called zero level. Other bins contain more pixel values. The number of XOR operations need to be performed on them to get the original pixel value is the number of its level. In this case there are five levels from zero to four. In software solution it means a cycle ("for") from zero to five and every cycle contains another cycle ("for") from one to n_{bl} where n_{bl} is the number of bins on the same level. If the window is chosen larger the number of level will increase with it. Using the already calculated pixel values the complexity of the whole system will increase also of course. However increasing of complexity will be slower then when only the *bin* values are used in the calculations.

Definitely there is a window size limit (WSL) for every FPGA chip. The WSL is the block size what the given FPGA can process in parallel. The WSL depends on the number of logic cells in the given chip and other resources. To enlarge the window size above to the WSL more FPGA chips must be applied [9].

The data source means image or other types of data pre-processing before the MOT is not necessary. Post-processing after the MOT can be any kind of lossless compression method to decrease the redundancy. Data sink can be a storage device (i.e.: PC). In the decoding process the data source is the file of projection lines. Pre-processing is necessary (uncompress the files). Post processing is not necessary here. The data sink in this case is the user's application.

5.2 Low Complexity System Design for Sliding Window Method

The sliding window method differs to the stepping window method in its basic. The sliding window method moves the window only with one row or column (depends on the direction of processing) forward. It means most of the pixels are common in two windows, which are neighbors of each other [9].

Another difference compared to the stepping window method that while at the stepping window method the MOT/IMOT computing is one single step, the MOT/IMOT computing has two parts at the sliding window method. First part is to calculate the final value of the given *bins*/pixels and calculate the next temporal value of the other *bins*/pixels in the window and second part is to move the data: write out the final values, move the temporal values into the corresponding registers and read in new data.

Figure 7 shows the symbolic logic circuitry of the sliding window method where $p_i = \{1, -1, 3, -3\}$ and $q_i = 1$. The virtual image size is defined by P and Q where Q = 4 and P = file size/Q. This means that the size of sliding window is chosen independently the file size. The size of sliding window is given by the hardware resources (number of logic circuits, memory, etc.) of the given FPGA.

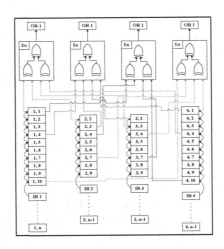

Fig. 7. Symbolic logic circuitry for sliding window method

Note if q is chosen larger (q=2) the logic circuitry and the size of sliding window (number of registers) will be multiplied by the new q value. When larger window size is required then the WSL uses more FPGA resources.

IMOT: The logic circuitry of the IMOT compared with the MOT shows that the reconstruction process is more complicated. There are single-pixel *bin* correspondences which result pixel values very simply but the bins of every other projection line need correction. It is necessary, because the *bin* value corrections generate new single-pixel *bin* correspondences and ensures the continuity of the reconstruction process.

The symbolic logic circuitry of the IMOT computing SLW co-processor is depicted in Figure 8. The picture does not show the total system only a part of it. In the image Pi (i=1..8) means the reconstructed pixel values, PLi (i=1..4) are the projection lines, while the rectangles above them represents the bins of the projection lines. In the image "tr" means temporary register which are necessary for the second step, the *bin* value correction. The outputs of Unit 1 are the reconstructed pixel values and it has an enable input. Unit 2 gives the *bin* values after the correction and works with the same enable signal as the Unit 1. In the image (Figure 8) the bit correction unit is depicted only for one projection line but the other three units are very similar. After the *bin* correction the new values are stored in temporary registers.

In the second phase the correct *bin* values from the temporary registers are written back to the original registers. The third phase is to slide the window. Move the values of the register forward by two places, and read in new data.

The MOT and IMOT functions are implemented as separate hardware co-processors of the main processor. This is possible in two ways. The main processor can be an off-chip or an on-chip solution.

The implemented algorithms are realized as separate co-processors and they work either in parallel or using run-time reconfiguration (This method was not tested yet) using relatively low working frequencies (100-300 MHz). This way can be obtained very high processing speeds.

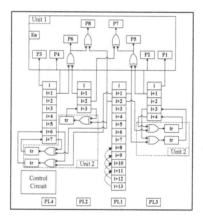

Fig. 8. Symbolic logic circuitry for IMOT (SLW)

6   Conclusions

The paper introduced from where the Mojette and Inverse Mojette transform originates. The Paper also outlined the different methods, how the Mojette transform and its Inverse is currently implemented in .Net environment and a computational experiment is described for one of the three methods integrated in MTTool. In the following part, the implementation of Mojette and inverse transformation in the embedded system using FPGA is introduced, by describing the dataflow in MOTIMOT via its logical model. In the software version (MTTool) more tests should be performed to get more accurate results, and by comparing them to the results of the hardware we should find an optimal way to perform the Mojette and for the more valuable Inverse transform. In the hardware version (MOTIMOT) finalizing the implementation of the co-processors as a whole with run-time reconfiguration is needed, concentrating mainly on the Mojette transform.

There was presented the MTTool software tool performance for several images. Also was presented two hardware implementation methods MOT and IMOT in FPGAs using parallel implementation of the Mojette algorithm.

Both algorithms (stepping window method and sliding window method) are capable to implement them on multiple FPGAs with a time-division multiplexer. This way the process time can be decreased for large images.

Acknowledgments. This work was supported by grants to projects: VEGA 1/0045/10, COST ICO 802 and CEI LPKT 26220120020.

References

[1] Guédon, J.-P., Normand, N.: The Mojette Transform: The First Ten Years. In: Andrès, É., Damiand, G., Lienhardt, P. (eds.) DGCI 2005. LNCS, vol. 3429, pp. 79–91. Springer, Heidelberg (2005)

[2] Guédon, J.-P., Normand, N.: Spline Mojette Transform Application in Tomography and Communication. In: EUSIPCO 2002 (September 2002)

[3] Guedon, J.P., Parrein, B., Normand, N.: Internet Distributed Image Databases. Int. Comp. Aided Eng. 8, 205–214 (2001)

[4] Parrein, B., Normand, N., Guédon, J.-P.: Multimedia Forward Error Correcting Codes For Wireless Lan. Annals of Telecommunications (3-4), 448–463 (March-April 2003)

[5] Normand, N., Guedon, J.P.: La transformee Mojette: une representation recordante pour l'image. In: Comptes Rendus Academie des Sciences de Paris. Theoretical Comp. SCI. Section, pp. 124–127 (1998)

[6] Katz, M.: Questions of Uniqueness and Resolution in Reconstruction from Projections. Springer, Berlin (1977)

[7] Autrusseau, F., Guedon, J.P.: Image Watermarking for Copyright Protection and Data Hiding via the Mojette Transform. In: Proceedings of SPIE, vol. 4675, pp. 378–386 (2002)

[8] Turán, J., Ovsenik, L., Benca, M., Turán Jr., J.: Implementation of CT and IHT Processors for Invariant Object Recognition System. Radioengineering 13(4), 65–71 (2004)

[9] Vásárhelyi, J., Szoboszlai, P., Turán, J., Serfőző, P.: Mojette Transform FPGA Implementation and Analysis. In: Proceedings of 10th International Symposium of Hungarian Researchers on Computational Intelligence and Informatics, Budapest, Hungary, November 12-14, pp. 351–363 (2009)

[10] Serfőző, P., Vásárhelyi, J., Szoboszlai, P., Turan, J.: Performance Requirements of the Mojette Transform for Internet-distributed Databases and Image Processing. In: IEEE OPTIM 2008, May 22-24, pp. 87–92 (2008)

Network Statistics in Function of Statistical Intrusion Detection

Petar Čisar[1] and Sanja Maravić Čisar[2]

[1] Telekom Srbija, Subotica, Serbia
  petarc@telekom.yu
[2] Subotica Tech, Subotica, Serbia
  sanjam@vts.su.ac.rs

Abstract. Intrusion detection is used to monitor and capture intrusions (attacks) into computer network systems which attempt to compromise their security. A lot of different approaches exist for statistical intrusion detection. One of them is behavioral analysis, thus in accordance with this, a model-based algorithm is presented. The paper also elaborates the way of determining the control limits of regular traffic using descriptive statistics. In addition, the paper deals with the statistical formulation of relation between the average and maximum value of network traffic and discuss the time factor in intrusion detection.

Keywords: network traffic curves, modelling, descriptive statistics, control limits, adaptive algorithm, intrusion detection.

1 Introduction

An Intrusion Detection System (IDS) generally detects unwanted manipulations to systems. The manipulations may take form of attacks by skilled malicious hackers or using automated tools. An IDS is required to detect all types of malicious network traffic and computer usage that can not be detected by a conventional firewall. Even the best packet–filtering can miss quite a lot of intrusions.

The performance of a network IDS can be more effective if it includes not only signature matching but also traffic analysis. By using traffic analysis, anomalous traffic is identified as a potential intrusion. Traffic analysis does not deal with the payload of a message, but its other characteristics such as source, destination, routing, lenght of the message, time it was sent, the frequency of the communication etc. Traffic payload is not always available for analysis–the traffic may be encrypted or it may simply be against policy to analyze packet payload.

An IDS may be categorized by its detection mechanism: anomaly-based, signature-based or hybrid (uses both of previous technologies).

I.J. Rudas et al. (Eds.): Computational Intelligence in Engineering, SCI 313, pp. 27–35.

Anomaly Detection Techniques:

- Protocol Anomaly Detection–refers to all expectations related to protocol format and protocol behaviour.
- Application Payload Anomaly–Application anomaly must be supported by detailed analysis of application protocols. Application anomaly also requires understanding of the apllication semantics in order to be effective.
- Statistical Anomaly–Statistical DDoS–To fully characterize the traffic behaviour in any network, various statistical measures are used to capture this behaviour. For example: there is a stable balance among different types of TCP packets in the absence of attacks. This balance can be learned and compared against short-term observations that will be affected by attack events. Additionally, the statistical algorithm must recognize the difference between the long-term (assumed normal) and the short-term observations to avoid generating false alarms on normal traffic variations.

With the theme of statistics and statistical anomaly detection have dealt the publications [2, 3, 4, 5, 6, 8, 10].

The main idea of this paper is directed towards statistical analysis of network traffic curves of larger users. Based on the identification of common characteristics of traffic curves, the general shape of traffic function will be determined, which represents the necessary condition for creation of algorithm for statistical detection of network anomalies.

2   Characteristics of Network Traffic

This chapter describes a graphical analysis of traffic curves. Based on the recognition of common characteristics, the model of traffic curve is formed. Samples of local

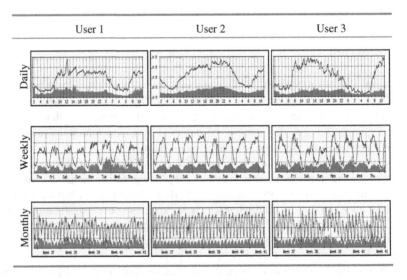

Fig. 2.1. Traffic curves of different users

maxima were taken from each segment of the curve, with help of which, by the application of descriptive statistics, the upper allowed values are determined. Two measurements were carried out in the interval of one month in order to recognize changes in the characteristic values of traffic.

The research uses daily, weekly and monthly traffic curves of several larger Internet users that derive from the popular network software MRTG, which is related to the period of one day, week and month. Without the loss of generality, the graphical presentation of curves from three users is given below, noting that the observed traffic curves of other users do not deviate significantly from the forms shown here.

Having in mind the previous figure, it can be created a general periodic curve trend (the curve of average value of traffic in observed time intervals) with period T=24 h, which results in the following figure [1]:

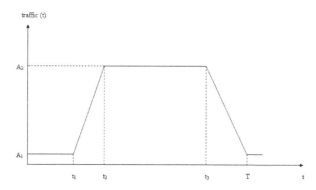

Fig. 2.2. Model traffic curve

In accordance with the previous figure, periodic traffic trend curve is defined as:

- for the interval $0-t_1$ (night traffic): $y(t) = A_1$

- for the interval t_1-t_2 (increase of morning traffic): $y(t) = A_1 + (A_2 - A_1) \cdot \dfrac{t-t_1}{t_2-t_1}$

- for the interval t_2-t_3 (daily traffic): $y(t) = A_2$

- the interval t_3-T (fall of night traffic): $y(t) = A_2 - (A_2 - A_1) \cdot \dfrac{t-t_3}{T-t_3}$

where A_1, A_2, t_1, t_2, t_3 and T represent values whose meaning is shown in the figure and vary from user to user.

In case of some users, for example, the user 3 in Figure 1, in days of weekend the fall of average daily traffic is about 25%. In this sense, the value of A_2 for a same user can not be considered as constant in all periods T.

In order to determine the range of expected values of traffic in a certain time during the day, 39 samples of local maxima were taken from the curve of traffic from all time segments: $0-t_1$, t_1-t_2, t_2-t_3 and t_3-T. Then, descriptive statistics is applied on them, with aim of calculating the lower and upper control limit. In this sense, the arithmetic mean and standard deviation of samples are calculated, and on the basis of them, with confidence interval around 99% (i.e. 3σ), the maximum and minimum of expected values of traffic are determined. Any value that falls outside the allowed interval, in statistical terms represents a network anomaly suspicious on attack. The user 1 is taken for an example and the following values are obtained:

Table 2.1. Samples and control limits

Sample	09-24 h (Mb/s)	24-03 h (Mb/s)	03-07 h (Mb/s)	07-09 h (Mb/s)
1	19	18	6	7
2	21.5	17.5	6.5	7.5
3	18.5	16	7	8
4	21	15.5	7.5	8.5
5	25	15	8	9
6	18	14.5	8.5	9.5
7	20.5	14	9	10
8	18.5	13.5	9.5	10.5
9	22	13	10	11
10	25	14	10.5	11.5
11	28	13	11	12
12	30	12	11	12.5
13	33.9	16	11.5	13
14	30	16	11.5	13.5
15	27	15.5	11	14
16	26	15	10.5	14.5
17	24	14.5	10	15
18	22	14	9.5	15.5
19	19	13.5	9	16
20	24	13	8	16.5
21	20	12.5	7	17
22	25	12	6.5	17.5
23	21	11.5	6	18
24	22	11	7	18.5
25	23	10.5	6.5	19
26	25	10	6	19.5
27	27	9.5	5	20
28	28	9	6	20.5
29	18.5	8.5	7	21
30	22	8	7.5	21.5
31	23	7.5	8	22
32	24	7	7.5	22.5
33	19	7.5	7	23
34	21	8	7.5	23
35	23	8.5	8	22

Table 2.1. (*continued*)

36	24	8	8.5	21
37	21	7.5	6	20
38	20	7	5.5	19
39	23.5	6	7	19
Avg	23.15128205	11.87179487	8.076923077	15.87179487
St.dev.	3.654188066	3.365298823	1.833670684	4.93206206
Max (99%)	34.11384625	21.96769134	13.57793513	30.66798105
Min (99%)	12.18871785	1.775898403	2.575911024	1.075608691

In the same way it is possible to determine the range of expected values for the traffic curve of any other user. In order to check the calculated values, two measurements were performed in range of a month and the following maxima were obtained:

Table 2.2. Maxima of two measurements

	Daily (Mb/s)	Weekly (Mb/s)	Monthly (Mb/s)
1st measurement	33.9	29.7	30.9
2nd measurement	33.1	33.4	32.4

Analizing the results from the previous table it can be concluded that nor in one case is not exceeded the calculated maximum value of traffic (in this case 34.1 Mb/s). This fact justifies the used method.

The research have also dealt with establishing the size and variation of characteristic values of the traffic in different time periods. In this regard, the observed values are the average and maximum traffic of several users, in daily, weekly and monthly periods. The results are as follows [1]:

Table 2.3. Differences in characteristic values of traffic

	Daily 1 [Mb/s]	Daily 2 [Mb/s]	Diff. [%]	Weekly 1 [Mb/s]	Weekly 2 [Mb/s]	Diff. [%]	Monthly 1 [Mb/s]	Monthly 2 [Mb/s]	Diff. [%]
User 1									
Max.	33.9	33.1	-2.4	29.7	33.4	12.4	9.7	9.8	1
Average	16.5	19.1	15.8	17.0	21.1	24.1	6.01	6.6	9.8
User 2									
Max.	3.94	3.63	-7.8	3.98	3.68	-7.5	48.2	49.2	2

Table 2.3. (*continued*)

Average	2.35	2.09	-11	2.28	2.09	-8.3	30.9	30	-3
User 3									
Max.	9.31	10.0	7.4	9.71	9.99	2.9	9.9	9.7	-2
Average	5.71	6.01	5.2	5.63	6.64	17.9	5.4	4.9	-9.2
User 4									
Max.	9.69	9.99	3.1	10.0	9.91	-0.9	10	10	0
Average	4.96	5.14	3.6	5.2	4.94	-5	7.4	7.6	2.7
User 5									
Max.	48.2	46.3	-3.9	48.5	45.2	-6.8	1.8	1.8	0
Average	29	24.4	-15.9	30.4	26.4	-13.1	0.14	0.14	0
User 6									
Max.	10.1	10.1	0	10.0	10.0	0	3.94	3.66	-7.1
Average	7.78	7.95	2.2	7.43	8.14	9.6	1.9	2.03	6.8
User 7									
Max.	3.98	3.97	-0.02	3.94	3.99	1.2	3.9	3.9	0
Average	1.74	1.79	2.9	1.88	1.99	5.9	1.9	2	5.2

By comparison of data from table it can be concluded that the changes in maxima and average values of traffic are relatively small–the average value of difference in maximum traffic is 3.26% while in average value is 8.44%.

3 Statistics of Maxima and Average Values

This chapter deals with determining the statistical relation between average and maximum value of traffic, analyzing this situation with multiple users in different time periods.

For each user, daily, weekly and monthly traffic were observed, for which the maximum and average values were determined by "MRTG" software. In this way, the following table is created:

Table 3.1. Max/Avg values for different users

		Avg (Mb/s)	Max(Mb/s)	Max/Avg	St. dev.
1	Daily	16.5	33.9	2.054545	-
	Weekly	17	29.7	1.747059	0.163
	Monthly	17.1	30.9	1.807018	-
		Avg (Mb/s)	Max(Mb/s)	Max/Avg	
2	Daily	29	48.2	1.662069	-
	Weekly	30.4	48.5	1.595395	0.052
	Monthly	30.9	48.2	1.559871	-

Table 3.1. (*continued*)

		Avg(kb/s)	Max(kb/s)	Max/Avg	
3	Daily	4955.6	9692.7	1.955908	-
	Weekly	5202.8	9992.8	1.920658	0.061
	Monthly	5374.4	9873.4	1.837117	-
		Avg(kb/s)	Max(kb/s)	Max/Avg	
4	Daily	2352.4	3938.3	1.674163	-
	Weekly	2285.8	3977	1.739872	0.217
	Monthly	1898.2	3945.2	2.07839	-
		Avg (Mb/s)	Max(Mb/s)	Max/Avg	
5	Daily	7.78	10.1	1.298201	-
	Weekly	7.433	10	1.345352	0.031
	Monthly	7.362	9.99	1.356968	-
		Avg(kb/s)	Max(kb/s)	Max/Avg	
6	Daily	5712.3	9306.7	1.629239	-
	Weekly	5626.9	9713	1.726172	0.056
	Monthly	6012.8	9785.4	1.627428	-
		Avg(kb/s)	Max(kb/s)	Max/Avg	
7	Daily	1737.7	3978.6	2.289578	-
	Weekly	1876.6	3940.1	2.099595	0.145
	Monthly	1970.4	3949.2	2.004263	-
			Mean (Max/Avg)	1.76	
			St. dev. (Max/Avg)	0.26	

Applying descriptive statistics on partial standard deviations of Max/Avg values for each user, the following results were obtained:

Table 3.2. Descriptive statistics for standard deviations

Mean	0.103668114
Standard Error	0.026754408
Median	0.061009887
Mode	-
Standard Deviation	0.070785509
Sample Variance	0.005010588
Kurtosis	-1.254753055
Skewness	0.677526024
Range	0.185791754
Minimum	0.03112295

Table 3.2. (*continued*)

Maximum	0.216914704
Sum	0.725676796
Count	7
Confidence Level (95. 0%)	0.065465677

On the basis of very small value of standard deviation (0.07) from the previous table, the authors are of the opinion that for the traffic of larger users the following general relation can be formulated:

$$traffic_{max} \approx c \cdot x \cdot traffic_{avg} \qquad (3.1)$$

where c is the constant specific for some particular traffic and represents its network characteristics. About 95% of all values of c belongs to interval $1.76 \pm 2 \times 0.26$.

The previously formulated relation gives the possibility of calculating the upper (allowed) traffic limit if the average value of traffic is known. In practice, there are numerous software for network monitoring, which have the possibility of continuous monitoring the average traffic. In this way, in certain fixed period of time (e.g. 10 seconds) the upper threshold value can be updated, which provides adaptivity in relation to the current state of traffic.

With aim of decreasing the number of false alarms, it can be suggested that, similar to AT–k intrusion detection algorithm, an attack alarm is activated when it comes to k consecutive (i.e. k ΔT, where ΔT represents a single measuring interval) exceedings of maximum value of traffic within a measurement period T.

4 Time Factor in Intrusion Detection

This chapter briefly elaborates the current views on the moment of the very beginning of network attack and presents information about necessary reaction time of intrusion detection system.

Considering the variety of attacks, it is rather difficult to precisely define the starting part of the attack timeline. Some attacks are immediately recognizable, perhaps taking the form of one or more packets operating over a short time period Δt– e.g. less than one second [7]. Others are active for a much longer period (e.g. hours, days or even weeks) and may not even be identified as attacks until a vast collection of event records are considered in aggregate. Thus, while every attack has a definite beginning, this starting point is not always discernible at the time of occurence.

The main idea of this paper is the detection of such type of attacks that are recognizable in real time (real-time detection) – true time zero plus some arbitrary small time slice beyond that – i.e. less than a few seconds [7]. Real-time, according to industry definitions, can be expressed like time interval 5 s-5 min.

Reaction time of security systems on the appearance of attacks is changing over time. So in the eighties and nineties of the last century the reaction time was about twenty days, from 2000-2002 about two hours, while from 2003 and later this time is needed to be less than 10 seconds.

5 Conclusions

In research proposed way of determining the control limits of network traffic is checked in case of different users, in different time periods and in case of time-distant measurements. No false alarm was generated. Algorithm based on this model has fixed and variable parameters. If the periodically update of variable parameter is enabled in a short enough time intervals following the actual traffic, this algorithm gets the feature of adaptivity, which further reduces the possibility of generating false alarms. By adequate application of descriptive statistics on the maximum and average values of network traffic, the general relation that connects these two values is formulated, which may be applicable in statistical anomaly detection based on threshold.

References

[1] Čisar, P., Maravić Čisar, S.: Model-based Algorithm for Statistical Intrusion Detection. In: Proceedings of the 10th International Symposium of Hungarian Researches on Computational Intelligence and Informatics, CINTI 2009, Budapest, pp. 625–631 (2009) ISBN 978-963-7154-96-6

[2] Engineering Statistics Handbook–Single Exponential Smoothing, http://www.itl.nist.gov/div898/handbook/pmc/section4/pmc431.htm (accessed January 6, 2010)

[3] Gong, F.: Deciphering Detection Techniques: Part II Anomaly-based Intrusion Detection, White Paper, McAfee Security (2003)

[4] Montgomery, D.: Introduction to Statistical Quality Control, 5th edn. Wiley, New York (2005)

[5] Moore, D., Voelker, G., Savage, S.: Inferring Internet Denial-of-Service Activity. In: Proceedings of the 2001 USENIX Security Symposium, pp. 9–22 (2001), http://www.caida.org/publications/papers/2001/BackScatter/usenixsecurity01.pdf (accessed January 6, 2010)

[6] Reid, R.D.: Operations Management, 4th edn. Wiley, New York (2011), http://www.wiley.com/college/reid/0471347248/samplechapter/ch06.pdf (Accessed January 6, 2010)

[7] Roesch, M.: Search Security. Next-generation intrusion prevention: Time zero (during the attack) (retrieved November 26, 2007) (2005), http://searchsecurity.techtarget.com/ti/0,289483,sid_gci1121310,00.html (accessed September 8, 2008)

[8] SANS Intrusion Detection FAQ: Can You Explain Traffic Analysis and Anomaly Detection? http://www.sans.org/resources/idfaq/anomaly_detection.php (accessed January 6, 2010)

[9] Siris, V., Papagalou, F.: Application of Anomaly Detection Algorithms for Detecting SYN Flooding Attacks. Computer Communications 29(9), 1433–1442 (2006)

[10] Sorensen, S.: Competitive Overview of Statistical Anomaly Detection, White Paper, Juniper Networks (2004)

Congestion Forecast – A Societal Opportunity

Roland L. Kovács[1], Dr. László Nádai[2], and Andrea Kovács-Tóth[3]

[1] ABS Systems, Knorr-Bremse Fékrendszerek Ltd.
Szegedi út 49, H-6000 Kecskemét, Hungary
roland.kovacs@knorr-bremse.com
[2] Computer and Automation Research Institute, Hungarian Academy of Sciences
Kende út 13-17, H-1111 Budapest, Hungary
nadai@sztaki.hu
[3] Nagyváradi utca 25, H-1044 Budapest, Hungary
andrea@kovacs-toth.com

Abstract. Information technological, economical, financial, legal and institutional aspects influence congestion forecast systems system setups. First, several solutions for the individual problem areas are developed and evaluated. Then, alternatives for congestion forecast systems are defined as coherent solution sets of the depicted problem areas. The information technological key issue is data quality. Successful introduction of such a system in terms of economy and finances deeply depends on the technological service provided. Legal and institutional solutions target in the first line data privacy requirements. Finally, the most attractive solution is outlined.

1 Introduction

The psychological and economical benefits of congestion forecast systems seem to be obvious. The basic goal of a congestion forecast system is to help users to avoid congestions. Individual and societal benefits can be achieved if this goal is reached. Individual benefits include decreased travelling time and savings on fuel consumption. Societal benefits would be reduced emission of motor vehicles and reduced traffic. By reducing traffic and stress a positive effect on the number of accidents can be expected. Overall, quality of life could be improved particularly in bigger cities.

2 Technological Backgrounds

2.1 Traffic Congestion

As strict mathematical models have a poor correlation to actual observed traffic flows, a simple definition may suffice in this context. Traffic congestion is a condition on networks that occurs as use increases, and is characterized by slower speeds, longer trip times, and increased queuing. In case, traffic flows continuously at a given point or interval of time within a selected segment of the road network, then there is no

I.J. Rudas et al. (Eds.): Computational Intelligence in Engineering, SCI 313, pp. 37–45.
springerlink.com

congestion. As average speed decreases then below a certain thresholds different levels of congestion can be observed.

In situations where one of the two extreme situations (no congestions, or complete congestion) can be observed for typical time periods predictably, there is no real need for congestion forecast. Just as there is no real need for rain forecast in the rain forest or the desert. On the other hand, there is need for congestion forecast for all the situations in between. However, this need may differ on state, regional or city level.

2.2 Congestion Forecast Systems

The process of congestion forecast features three key participants. The data management institution plays the central role. The central institution is responsible for storage, processing, deletion, and maintenance of data. Data acquisition and provision to the data management institution is done by the input institution. The output institution exploits the necessary information by sales and data forwarding via service providers to the final consumers.

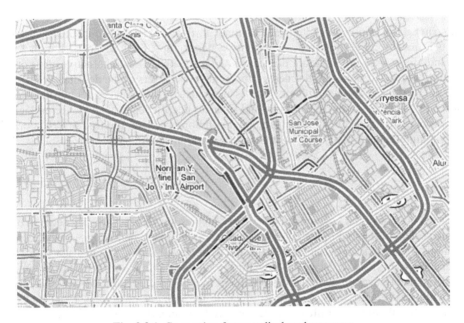

Fig. 2.2.1. Congestion forecast displayed on a map

Congestion forecast systems can be sorted into three groups based on the source of data. These are person-dependent, automated and – as a combination of the two – hybrid systems.

A person-dependent system treats congestions as unexpected events by the cooperative data management of different institutions (ÚTINFORM, FŐVINFORM, police, fire department, dispatcher centres, etc.). The major problem of these type of systems is that congestion spots are reached with significant delays and the institutions can provide data only if it has been reported to them. Due to lack of real-time and redundant information data cannot be synchronized and filtered easily which results in poor data quality. Further disadvantages are the costs related the acquisition of input data. The only advantage is that start-up investments are relatively low.

Automated systems provide road load information by the use of telematic systems that can be built into passenger cars or other vehicles. Automated systems that provide confident congestion data would be based on GPS systems. During a congestion event vehicles equipped with GPS would provide slow motion data. When congestion is cleared the vehicles speed up, this is detected by the data management system and forwarded to the contracted service providers. In case several vehicles provide motion data congestion situation can be identified more accurately. Also, the number of the data providers has to be large enough to cover representatively the important segments of the road system. Automated systems that provide distortion-free data for a large road network may only be realized with a large number of individuals who provide data. Drawbacks are high investment costs as wide-spread deployment of telematic systems is needed. On the other hand, multifunctional usage that goes beyond congestion forecast could justify these costs.

Of course, hybrid systems can be established as well. These systems could extend data provided by automated systems, e.g. road maintenance data, traffic weather data would be valuable. Road users could avoid critical road sections, thus avoiding the development of congestions. Another option is the usage of mobile phones that are much wider spread than telematic units.

Beyond information technological aspects it is important to integrate supply and demand on a system level and to manage this connection appropriately.

3 Stakeholders

The process of data transmission aiming congestion avoidance can be divided into two phases that is connected by the agreements about the forwarding of information. First, microscopic traffic data has to be supplied to a central data management facility. Second, macroscopic traffic data is transmitted to mobile devices of individual participants of the transportation system.

The information for congestion avoidance is or may be provided – via different technologies – by the participant of the vehicular traffic themselves directly to the data management centre. Thus, the congestion database may be built up based on live real-time data of a small organized group of traffic participants or even of a larger group. It is even possible to make data supply obligatory.

The prerequisite for sales of the service - depending on the state of technical developments – is the creation, management and exploitation of relevant data records, which includes the following tasks: data collection, implementation and management of the databank, data processing and use of information obtained from the databank. These activities can be assigned to data management and data processing organizations. The data management centre is responsible for lawful execution of data processing operations.

The primary end-users are the drivers who receive road traffic information. The users may be individuals or legal persons and owners or employees of unincorporated companies.

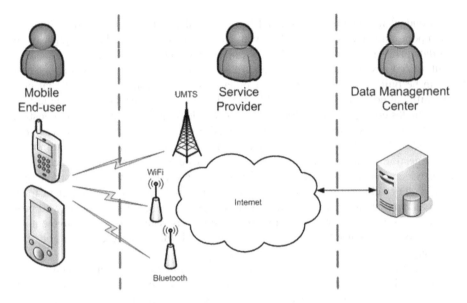

Fig. 3.1. Congestion forecast stakeholders

The transmission of congestion forecast information can be accomplished by the data management centre, or an intermediary service. The service provider may be the mobile network operator company which has a contract to the data management centre and to the final user. Theoretically, the service provider could be the data manager as well, but the mobile network operator may attain competitive advantage as data provided to other mobile network operators could be analyzed.

4   Economic Context

In addition to the information technological aspects the economical view on the systemic link between supply and demand and the way it is managed is a strategic key element.

4.1   The Product

The product is data on road traffic congestion. An essential condition for the marketing of the product is that the information distortion-free. The database is relatively undistorted, if the verification process works on a large amount of data, and targeted traffic segment is covered continuously.

The product – examined in space and time dimensions – can form different sets of products. The information may cover the entire public and local road network, only the road network, only the larger cities or the network of small and medium-sized cities. The period of use of congestion information can change, so fees may be charged on yearly, monthly, daily base.

4.2 Market Forecast and Price Sensitivity – The Supply Side

For marketing purposes supply and demand has to be evaluated. The problems of supply are lack of trust, vulnerability of personal data, low number of vehicles equipped with on-board GPS, low connection willingness, therefore, the low reliability of the product information, and hence its relatively high cost.

The biggest initial problem is product quality. The relatively distortion-free product would require large number of vehicles with GPS installed. Reliable and complete traffic input data is only possible with the intervention of the public sector. If data management is done by the private sector than data management will comply with their interests and therefore data loss may occur.

The key solution lies in the multi-purpose use of GPS, which serves public and private interest at the same time. The payment of taxes imposed on motor vehicles can be combined with the use, i.e. additional road toll is paid over a certain amount of usage. If the idea was supported by the government longer-term, the necessary technical conditions must be created with on-board GPS systems.

In addition, the cooperation between insurance companies dealing with motor vehicle and the government to support the purchasing of congestion-warning systems may have positive impacts. GPS-based vehicle tracking could significantly reduce insurance payment due to theft and there may be a discount on insurance fees for vehicles equipped with GPS-based tracking systems. Cumulative advantage could be obtained by many small and medium-sized enterprises through maintenance works and sales of replacement parts. This in turn would extend the options of the banking sector. Additional advantages would be an alternative EU-wide usage-proportional road toll payment option for the passenger cars.

On this basis, the security of the vehicle fleet could be improved, toll payment would depend on the actual road usage, and the acquisition of automated, objective, undistorted congestion data would become possible.

Optimized fuel consumption would reduce the environmental pollution. The additional tasks of the mobile network operators would lead to revenue increase that would result in tax increase that leads to consolidated macroeconomic revenue. The participation of SMEs would increase employment.

4.3 Price Structure of the Product – The Demand Side

Demand analysts say Hungarian population is very price- and cost-conscious. Most important task is to create consumer confidence. First, potential buyers are going to test the product. The majority would buy congestion forecast information for a smaller area. It is not likely that annual subscriptions will be purchased for the whole national area. At national level, there will be demand for occasional connections only.

The mobile network operators have to carry out very serious business-psychological demand analyzes to find out how to contract with the data manager in order to compensate for the development costs and the operational services fees.

The data service in the major cities is not required continuously by road users, usage is mostly related to get to the workplace and back home. Thus, congestion information is claimed just in a fraction of the day. The majority of congestion areas are already well known as they appear periodically. Most drivers are less interactive and use the usual route and do not change. Country-wide demand covering the whole year and all time of day shall not be a real demand due to relatively high costs.

5   Legal and Institutional Aspects

In the three-partner congestion forecasting process the participants are linked by a contractual relationship. The contract requires that the mobile network operator license permits this type of service, and that the available network covers user demand areas.

The reliability of the information has to be guaranteed by the data transmission contract between service provider and data manager. The key object of the end-users contract is the congestion information, and therefore the alternative routes suggested. The data processing methods and the spread of automated information acquisition systems will minimize the distortion of information.

The handling of traffic data is a very sensitive area of the implementation of the congestion-warning service. Data protection is the responsibility both of the service provider and the data manager. The data must be protected against unauthorized access, alteration, transmission, publication, deletion or destruction, accidental destruction and damage. Specific security arrangements must be taken by the data management centre and by the service provider to ensure technical protection of personal data. According to the statutory provisions on the protection of personal data, the individual user must declare that the service provider is empowered to use personal data for temporary service records and for the temporary transfer to the central database. Regarding the new service it is worth mentioning that the mobile network operator companies have to inform the affected customers about all details of handling of personal data.

Due to the problems of processing of personal data the greatest emphasis should be placed on the data management centre. The data management centre can be established by private organizations and the public sector. The SWOT analysis below shows the possible alternatives for institutional background and their strengths, weaknesses, opportunities and threats.

To the establishment of institutional background it is necessary to take the following aspects into account: is the collected road traffic information personal or public data; under what conditions the data can be treated, including the questions: is the supplying of data voluntary or obligatory, can the handling of data be considered as a state responsibility. It is important to know the fields of activity, the legal status and the possible types of organizations who are involved in data collection and data processing. It is necessary to take notice of the scale of developments and the very significant resource needs.

As information related to public interest is concerned there are multiple reasons to establish a state authority. However, public interest may arise indirectly. In the present case, it is proposed to put the focus on the data management centre. Because of data privacy problems and trust reasons, the system cannot be spread widely without the assistance of state interaction.

Table 5.1. SWOT analysis of data management done by the public sector

STRENGTHS	WEAKNESSES
Ensures protection of data privacy	Bureaucratic organisational structure
Provides financial background for start-up costs	Low operational efficiency
May enforce technical standards	Lack of interest in profit
OPPORTUNITIES	**THREATS**
Data of public interest may generate micro and macro economic profit	Long response times between public sector and users
Environmental impact may decrease	Changes of user requirements may not align with public interest
Number of accidents may decrease	
Due to increased traffic speed, larger cities may become more viable	

Table 5.2. SWOT analysis of data management done by the private sector

STRENGTHS	WEAKNESSES
Advanced technology is available	Distrust in respect to data privacy may arise
Due to competition operational costs are lower	Transparency is not fully ensured
Responsiveness to user demands ensures usability of the system	High development and deployment costs may cause problems
OPPORTUNITIES	**THREATS**
System may be developed fast due to private interest	Low market penetration due to distrust
Prosperous co-operation may evolve with e.g. insurance companies	Data of public interest may be lost

Various legal entities and corporations may be taken into consideration: governmental organizations, non-governmental organizations, public benefit organizations, non-profit companies and profit-oriented companies. Governmental organization may be established, if the activities listed in the public interest function have to be fulfilled.

The revenue of governmental organizations is typically support by the budget and only in exceptional cases can target funding be applied. By state resources can be achieved the initial investment and provided the necessary improvements cover.

Under statutory provision it is possible that a company is founded by the ceased governmental organization for the same activity. The companies are in general created for profit-oriented economic activity, but they can operate on non-profit basis, too. A non-governmental (state-owned) company established by a central government organization has in most cases the „public" or „increased public" qualification. Companies with those qualifications get advantages like tax relief, governmental support, capital injection and other services (e.g. infrastructure). The non-profit company form seems to be favorable – with regards to operating conditions – to the governmental organization, although in this case economic restrictions take effect because of being state property.

This solution has also disadvantages. The sales of congestion forecast service by the non-profit company is more difficult than by a profit-oriented company due to the lack of profit-interest. The profit from public activities can not be taken out of the company, it can be spent only on the expansion and development of its main activity. The state-owned entities – due to the state capital – gain a significant competitive advantage over other market operators on services market. However, the profitability and efficiency of a governmental company can not reach the ones of a market operator.

6 Concluding Remarks

Not just because of the financial crisis, but also because of data privacy concerns, the private sector is handled with some distrust by the society. Such distrust may cause low numbers of participants in automatic systems that in turn would cause low quality congestion data comparable to person dependent systems. Profitability requirements of the private sector may increase costs to be paid by end-users.

Decreased environmental impact and lower fuel consumption are the most obvious micro- and macroeconomic benefits. Others are the options to implement road usage dependent toll payment or theft protections system that could lower insurance costs. Further benefit would be that new opportunities and market were generated for SMEs. However, data privacy protection is the most sensitive issue regarding congestion forecast systems.

As outlined, there are notable differences in the services performed by the private and public sector. While private data management centres would be organized along private interest, public interest may be represented best by a public institution. In the latter case it is ensured that public tasks can be accomplished according public interests from public funds as defined by proper legislation. That way that system could show huge societal benefits. Of course it cannot be disclosed that with better economics a framework for private owned companies that represent public interest is going to be established.

Gradual introduction of congestion-warning information systems shall generate important macro-economic benefits and savings. The congestion forecast project's basic problem is related to the handling of personal data. The solution is to create confidence in users that depends on fair and credible data handing of the data manager and the mobile network operator.

Acknowledgments. This research was supported by the National Office for Research and Technology (NKTH) in Hungary.

References

[1] Kevevári, B., et al.: Közlekedési torlódás-előrejelzési projekt megvalósításának közgazdasági, pénzügyi, jogi, szervezeti, stratégiai elágazási pontjai, Pénzügykutató Zrt (2009)
[2] Kovács, R., et al.: Congestion Forecast Strategies. In: 10th International Symposium of Hungarian Researchers on Computational Intelligence and InformaticsSymposium Proceedings, Budapest, Hungary, November 12-14, pp. 167–172 (2009)

[3] Maibach, M., et al.: Handbook on Estimation of External Cost in the Transport Sector. In: Internalisation Measures and Policies for All external Cost of Transport (IMPACT), Delft, CE (2007)

[4] Jakob, A., et al.: Transport Cost Analysis: a Case Study of the Total Costs of Private and Public Transport in Auckland. Environmental Science & Policy 9, 55–66 (2006)

[5] de la Fuente Layos, L.A.: Short Distance Passenger Mobility in Europe. Statistics in focus, Transport, Eurostat 5, 8 (2005)

[6] de Blaeij, A., et al.: The Value of Statistical Life in Road Safety: a Meta-Analysis. Accident Analysis and Prevention 35, 973–986 (2003)

[7] Gibbons, E., O'Mahony, M.: External Cost Internalisation of Urban Transport: a Case Study of Dublin. Journal of Environmental Management 64, 401–410 (2002)

3D-Based Internet Communication and Control

Péter Korondi[1], Péter Baranyi[2], Hideki Hashimoto[3], and Bjørn Solvang[4]

[1] Computer and Automation Research Institute, H-1111 Budapest, Hungary
    also with Department of Mechatronics, Optics and Applied Informatics,
    Budapest University of Technology and Economics, H-1521 Budapest, Hungary
    `korondi@sztaki.hu`
[2] Computer and Automation Research Institute, H-1111 Budapest, Hungary
    `baranyi@sztaki.hu`
[3] Institute of Industrial Science, The University of Tokyo, 4-6-1 Komaba Meguro-Ku,
    Tokyo 153-8505, Japan
    `hashimoto@iis.u-tokyo.ac.jp`
[4] Department of Industrial Engineering, Narvik University College, 8514 Narvik, Norway
    `bjs@hin.no`

Abstract. Nowadays the main center of Research and Development are universities and academic institutions. The development of the Internet and network technologies made possible for research institutes to improve their cooperation and communication. The aim of this paper is to present the foundations of a uniform system for international collaboration, based on 3D Internet. The aim of the network is to create a cooperative and testable connection between mechatronical and industrial robot systems even in continental distances. In this paper we introduce a virtual laboratory which provides a suitable environment for distant institutions' laboratories for a close collaboration.

1 Introduction

Universities and institutes experiment with researching new technologies and improving existent technologies, such as the dynamically improving area of robotics and cognitive informatics devices. The cooperation of R & D institutes is problematic, because large distances make collaborative work difficult (e.g. transferring industrial robots from one place to another).

Using a virtual 3D laboratory can solve this problem. It connects research groups and distant laboratories over the Internet. All groups can share their own equipment and use the others' without actually being in the same room. This set-up provides an effective environment to test devices and algorithms separately. The industrial robot doesn't have to be leased and transferred into the laboratory when only testing its features, meaning better time and cost efficiency. Algorithms can be written – like creating a path finding algorithm for a welding robot to move and follow the surface – and be tested far from the robots.

I.J. Rudas et al. (Eds.): Computational Intelligence in Engineering, SCI 313, pp. 47–60.
springerlink.com   © Springer-Verlag Berlin Heidelberg 2010

Programming a robot is a very difficult and time-consuming task, not to mention the requirement for specially trained programmers and operators. On the contrary, the virtual laboratory provides an efficient way to teach a robot without much code writing. Instead of telling the robot joints about every movement element, the system is capable to recognize the gesture of the operator and send a high-level command directly to the robot which has the same effect [1].

Our daily activities are more and more linked to the "digital social behaviour". Investors in industry are highly aware of these trends and allocate more financial means to develop personal-fit products to reach the level when these products – besides remain functional – become indispensable partners for their owners and – as it can be clearly marked in the case of new generation mobiles – express even their cultural and social affiliation. The more complicated it is to use a device, the less it can be used in everyday life since we are not willing to learn all functions. Therefore the demand of the users that they would like to „discuss" compound needs with their personal informatics in an easy, user friendly way is fully understandable [4].

Man live in a 3D world. All our knowledge is in 3D and we use 3D in our non- and para-verbal communication. No wonder that it is a natural need from the customers that they would like to communicate with their personal informatics in the very same way; in 3D. That is why – among others – internet, the appearance of all allocated and collective knowledge of man, should also be in 3D.

The three-dimensional opportunities of informatics help significantly the communication and the representation of the information [11, 10]. These are the reasons why every day a new 3D application shows up: For example in Japan 3D TV broadcast is already available and in Europe it will start soon, 3D televisions and monitors are commercially available for an affordable price and numbers of 3D display devices based on new technologies are born every day. Cell phones are also available with 3D interface and by moving them in 3D it creates a link between our tools for informatics and their 3D content [8]. All of these are connected to each other through the internet. Internet broke into our daily life incredibly fast and so it is predictable that 3D internet will have the same success. The largest internet software companies announced that soon they will publish their 3D internet interfaces. These improvements are about to change the traditional keyboard and mouse interfaces [2], and require a more effective communication channel for data input, which is examined within the framework of comprehensive cognitive research. 3D visualization and cognitive infocommunication create 3D media research directions, which is actually the summarization of 3D representation and content handling.

Present networks, communications and content handing devices differ radically from those in future 3D internet. Media network is a technology that allows anyone to create, edit, use and enjoy any media content, wherever he/she is located. The content will not only contain sound and picture as in present phone and television services but will also offer a wide range of interactive services in the field of informatics education and entertainment thus creating new business opportunities. Currently most media content is used for broadcasting or producing music/movies. In Europe broadcasting is right in the state of changing from analog to digital technology. Digital broadcasting is more efficient in using the spectrum and in addition it offers the possibility for integrating data services and interactivity. In the case of digital broadcasting it can be observed that through recording or trough the service offered by the supplier, the content itself becomes available nearly any time for the user and gets more and more interactive. Briefly, future media network can be defined as a service that can be reached easily by anybody and anywhere for professional and leisure purposes.

Of course technically several levels of complexity lie behind this simplicity but users do not need to know this complexity. To make it possible the following fundamental things have to be changed:

- Media needs to become part of the network, just opposite the today's practice where it just something to be delivered from A to B.
- Content can come from anyone and an intelligent, user-friendly indexing engine has to generate the matching metadata.
- Intuitive and multimodal input interfaces need to offer an interconnection to and in the media environment providing a more natural interaction than what we have today.
- Display of the content should be unnoticeably adapted to the user, to the environment and to the display capabilities.

2 The Basic Definitions for the Cognitive Info-Communication and for the 3D Internet

The classic information-technology is composed of 3 pillars (see Fig. 1 left).

- *The media* is to create and manipulate the information content.
- *The communication* function is to transmit the information.
- *The informatics* task is to process the information.

Today, the boundaries between the three pillars are becoming increasingly blurred. That is called the convergence theorem in the IT literature. Thus the intermediate areas are becoming in the focus of attention.

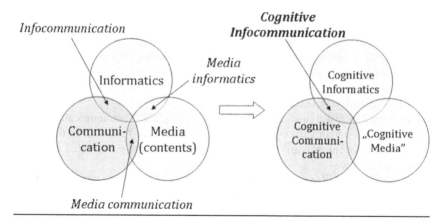

Fig. 1. Divisions of the traditional and cognitive information-technology - **Left**: The three pillars of traditional information-technology - **Right**: The location of cognitive info communication

- *The media communication* is responsible to deliver the information to the broad masses.
- *The media informatics* uses the strength of the informatics to build up the interactive media.
- *The info communication* handles the communication both between the people and the information-technology tools and the communication between the information-technology tools.

2.1 Cognitive Info-Communication

The information-technology triplet can be found in every corner of the cognitive research field. Cognitive science, or in other words acquaintance science, was developed in the fifties as a branch of the interdisciplinary sciences, trying to understand how the human sensing works and examining the sensing connection with brain activities and human intelligence. This paper focuses on the cognitive info-communication whose place is shown in Fig. 1 right. The previous definitions can be supplemented by the following interdisciplinary sciences:

- *The cognitive communication* is closer to cognitive sciences and its task is to analyze how information is sent through the transmission channels. These channels include cognitive linguistics and other non- and para-verbal channels introduced by other cognitive sciences, including the cases, when our senses are not used in the usual way, for example, when the visually disabled use their hands to see.
- *The cognitive informatics* should belong to the informatics branch. It investigates the internal information processing mechanisms of the human brain and their engineering applications in computing.
- *The cognitive info communication* handles the communication channels between people and information-technology tools and also the communications which are based on cognitive informatics processes.

The definition of cognitive info-communication is used in greater extent. In Fig. 2 an example is shown for cognitive info-communication. There is machine communication between the low level controller and a robot or intelligent tool. The link between a robot and the high level intelligent control is realized at the info-communication level. When the operator gives direct order to the robot that is also called info-communication. The cognitive info-communication level is only reached if the entire communication process – from the natural intelligence, generated by the human brain to the controlled task – is examined as a whole.

It is necessary to distinguish two cases. In one people are communicating between each other (at both ends of the communication channels are humans). In this case the task of the cognitive information device is to deliver non- and para-verbal information in the best possible way.

It is a completely different situation, when we communicate with a machine or with an artificial intelligence. Not long ago the programming language of machinery was completely abstract and only a few professional had the privilege to understand it. Today not only a small group of professionals are forced to communicate with machines, thus there is a strong demand to develop cognitive information channels. Here we have to emphasize the customization of personal tools for informatics to fit for

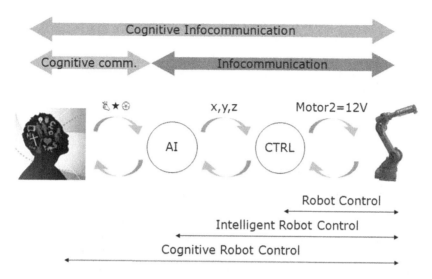

Fig. 2. The cognitive info-communication

personal needs. For users the personal tools for informatics are almost like "partners", with whom the communication should be carried out in the same way as with another human being.

The most important condition of this is that we should be able to talk to our device; moreover the para- and non-verbal channels should preferably work in both directions and in 3D because that is our natural environment. With this the real challenge is when the artificial intelligent device communicates with a human because the "feelings" of the artificial system – voltage, current, torque, angular momentum, consumption, etc. – are completely different from those of the humans. All these feelings have been transformed to human senses in such a way that either the sensory organs with the appropriate resolutions and speeds are attached together or their combinations, since in many cases we can't even break down which senses determined our action. For example, a racing driver autopilot manoeuvres according to the engine torque, speed and many other parameters that are hardly understandable for people while humans drive according to the shifting of the landscape, the sound of the engine and many divergent acceleration measurement sensors in our body. When we drive a car in a virtual space and control an autopilot, the "senses" of the autopilot have to be transformed to human senses using cognitive info-communication tools.

2.2 3D Internet

"3D Internet" is such content service, which is taking advantage of the opportunities of the Internet to give the user a stereoscopic 3-dimensional viewing experience, or attached multi-media (interactive) content. To be able to carry out this, basically new input and display hardware and software tools are needed. From these the key element is the stereoscopic visualization. These can be grouped in a number of different ways. To give a detailed description of these technologies would go beyond the limits of this article, therefore we made a summary in tabular form (see Table 1).

Table 1. Summary of 3D stereo technologies

Technologies which require glasses	Technologies which don't require glasses
Passive glasses	Optical filter based on parallel obstacles
- anaglyph	
- polar-filter	
- infitec	
Active glasses	

It is important to emphasize here that compared to the many conventional display devices significantly different new technologies are developed in this field (one promising domestic example is holographic TV). In home usage the most widely used technology is anaglyph technique because it only requires a simple screen and even a home-made anaglyph (colour filter) glasses is sufficient. This technology also appeared on YouTube. On the file sharing portals many films can be found whose stereo has been made by anaglyph technique. The first stereo cinemas used polar-filter technology, so this tradition has a history of several decades but presently modern cinemas have switched to stereo infitec technology. In both cases an appropriate optical filter is placed in front of the projector, which makes this technology too complicated to use in the case of monitors. On the other hand you can buy a 120-Hz monitor for a reasonable prize, which is capable to display images with 60-60 Hz alternating frequency, projecting the stereo image separately for the right and left eye. The monitor comes with an accessory, active shutter glasses, which is synchronized with the monitor and able to split the images for the left and right eye. Although it is likely that future of 3D screens would work without glasses. Such displays have appeared already on the market but their prices are over twice as much as the ones which use glasses.

2.3 Telemanipulation and Monitoring Based on the 3D Internet

When a cognitive info-communication instrument is used to give an online command through the 3D internet to a robot to perform a task, then we came to the 3D Internet-based telemanipulation. This also includes monitoring. The 3D visualization of the monitored information content of the human cognitive processes, perceptual abilities and as well as the speed and the importance of the information, is a separate science.

2.4 Intelligent Space, as the Forerunner of the 3D Internet-Based Cognitive Info-Communications

The "Intelligent Space" is an extension of the 3D virtual reality equipped with intelligence. In this sense it goes beyond the boundaries of 3D Internet but in current case it can be viewed as the preliminary version of it and it is considered to be an important field of application because it has determinative role in the intelligent space concept, where distributed intelligent devices are connected through the Internet.

A defined area (e.g. a room) can be considered as an intelligent space, if it is equipped with distributed actuators, sensors and robots who jointly "understand" and monitor the actions taken place in the virtual space and thus able to influence the events or help the humans staying in the virtual space.

The "Intelligent Space" is the first definition and is the trademark of the first implemented system called iSpace, which burst into the public awareness in the nineties as the result of Hideki Hashimoto, professor of Tokyo University, work [6, 5]. The tools for the 3D Internet and cognitive info-communication are ideal for iSpace-s.

3 The Research Related to the Intelligent Space

Several events took place parallel in the last 15-20 years, at the beginning maybe a little isolated from each other, which finally ended up in the definition of cognitive info-communication by SZTAKI. SZTAKI gained significant competence in the 3D Internet and other related terms referred in the previous parts. In this section, the intelligent space related research will be highlighted from the point of the parallel events mentioned above.

The concept of iSpace and initial toolbox were born in the early 90' in the laboratory of Professor Hashimoto at the Tokyo University.

The Department of Electro Technique (ET) at BME got involved soon in the research. At first the University of Tokyo made it possible for a Hungarian researcher to be present continuously at the project later on even the iSpace project management was carried out partly by a researcher from ET BME. At the same time in close cooperation with SZTAKI researchers at the Tokyo Institute of Technology and later at Gifu County Research Center gained experience as intelligent system laboratory leaders and carried out experiments in a complete, 6 sided 3D virtual room. This room is called CAVE (Cave Automatic Virtual Environments). The results of cooperation were promising so SZTAKI initiated and with the support of the Department of Electro Technique organized the IISL (Integrated Intelligent System Laboratory, Hungarian, Japanese, more: http://www.iisl-lab.org/) formation.

In the framework of IISL a virtual laboratory was designed and created by the Hashimoto laboratory at University of Tokyo. A simplified 3D model of it from the late 90's can be seen in Fig. 3, where a virtual robot could be controlled.

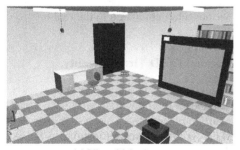

(a) Virtual laboratory   (b) Real laboratory

Fig. 3. Virtual and real laboratory

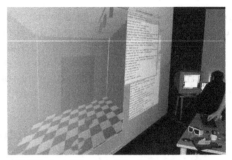

(a) 3D projection screen   (b) Projected image of the Tokyo laboratory

Fig. 4. ELTE 3D Center for Visualization

Initially, this 3D model was projected onto the 2D screen where it could be observed [3]. Later, when the Research Center for Visualization from ELTE University has also joined the project, the 3D model could be seen using special stereo glasses (see Fig. 4) [12, 13]. With the help of Japanese and Norwegian support the BME Mogi-ELTE-MTA SZTAKI cooperation has created a number of demonstrations, which can be considered as 3D Internet-based cognitive info-communication application.

4   Internet-Based Robot Control with the Help of 3D Virtual Space

The demonstration objective was to show the nearly one decade long joint research results, the Hungarian and Japanese hardware and software tools to show how to integrate the internet and the stereo 3D visualization service. The operator residing in Budapest (Hungary) controlled the virtual robot with a virtual joystick in the 3D virtual laboratory. The motion description of the robot was sent through the Internet to the Tokyo University iSpace, which was directed according to the instruction. The two robots were not in direct contact with each other, it was the iSpace that created the connection between the virtual 3D and the real model. One of the most challenging problems was the time-delay caused by the internet and transposed data packets sent through the internet. The average time-delay was measured continuously by a predicting algorithm to estimate the position of the virtual robot compared to the real one robot [13].

4.1   The Physically Distant Objects Cooperation in a 3D Virtual Reality

The Virtual Collaborating Arena (VIRCA) is making a virtual laboratory that gives a way to the team of researchers or industrial engineers to collaborate with each other control a physical device remotely in easy, reality environment. This system was developed in Cognitive Informatics Group of MTA SZTAKI research institute. VIRCA is a software backbone for virtual robot collaboration, cognitive infocommunication and telemanipulation. The basic concept of VIRCA is that it is composed of distributed components. Such components can be registered into a global naming service, which in turn allows the user to connect the component the way it is necessary for his or her purpose.

In the virtual laboratory every robot, device and the operator has its own 3D representation. This means that the operator can see the lab in 3D and feel that he/she is actually inside the virtual world. This is important, because it makes task solving much more user-friendly and he/she could enter into the spirit of a real lab. Also, with cognitive devices – like a vibrational glove or speech synthesis – it is possible to give not only visual feedback of the task execution but to provide various ways of two-way interaction. Robots and devices can communicate with each other directly or through the virtual world. Direct connection can occur when no command reinterpretation is needed and the two devices have the same communication interface. For example a simple joystick can send the handle's position information and the robot can move towards the designated direction. When the way of communication is more complex, the intermediation of the virtual world is indispensable. For example, if a robot has to know the position of a given object, it must query for it from the 3D world. The basic concept of the VIRCA is exemplified on Fig. 5.

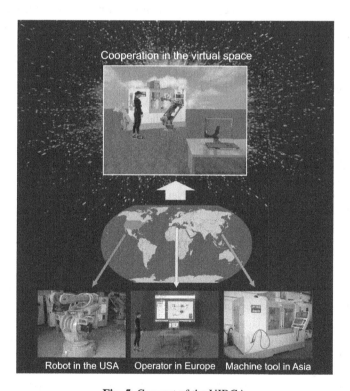

Fig. 5. Concept of the VIRCA

The demonstration shows how the physically distant devices can virtually cooperate in a common virtual space, for example to move a virtual ball (http://dfs.iis.u-tokyo.ac.jp/~barna/VIRCA/). In the impressive part of demonstration, a mobile robot and a robotic arm are shown, which are actually 1 km away from each other. Both robots are animated in a common 3D virtual reality room, where a virtual ball was situated as well. First, the mobile robot is moved so that its animated image in the virtual space brings the ball close to the animated image of the virtual robot, then the robotic arm reaches out in a way that it can reach the animated image of the virtual ball. The animated robotic arm held the virtual ball and put it a little bit further and in the same time the real robot was also moving grabbing the air.

To thoroughly investigate, plan and develop additional features for the aspects of 3D Internet network a larger 3D virtual network is needed. Therefore, Mogi from BME and MTA SZTAKI has established a consortium to build a large 3DICC (3D-based Internet Communication & Control) laboratory based on the professional expertise gained within the IISL framework and on large founding of applications. The first Hungarian 3D CAVE (Cave Automatic Virtual Environments) will be built in the laboratory as well. Combining CAVE with projectors of ELTE's 3D Visualization Center a unique system will be developed in Europe, containing two large 3D virtual spaces which are linked through high-speed network and related cognitive information communication channels. The evolving system will be able to connect to the international virtual 3D and communication network. The physical proximity of the two 3D systems makes it possible to investigate and test the aspects of the 3D internet network.

VIRCA is the base for the future iSpace Laboratory Network, which can be used to connect 3D internet research laboratories and so can open a new dimension in the remote laboratories (either research or for industrial purposes) cooperation. In advance nearly twenty research groups have indicated their intention to join the network from Europe, Asia and America. The majority of them are institutes or universities, who are respective leaders in their own countries. Here we highlight the Japanese research teams founded by the Japanese government, which are developing communication standards for the new generation robots. These standards will be shared with the members through the iSpace Laboratory Network in the development stage, thus the iSpace Laboratory Network plays an important role in the development and dissemination of these standards worldwide.

The final version of iSpace Laboratory Network will have real robots and their animated 3D virtual models, as well as tracking systems. Suppose that a small enterprise in Norway (which is a member of the iSpace Laboratory Network) wants to solve a task with robotic manufacturing cell, they intend to use a Hungarian specialist (who also has access to the iSpace Laboratory Network) – with lower hourly wage – to program the robot. The Hungarian expert puts on a motion tracking data suit and enters into the iSpace Laboratory Network (see Fig. 6). There, he decides the suitable process to be carried out and selects the robots, machining tools through animated icons. Connecting the icons graphically and choosing the necessary data traffic, he can create an actual link. Then the remote expert performs the processes virtually. Finally, based on the already completed virtual task, the real program is automatically generated and used as an input for the real robot to perform the desired task in the real world. This is just one industrial example, but presumably iSpace can be useful for computer games and also for establishing human contact. There is a lot of potential in a 3D laboratory network.

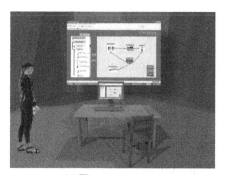

(a) The remote expert

(b) The 3D virtual reality seen by the expert

Fig. 6. Concept of iSpace Laboratory Network

4.1.1 Experiment

In the following the cooperation of physically distant objects will be demonstrated. An NXT Lego Robot and a KUKA KR6 industrial robot was connected through the services of 3D Virtual laboratory. The cooperation consists of manipulating a ball in the virtual world but both robots perform the actual movement commands. The control can be monitored by a live web-camera connection. The user only gives high level commands, like: this robot place this ball to that position.

Initially, only a virtual ball is placed into the virtual room. The NXT mobile robot and the KUKA robot is connected afterwards. The operator selects the NXT and the virtual ball and clicks on the floor so that the NXT should take the ball to it's new position in the room. Then the operator selects the KUKA robot and the virtual ball, and clicks somewhere else on the floor so the KUKA grabs the ball, and takes it to its new selected position. A snapshot from the experiment can be seen in Fig. 7.

Fig. 7. Results for the collaboration experiment

During the test an interesting phenomenon occurred. The KUKA robot grabbed the virtual ball and waited for the position just like the NXT mobile robot. So after giving the final command to place the ball on the floor the NXT mobile robot moves just under the virtual ball and the KUKA robot drops the ball on top of it. Physics is enabled so the ball bounces off the NXT as expected.

5 Robot Communication and Programming Based on 3D Internet

This is actually the long-term objective of three large integrated projects, thus it may be suitable for giving a summary on the ongoing professional research. This is a large scale goal, which can only be achieved through broad international cooperation. The project specifically focuses on industrial applications in order to increase the competitiveness of small and medium-sized companies with introducing the use of new robotic programming paradigm.

Motivation is doubled because in case of small series production the frequent changing of the robot programs makes the conventional methods cost so high that it becomes uncompetitive [7]. Further problem is that programming a robot requires skills, which are often not present at the small and medium-sized enterprises. However, if the owner can communicate with the robotic processes without specific informatics knowledge and can "explain" the essence of the change, as he would do with a colleague and in case the programs are generated largely automatically than the transition period time, price, expert demand will decrease significantly. This project focuses on the preparation of such systems. If we are seeking 100% automation the safety preparations for all extreme cases will significantly increase the costs again. The solution is to involve the human intelligence at the supervisory level (brain in the loop), when processing the tasks. This again requires competence in the field of cognitive info-communication. If in this process the decisions made by human brain are supported by an advanced intelligent communication and control system and a process that is automated upto a very high level efficiency will increase significantly. Artificial systems are lack of situation awareness, global overview and intuition. That makes them unsuitable for handling complex, flexible autonomous processes. This weakness can eliminated if a human intelligence participate at the highest level of control and it is backed by a very efficient communication system. The common demonstration of BME Mogi-MTA SZTAKI NUC (NUC-University of Narvik, Norway) shows a work-piece, which has manufacturing defects on the surface [7]. These errors are often removed by polishing, grinding with manual labour individually. Performing this type of action involves high health risk (harmful substance inhalation, eye damage, etc.), thus the automatization of these processes by industrial robots can be required in many areas. However, detection the manufacturing defects automatically is rather complicated.

In the developed inspection system the operator can show it on either the real work-piece or on the virtual model which parts need polishing. Several demonstrations were made to show this process. There was a demonstration, when the operator was moving in a motion tracking suit and the robot received the command sent by the operator.

There was also a case when the robot received the commands based on visual information, but in neither case was the data coming from the operator sufficiently accurate by itself. Conversely, if the incorrect information coming from operator was compared with the work piece's CAD model, then the defects could be identified clearly, and then the grinding path could be generated for the robot (see Fig. 8) [9].

(a) work-piece

(b) assigning the path for the robot

(c) identifying the path for the robot

(d) 3D animation of the process

(e) carrying out the process in real conditions

Fig. 8. Generation of robot grinding path

6 Conclusions

The possible future of 3D internet and cognitive info-communication was presented. The highlighted currently running projects enforce our vision of these concepts. Usage of 3D internet in distant telemanipulation of mobile robots and advanced visualization of spaces (iSpace) was also demonstrated. Through an experiment, the usability of such a distributed virtual environment is highlighted.

Acknowledgments. The research was supported by HUNOROB project (HU0045, 0045/NA/2006-2/ÖP-9), a grant from Iceland, Liechtenstein and Norway through the EEA Financial Mechanism and the Hungarian National Development Agency and the National Science Research Fund (OTKA K62836).

References

[1] Baranyi, P., Solvang, B., Hashimoto, H., Korondi, P.: 3D Internet for Cognitive Info-Communication. In: Proc. of 10th International Symposium of Hungarian Researchers on Computational Intelligence and Informatics, pp. 229–243 (2009)

[2] Bowman, D.A., McMahan, R.P.: Virtual Reality: How much Immersion is Enough? Computer 40(7), 36–43 (2007)

[3] Brscic, D., Hashimoto, H.: Model-based Robot Localization Using Onboard and Distributed Laser Range Finders. In: Proc. of IROS, pp. 1154–1159 (2008)

[4] Jacko, J.A., Sears, A.: The Human-Computer Interaction Handbook: Fundamentals, Evolving Technologies and Emerging Applications, 2nd edn. CRC Press, Boca Raton (2008)

[5] Korondi, P., Hashimoto, H.: Intelligent Space, as an Integrated Intelligent System. In: Proc. of International Conference on Electrical Drives and Power Electronics, pp. 24–31 (2003) (keynote paper)

[6] Lee, J.H., Hashimoto, H.: Intelligent Space Concept and Contents. Advanced Robotics 16(3), 265–280 (2002)

[7] Solvang, B., Sziebig, G., Korondi, P.: Robot Manipulators. In: Robot Programming in Machining Operations, pp. 479–496. I-Tech Education and Publishing (2008)

[8] Sziebig, G.: Achieving Total Immersion: Technology Trends Behind Augmented Reality - a Survey. In: Proc. of Simulation, Modelling and Optimization, pp. 458–463 (2009)

[9] Sziebig, G., Zanaty, P.: Visual Programming of Robots. In: Proc. of the Automation and Applied Computer Science Workshop (AACS 2008), pp. 1–12 (2008)

[10] Tamás, P., et al.: 3D Measuring of the Human Body by Robots. In: Proc. of 5th International Conference Innovation and Modelling of Clothing Engineering Processes IMCEP 2007, pp. 109–115 (2007)

[11] Wenzel, K., Antal, A., Molnár, J., Tóth, B., Tamás, P.: New Optical Equipment in 3D Surface Measuring. Journal of Automation. Mobile Robotics & Intelligent Systems 3(4), 29–32 (2009)

[12] Zanaty, P., Brscic, D., Frei, Z.: 3D Visualization for Intelligent Space: Timedelay Compensation in a Remote-controlled Environment. In: Proc. of Conference on Human System Interactions, pp. 802–807 (2008)

[13] Zanaty, P., Sziebig, G., Korondi, P., Frei, Z.: 3D Virtual Model for Intelligent Space. In: Proc. of 9th International Symposium of Hungarian Researchers on Computational Intelligence and Informatics, pp. 151–162 (2008)

Global, Camera-Based Localization and Prediction of Future Positions in Mobile Robot Navigation

István Engedy and Gábor Horváth

Department of Measurement and Information Systems
Budapest University of Technology and Economics
Magyar tudósok körútja 2, H-1117 Budapest, Hungary
engedy@mit.bme.hu, horvath@mit.bme.hu

Abstract. Localization is a main problem in most mobile robot navigation systems. One of its functions besides determining the position of the robot is to sense the surrounding environment. There are different solutions for this problem with a wide scale of complexity from local collision avoidance to global positioning systems. In this paper we will present a camera- and image processing-based global indoor localization system for the localization of a mobile robot and all other objects in its surrounding environment. We will present how the system is able to locate different objects marked with simple marker shapes, and any other objects in the working area. The presented method is also able to keep track of moving objects, and predict their future positions. We will show the basic steps of this image processing algorithm, including how Fourier transformation is used to determine the main localization parameters.

1 Introduction

The field of mobile robotics is a developing industry nowadays. There are numerous applications of mobile robots. They are sortable based on their operation environment: there are land, aerial, water and underwater mobile robots. They can differ in the level of autonomy they have during operation. Some mobile robots need no human interaction or supervision even in unknown environments, while others are only able to follow a previously painted line or curve on the ground, and there are remote controlled robots as well, which always need human supervision. In the sequel, we narrow the scope of this paper to the discourse of an autonomous land mobile robot, which is able to navigate in the specified environment.

Mobile robot navigation systems have a number of well separable subtasks. Each has also numerous different solutions of different approaches. One of the most important functions of a navigation system is to sense the surrounding environment and to determine the location of the robot, which is the localization of the robot and the surroundings. If the entire working area should be known, we were talking about mapping. The other important task is motion planning. This can be divided into at least two separate subtasks, the path planning, that is planning a route from one point to

I.J. Rudas et al. (Eds.): Computational Intelligence in Engineering, SCI 313, pp. 61–73.
springerlink.com

another, and planning the movement itself, how to actually follow the path. The latter, in some real-world system, can be rather difficult due to the kinematic constraints of the robot [1].

1.1 Localization

Localization is determining the position of the robot and other items in a reference system. It is separable to three different problems: tracking of a known object, finding an unknown object, and kidnapping of the robot, which is to take it to an unknown place. To solve these problems, sensors that provide information of the physical reality are needed for the robot. The various sensors limit the type of the localization techniques.

There are sensors that operate only in a short range. Robots using such sensors can observe only the immediate environment. Collision avoidance can be solved more or less easily, but for the task of mapping, the robot must walk through the entire working area. Different sensors can be used in this task e.g. tactile sensors like sensor mustache, or bumper, or sensors such as laser, ultrasonic, or radar distance sensors. A camera mounted on the robot may also belong to this category; the robot has the ability to monitor the environment: find and identify objects and obstacles.

Long-range distance sensors typically require some other device, or the sensor is not mounted on the robot. The GPS system is such a sensor, which requires the signals of GPS satellites to calculate its own position. A robot can be easily localized using this sensor, but the surrounding environment cannot be observed. Another solution would be the application of camera or cameras mounted at fix positions, which monitor the entire workspace, so the robot can see all the obstacles in the workspace, thereby providing an accurate map for the robot. In this case localization would be carried out by using image processing algorithms. Disadvantages of these methods are that they need to be deployed in the specific environment. The satellite signal is detected only under the open sky, while the cameras cannot see through walls [3].

There are a number of localization methods, which use the above-mentioned sensors. They have in common the need of some recognition algorithm, which is based on the sensor signals, by which it is able to locate the robot and any objects in the workspace. In cases when only the immediate surroundings of the robot is detected by the sensors, the exact location of the robot can be determined by the surrounding environment and an a priori known map. Often there are ambiguous solutions (for example, there are several similar intersections in a maze), so it is necessary to process not only the currently measured, but the previously measured sensor data, so the more information may make the localization of the robot unambiguous. Such methods include the Monte Carlo localization, where the possible locations are represented by particles, or a Kalman filter solution, where recursive state estimation is used in simultaneous localization and mapping (SLAM) [4].

When the entire workspace is known, for example, through camera surveillance, there could be also cases when the localization of the robot or other object is ambiguous, such a case might be the robot soccer, where robots are very similar, and they cannot be distinguished from one another merely by an image. The solution for this problem may be to distinguish the robots from each other with markers.

In this paper we will present the localization part of a mobile robot navigation system. The navigation system is based on a regularized version of the backpropagation through time method, and its main purpose is to navigate a robot car among moving obstacles [2]. It is tested on a real car-like robot, and as a real-life experiment, all parts of the navigation system has to be functioning in a real-world system, including the localization system.

2 Localization System Architecture

As it was shown in the introduction, a number of solutions exist for the problem of localization. One of them is to observe the entire workspace by a camera, and use an image processing algorithm to obtain important information for the robot from camera image. Such a system is described and implemented in this paper, where only a single camera is watching vertically down from the ceiling to the workspace. The workspace, in which the robot can move, corresponds to the field of vision of the camera. This camera system is responsible for the localization of the robot, the target, and the mapping of obstacles.

Such a camera-based localization system has many advantages, including the fact that as the entire workspace is seen, not only the nearby environment of the robots will be known, but the whole workspace too, including the locations of all objects. This is particularly important in a rapidly changing environment, because the current state of the terrain is always taken into account during path planning, thus there is much less danger of collision with the obstacles.

The disadvantages of such a system are that it is unable to locate objects that cannot be seen directly by the camera; which renders this method unusable in most real-world applications. The problems of mounting of the camera also reduce the number of possible fields of application.

Despite these disadvantages, this method is able to perform remarkably well in some indoor applications such as robot soccer. Also it is useful for laboratory experiments, like testing new methods for other navigation subtasks, which require localization.

2.1 Preprocessing of Camera Images

The system is built up from a webcam and an image processing algorithm. A webcam is mounted on the ceiling, monitoring an area on the floor (Fig. 1). The camera provides colored QVGA resolution images to the computer at 15 frames per second. Those are acquired through the DirectShow multimedia framework in the implemented application. The images are processed in real time by the algorithm, which is responsible of locating all the objects in the working area, and recognizing some special objects, like the robot itself, or an object determining the target position. These special objects are marked with markers. The possible markers are given to the image processing algorithm, so it is able to recognize them. All other objects, which cannot be matched with any markers, are considered as barriers.

Image processing takes place in short as follows (Fig. 2). The image is noise-filtered after it has been captured. In the next step a previously recorded background image is subtracted from the image to determine the shapes of the objects laying on

Fig. 1. The situation of the camera and the working area

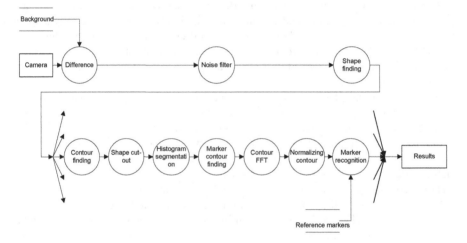

Fig. 2. The data flow of the image processing

the workspace. Then another noise filtering comes, which is followed by the contour-finding algorithm, and as a result, the contour of each shape (short of the shape of the object) is determined so all shapes are separated by cutting them out from the original image. This makes possible to process them independently. Another contour-finding step is made on the binary version of the cut-out images where binarization is based on the intensity values. The contour of this marker then processed further. The size, location and orientation of the marker are calculated based on the Fourier-transform of the contour. Then the contour is normalized using these parameters, so it can be compared to the pre-defined contours of the reference markers, to check if it is a pre-defined type of marker.

2.1.1 Differentiation from the Background Image and Noise Filtering
To find objects in the working area a simple subtraction is made to detect changes in the image. The actual camera image is compared to a pre-recorded background image, which was captured when there were no objects on the working area. Any deviation from this picture could mean an object, as the camera and the floor compared to each other's position do not change. The only errors could origin from the changes of the

lighting, which cannot be considered to be static. To over-come this problem, not the RGB values are subtracted from the actual image and the background image, but first the RGB values of each pixels are converted to HSL (hue, saturation and lightness) representation, and the lightness value is dis-carded. This way the effect of different lighting can be eliminated or at least reduced, as the saturation and hue of the objects are independent of the intensity of the lighting, especially when artificial lighting is applied, like in indoor environment.

It should be noted that due to the subtraction objects that have similar color as the background, or which are transparent, can only vaguely found by the system, if it can be found at all, so using such objects in the workspace should be avoided. For an object, seen by the camera the same as the background, there is no way for the image processing algorithm to detect, so the object is invisible to the system.

The signal of the camera is of course noisy. The noise would make the subtraction based object finding less accurate. For denoising, median filtering is used; the binarization of the image is done by applying a threshold value. If the sum of the differences between the actual camera image and the background in a 3 by 3-pixels window is larger than a predefined threshold value, the pixel in the middle of the window is considered as part of an object, so it will be a binary value 1 on the result image, in any other case it will be the value of 0.

Despite the median filtering and the threshold, there could still be some point defects on the image, mainly along the boundary of the objects because of shadows, or in places where the object has the same color as the background. Even different lighting conditions could cause such errors. Due to these problems additional filtering, smoothing is needed. The morphological operations, opening and closing are used to solve this problem in 3 by 3 windows (Fig. 3). Opening (erosion, dilation) clears all the point defects, and then closing (dilation, erosion) will patch the gaps at the edge of the object, and the point defects in objects. (Dilatation: value of the pixel will be 1 if there is at least one pixel of value 1 in a 3 by 3 neighborhood, erosion: the pixel value will be 0 if there is at least one pixel of value 0 in a 3 by 3 neighborhood.)

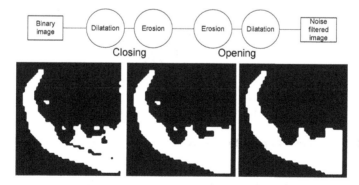

Fig. 3. Denoising the difference image, using closing and opening operations. Upper figure: the process of noise filtering; Lower figures: before closing and opening; after closing; after closing and opening.

2.1.2 Localization of the Shapes

After all the shapes of the objects are available on the image, the exact position, size, shape and orientation of these objects should be determined. To do this, we use a contour finder algorithm, which not only displays the binary image of the contour of the shapes, but also gives a list of the coordinates of the sequential contour points. Since there are several shapes on the picture, there will be multiple contours. These are returned in a list of the coordinate lists.

There are two parts of this algorithm: the shape finder, and the contour following algorithms. The task of the shape finder is to find each shape on the previously denoised binary image. To do this, the whole image is scanned row by row, and once a pixel of value 1 is found, the contour following algorithm starts, as it is sure that the point just found is part of a contour. After the contour following algorithm is finished, the shape is simply removed from the image, ensuring that each shape is processed exactly once.

2.1.3 The Contour Following Algorithm

The contour following algorithm receives as input the image itself, and one point of the contour. It starts from that point to follow the edge of the shape. It is known in which direction this contour point was found. This direction is important to walk clockwise around the shape. The point surely is found in the upper left "corner" of the shape and this will be the first tested point. The algorithm does the following at each test points: it tests all eight neighboring points, clockwise, starting from the last found contour point (Fig. 4). If it finds a pixel with the value of 1, then it goes to this pixel which will be a contour point too. The algorithm will stop once it reaches the first contour point again (Fig. 5).

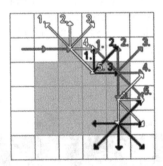

Fig. 4. The process of contour following

2.2 The Localization of Special Objects

The image processing algorithm is expected to recognize certain objects. Such an object would be the robot itself, another is the target object. The recognition could be carried out by recognizing merely the object, or it can be done with the help of some kind of marking on the object, and the recognition of this marker. The image processing system of our localization method uses the latter.

```
function CONTOUR_FOLLOW (image, column, row) returns contour
  DIRECTIONS = {{-1,-1}, { 0,-1}, { 1,-1}, { 1, 0},
              { 1, 1}, { 0, 1}, {-1, 1}, {-1, 0}}
  x = column
  y = row
  dirindex = 0
  dir = DIRECTIONS[dirindex]
  repeat
    contour.ADD(POINT(x, y))
    while image.value[x + dir.x, y + dir.y] == 0
      dirindex = (dirindex + 1) mod 8
      dir = DIRECTIONS[dirindex]
    end while
    x = x + dir.x
    y = y + dir.y
    dirindex = (dirindex + 5) mod 8
  until x == column and y == row
  return contour
end function
```

Fig. 5. The algorithm of contour following

Any object that the system must recognize, should have a rather simple shape on it, called marker. There are lots of localization solutions based on marker recognition. Some of them are active markers, they use additional hardware components, like lights that might be blinking at a known frequency, thus making it easier to recognize. There are other, passive markers too, which are different colored and shaped blobs that can be drawn on a single piece of paper. Some of these use the color information to recognize something on the camera image; others recognize the shapes for the same purpose.

In our case the marker is a dark shape on a light background, and sharp in outline, and it is found by the recognition of is shape. A further requirement of the markers is that they must have only one axis of symmetry, to ensure that their direction can be determined definitely. However, it is good if it really has one axis of symmetry, we will show the reason for this later. Such potential markers are visible in the figure below (Fig. 6).

However not all objects are needed to be marked with a marker. Those shapes, on which the system cannot find a marker, are automatically classified as obstacles.

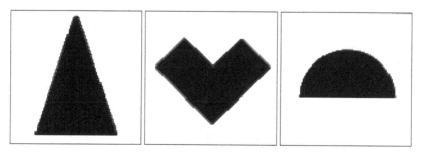

Fig. 6. Possible markers. All has exactly one axis of symmetry.

2.2.1 Extracting the Markers from the Shapes

The image processing steps so far has extracted the contours of each shape. In order to determine which marker can be seen on the shapes, the algorithm uses the original camera image to cut out the shapes using their contours. The process of cutting out is done with a use of a mask, which can be easily created using the contour of the examined shape.

The resulting cut-out shape is still in RGB space, so it has to be segmented to the marker and the background, this way getting a binary image. It has to be done so that on the resulting binary image the marker has a value of 1 and the background has a value of 0. This can be achieved easily if the image is segmented according to its intensity. To do this, first the algorithm calculates the image intensity histogram. Then using this histogram, the algorithm finds the threshold value, which separates the light and dark points from each other the best. The algorithm is iteratively trying to fit two Gaussian functions to the light and dark pixels on the histogram. The threshold will be the value which is half-way from the centers of the Gaussian distributions. Using this threshold, the image is easily segmented by the intensity of its pixels.

After segmentation of the cut-out image is done, it is necessary to perform the opening and closing operations again, as it was done at the initial processing of the camera image. If there is a marker of the cut-out image, then that must be the only shape visible on the resulting binary image. To recognize it, first the contour of the marker must be found. Based on the contour it is much easier to recognize the marker. The contour finding works just like as it was described above.

Fig. 7. The cut-out image, the segmentation of that image, and the contour finding of the shape

2.2.2 Determining the Type, Position, Size and Orientation of the Marker

The markers are not only needed to be recognized, but it is also necessary to determine its location on the image, its orientation and its size. Furthermore, this information makes easier the recognition of the marker. The recognition is to compare the marker to a set of known markers, and select the best match. However the marker shape from the camera image is scaled, rotated and translated on the image, not to mention the quality loss due to the noise on the camera image. To overcome this problem, the marker shape must be normalized, which means it has to be translated to the origo, rescaled, and rotated to face a predefined direction. This might look a hard task with a yet unknown marker shape. If the marker is not recognized yet, it cannot be told, which part of it is its top, and which is its bottom, thus it cannot be rotated to the proper direction.

There is a simple way of calculating all these parameters we need to normalize the marker shape, based on its contour, without recognizing it. The points of the contour are used as complex numbers ($c_k = x_k + jy_k$) where j is the imaginary unit. This way the contour can be Fourier-transformed, using the FFT algorithm. To have the same number of Fourier-coefficients for each marker, the contour must be resampled using fixed number of samples, in this case $N = 32$ samples.

The first Fourier-coefficient, F_o is the center of gravity of the shape, so it is the position of the object (1). The absolute value of the second Fourier coefficient, F_1 determines the size of the shape, and its square is in relation with the area (2). Furthermore the orientation of the shape (3) and the phase of the sense of rotation (4) can be determined from the phase of the second and last Fourier coefficient. The shape can be rotated by multiplying the coefficients with $e^{j \cdot \varphi_1}$ (5), and the phase of the sense of rotation can be shifted by multiplying the coefficients with $e^{i \cdot j \cdot \varphi_2}$ (6), where j is the imaginary unit.

$$F_0 = \frac{1}{N} \sum_{k=0}^{N-1} c_k ; F_0^{'} = F_0 + \Delta \tag{1}$$

$$\forall i > 0 \longrightarrow F_i^{'} = F_i \cdot \lambda \tag{2}$$

$$\varphi_1 = \arg(F_1) + \arg(F_{N-1}) \tag{3}$$

$$\varphi_2 = \arg(F_1) - \arg(F_{N-1}) \tag{4}$$

$$\forall i \longrightarrow F_i^{'} = F_i \cdot e^{j \cdot \varphi_1} \tag{5}$$

$$\forall i \longrightarrow F_i^{'} = F_i \cdot e^{i \cdot j \cdot \varphi_2} \tag{6}$$

2.2.3 Normalizing the Contours of the Markers

The contour can be translated, rotated and scaled by changing these values. By knowing these parameters, the contour can be normalized, that is setting these values to predefined constants. The center of gravity of the contour is translated to the origo, by changing only one the first Fourier coefficient to zero. The area of the shape can be normalized easily both in the contour point representation and in the Fourier-spectrum too. In the Fourier-spectrum, all the coefficients must be divided by the absolute value of the first coefficient to normalize the area of the shape. The shape can be rotated so that the longest diameter will be parallel to the imaginary unit vector. To do this, the orientation must be rotated by the calculated $-\varphi_1$, and the phase of the sense of rotation must be shifted by $-\varphi_2$. The shape can be normalized to identical contours applying inverse FFT to the altered coefficients, whatever direction was it previously aligned, whatever position it was translated on the image, and whatever size it was scaled (Fig. 8).

There is however, a defect in this method: there is a 180-degree rotation uncertainty, so it cannot be determined if the normalized contour is oriented properly, or it is rotated by 180 degrees. This ambiguity is resolved during the marker recognition,

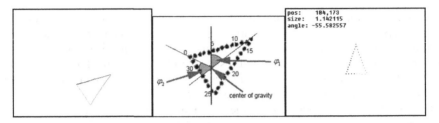

Fig. 8. The contour of the marker, its center of gravity, its orientation, its phase of sense of rotation (from dark to light), and the resampled and normalized contour

by comparing the normalized contour with the reference markers, and also with their 180-degree rotated counterparts. The phase of the sense of rotation should be changed with integer multiples of $2\pi/N$, otherwise the shape of the marker is distorted.

The similar contours are transformed to numerically similar contours using the normalizing method above; the marker recognition can be carried out pretty easily. The unknown normalized contour is compared with all the normalized contours of the reference markers, and the most similar marker is selected as the recognized marker. If the difference of the unknown marker and the most similar reference marker excess a specified threshold value, the contour is considered as an unrecognized shape.

The contours are compared by the quadratic distance of their points. This is possible because the contours have the same number of sample points as they were resampled during the normalization, and the phase of sense of rotation (φ_2 in Fig. 8) is also identical. The contours can be compared using this method with sufficient accuracy even in noisy cases; moreover, it takes about an order of magnitude less computing time, as if the bitmap of the markers would have been compared. It is also better than a comparison method which would calculate numerous features from the images, because this method not only recognizes the markers, but it also determines their location, orientation and size, which are also needed during localization.

3 Obstacle Position Prediction

Until this point the algorithm was processing each camera frame separately. However there are situations, where not just the present situation of the workspace and the location of the robot need to be known, but also previous states, or even future estimations as well.

Such a situation could be the avoidance of fast moving obstacles, which would be hard or even impossible to avoid if the navigation system only resting on the present state of the world. Luckily, our localization system provides all relevant information of the workspace globally, not just at the surroundings of the robot, this way the fast moving obstacles can be observed long before it gets near the robot, thus allowing the robot to change its course in time.

To make the navigation method able to avoid such obstacles, the prospective positions of these must be known. This is only possible by prediction. There are many ways to predict the future positions of the obstacles, but all of them needs that we know some of the previous samples of the position-time function of the obstacle. This problem can be taken as an extrapolation problem, so most of these methods are known extrapolation methods. The easiest way is to use the sample points to fit a polynomial onto them. This way the future positions could be extrapolated.

3.1 Artificial Neural Network-Based Prediction

There are also soft computing methods to predict a time series. To predict the prospective positions of the obstacles, we use dynamic artificial neural networks (ANN), which are trained with training samples generated from the previous positions of the obstacles. One training sample consists of a target point, which is one point of the time series, and some input points, which are the previous points in the time series, in the proper order. The training of the ANN is done the usual way, using the back-propagation algorithm.

Time series prediction is considered as a modeling task when a dynamic neural network is used to model the system generating the time series, in this case, the obstacles and all their physical constraints. The complexity of the system determines how many previous sample points should be used in the input of the ANN. Also the sampling rate is determined by the system too. In this case the sampling rate was predetermined by the hardware elements, the sampling rate of the webcam. The complexity of the movement of an obstacle is a harder question to answer. If we narrow the scope of obstacles only to passive objects, which does not have any logic or intelligence and move only by their inertia, the complexity of their movement would be low. If we assume that the obstacles could have some inner intelligence, the prediction could be even impossible, because the obstacle might decide to alter its route. Therefore we did not want to predict all possible obstacles. We have elaborated a method that is able to predict the movement of objects similar in its moving pattern as the robot. This makes possible to use this method in collaboration environments, like robot soccer.

3.1.1 Obstacle Identification

There might be more obstacles. This rises an important question. How do we tell, which obstacle is which in two consequent camera image? The answer is quite simple: as we are predicting the positions of the obstacles in the future, we have a prediction, where the obstacle might be on the next camera image. If there is one in a range where we expect to be, that is regarded as the same obstacle.

If there is nothing in the range of our expectation, the predicted position is regarded as the newest location of the obstacle, because this must be a missing data due to some errors in the image processing algorithm. However this missing data substitution doesn't work forever, so after a couple of consecutive misses, the obstacle is regarded as obsolete, and it is not maintained any further.

If there are more obstacles in the range, like in situations where two obstacles crosses each others' ways, we take the nearest one to our expectation. This doesn't cause any problems later, because the momentum of the obstacle is stored inside the neural network, so the next prediction of these multiple obstacles will be different, as they move in different directions.

3.1.2 Training of the Neural Networks

Because there might be more obstacles, one ANN is allocated to each obstacle. The inputs are some of the previous positions in order, and the target is the following position in the path of the obstacle. This can be seen on Fig. 9 in three steps, lines marked with different styles.

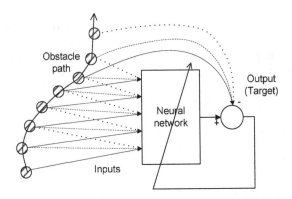

Fig. 9. The obstacle position predicting ANN

At the beginning of the navigation, there are insufficient samples, so more samples are needed to be collected, before the robot can start moving. Another problem is that when there are just enough samples, they make too few training point, to train an ANN. In such situation it is better to use fewer samples in the input of the ANN, and make more training points, and as more samples are collected, the length of the input of the ANN could be increased.

The predicted obstacle positions can be used as an input to predict more positions, so the prospective path of the obstacle can be predicted. This predicted path can be used in the other parts of the navigation system, especially in the path planning. This method makes it possible to avoid collision with obstacles in the future that are not close to the path of the robot at the beginning of the navigation, but they are moving towards the robot.

4   Results and Conclusion

The localization system was built in a real navigation system, so it has been tested in real experiments. The results have shown that the segmentation of the camera image using a background image works perfectly in practice, which means that all the shapes and only the shapes of objects in working area are determined. Low-light conditions degrade the image signal-to-noise ratio; this also causes some segmentation errors, which can be compensated by the following noise filtering step. Problem only occurs when an object has very similar color as the background, in such case it is not detected by the system.

The contour finder algorithm and the contour normalization using Fourier-transformation also work well. Every error of the contour finding can be originated from the error of the input of the contour finding, so the algorithm itself is performing well, only the erroneous input could lead to wrong results.

The segmentation of the cut-out image based on the intensity histogram gives good results if the brightness of the cut-out images is adequate. It is important that the light and dark parts are well separated from each other. Unfortunately in most cases poor lighting conditions are the origins of the errors during localization.

The contour recognition step is the origin of the largest uncertainty during the image processing. The detection and recognition of a marker is correct, there are very few misrecognized markers. The recognition of obstacles however, which does not have markers on them, is worse. There are a lot of false positive marker recognitions. To overcome this problem, the markers are filtered by their size parameter. In most false positive cases, the markers are much smaller or larger than the markers used on the special objects, like the robot or the target object, so these can be considered as obstacles too, despite there is a recognized "marker" on them. This improves the performance a lot.

After the obstacle position prediction part was installed, the quality of the obstacle avoidance improved greatly. Its only drawback is its quite big need of computational power. The prediction was tested with straight and circular movements. In both tests the system performed well, after 3-4 training samples were collected at the beginning.

In this paper we have presented a working localization system. Its weakness is that it can be applied only when we have a fixed position camera, but it can be used well in indoor applications, like in real experiments of the testing of other navigation system parts, where the localization is considered a solved problem, or in applications like robot soccer. This localization method has also some problems as mentioned above, but with further development, these problems can be resolved. The most important improvement is to use better classification method for the marker recognition, like artificial neural networks, or support vector machines.

Acknowledgments. The authors gratefully acknowledge the support of the Hungarian Fund for Scientific Research (OTKA), Grant #73496.

References

[1] Latombe, J.-C.: Robot Motion Planning. Kluwer, Boston (1991)

[2] Engedy, I., Horváth, G.: Artificial Neural Network based Mobile Robot Navigation. In: Proc. of the IEEE International Symposium on Intelligent Signal Processing, Budapest, Hungary, pp. 241–246 (August 2009)

[3] Russel, S., Norwig, P.: Artificial Intelligence a Modern Approach. Prentice Hall, Englewood Cliffs (1995), ISBN:0-13-080302-2

[4] Thrun, S., Burgard, W., Fox, D.: Probabilistic Robotics. MIT Press, Cambridge (2005) ISBN:0-262-20162-3

[5] Engedy, I., Horváth, G.: A Global, Camera-based Mobile Robot Localization. In: Proc. of the 10th International Symposium of Hungarian Researchers on Computational Intelligence and Informatics, Budapest, Hungary, November 12-14, pp. 217–228 (2009)

New Principles and Adequate Robust Control Methods for Artificial Pancreas

Levente Kovács[1], Zoltán Benyó[1], Adalbert Kovács[2], and Balázs Benyó[1]

[1] Dept. of Control Engineering and Information Technology, Budapest University of Technology and Economics, H-1117 Budapest, Magyar tudósok krt. 2, Hungary
`lkovacs@iit.bme.hu`
[2] Dept. of Mathematics, University "Politehnica" of Timisoara,
RO-300006 Timişoara Romania, P-ta Victoriei Nr. 2
`profdrkovacs@yahoo.com`

Abstract. The current work is a short review of the research results obtained by the Biomedical Engineering laboratory of the Control Engineering and Information Technology Department, Budapest University of Technology and Economics (BME) in the field of automatic control of Type I diabetes mellitus, and mainly is focused on the PhD dissertation written by the first author [1]. The topic focuses on modeling formalisms and implementation of robust control algorithms for optimal insulin dosage in case of Type I diabetes patients. Regarding modelling concepts, an extension of the modified Bergman minimal model, the analytical investigation of the high complexity Sorensen-model and model synthesis of a novel molecular-based model are described. From robust control methods implementation firstly, the minimax method is applied and is described that its limitations can be spanned using Gröbner basis. Secondly, the graphical interpretation of the H_{inf} method under *Mathematica* program is extended with disturbance rejection criteria. Thirdly, an LPV (Linear Parameter Varying) type robust control method is described for the high complexity Sorensen-model; hence it is possible to deal directly with the non-linear model itself. Finally, actual research tasks are summarized related to the previously mentioned topics and further research directions are formulated.

Keywords: diabetes mellitus, glucose-insulin control, minimal-model, Sorensen-model, modern robust control, LPV control, *Mathematica*.

1 Introduction

Diabetes mellitus is one of the most serious diseases which need to be artificially regulated. The statistics of the World Health Organization (WHO) predate an increase of adult diabetes population from 4% (in 2000, meaning 171 million people) to 5,4% (366 million worldwide) by the year 2030 [2]. This warns that diabetes could be the "disease of the future", especially in the developing countries (due to the stress and the unhealthy lifestyle).

I.J. Rudas et al. (Eds.): Computational Intelligence in Engineering, SCI 313, pp. 75–86.
springerlink.com   © Springer-Verlag Berlin Heidelberg 2010

Diabetes mellitus appears if for some reason the human body is unable to control the normal glucose-insulin interaction (e.g. the glucose concentration level is constantly out of the 70-110 ml/dL range). The consequences of diabetes are mostly long-term; among others, diabetes increases the risk of cardiovascular diseases, neuropathy and retinopathy [3]. This metabolic disorder was lethal until 1921 when Frederick G. Banting and Charles B. Best discovered the insulin. Today the life quality of diabetic patients can be enhanced though the disease is still lifelong.

In many biomedical systems, external controller provides the necessary input, because the human body could not ensure it. The outer control might be partially or fully automated. The self-regulation has several strict requirements, but once it has been designed it permits not only to facilitate the patient's life suffering from the disease, but also to optimize (if necessary) the amount of the used dosage.

From engineering point of view, the treatment of diabetes mellitus can be also represented by an outer control loop, to replace the partially or totally deficient blood-glucose-control system of the human body. However, the blood-glucose control is a difficult problem to be solved. One of the main reasons is that patients are extremely diverse in their dynamics and in addition their characteristics are time varying. Due to the inexistence of an outer control loop, replacing the partially or totally deficient blood-glucose-control system of the human body, patients are regulating their glucose level manually. Based on the measured glucose levels (obtained from extracted blood samples), they decide on their own what is the necessary insulin dosage to be injected. Although this process is supervised by doctors, mishandled situations often appear. Hyper- (deviation over the basal glucose level) and hypoglycemia (deviation under the basal glucose level) are both dangerous cases, but on short term the latter is more dangerous, leading for example to coma.

Starting from the 1960s lot of researchers have investigated the problem of the glucose-insulin interaction and control. The closed-loop glucose regulation as it was several times formulated [4, 5, 6], requires three components: glucose sensor, insulin pump, and a control algorithm, which based on the glucose measurements, is able to determine the necessary insulin dosage.

Our research work focused on the last component and analyzed robust control aspects of optimal insulin dosage for Type I diabetes patients.

To design an appropriate control, an adequate model is necessary. The mathematical model of a biological system, developed to investigate the physiological process underling a recorded response, always requires a trade off between the mathematical and the physiological guided choices. In the last decades several models appeared for Type I diabetes patients [7].

In the last decades several models appeared for Type I diabetes patients. The mostly used and also the simplest one proved to be the minimal model of Bergman [8] and its extension, the three-state minimal model [9]. However, the simplicity of the model proved to be its disadvantage too, while in its formulation a lot of components of the glucose-insulin interaction were neglected. Therefore, the model is valid only for Type I diabetes patients under intensive care. The dynamic characteristics of the model are created by artificially dosed glucose input. As a result, the model can simulate only a 3 hours period. Furthermore, it was demonstrated, that the model control possibilities are limited, while it is very sensitive to its parameters variance.

Henceforward, extensions of this minimal model have been proposed [10, 11, 12, 13], trying to capture the changes in patients' dynamics, particularly with respect to insulin sensitivity, Also with respect to the meal composition, minimal model extensions were created [14, 15]. Probably the most complex and general Type I diabetic model derived from the minimal-model is that presented in [16].

Beside the Bergman-model other more general, but more complicated models appeared in the literature [17, 18, 19]. The most complex one proved to be the 19th order Sorensen-model [18]. Even if the model describes in the most exact way the human blood glucose dynamics, its complexity made it to be rarely used in research problems. Nowadays, it is again more often investigated (due to its general validity).

Regarding the applied control strategies, the palette is very wide [7, 20]. Starting from classical control strategies (ex. PID control [21]) to soft-computing techniques (ex. neuro-fuzzy methods [22]), adaptive [13, 23, 24], model predictive [4, 25], or even robust H_{inf} control were already applied [5, 6]. However, due to the excessive sensitivity of the model parameters (the control methods were applied mostly on the Bergman minimal model), the designed controllers are true only for one (or in best way for few) patient(s).

As a result, investigations demonstrated [4, 6], that even if the best way to approach the problem is to consider the system model and the applied control technique together, if high level of performance is desired, a low complexity control (like PID) is not effective. Therefore, the literature has oriented in three directions: adaptive control, model predictive control (MPC) and modern robust control techniques.

The advantage of the first two control possibilities lies in the retuning possibility of the controller or predicting the next step even in its working conditions. However, their disadvantage appeared if the complexity of the diabetes model was grown. Robust control adjusted the disadvantages of the adaptive or MPC control techniques, but the designing steps are more difficult.

In this paper a review is given of the research results obtained by the Biomedical Engineering laboratory of the Control Engineering and Information Technology Department, Budapest University of Technology and Economics (BME) in the field of automatic control of Type I diabetes mellitus. The topic focuses on modeling formalisms and implementation of robust control algorithms for optimal insulin dosage in case of Type I diabetes patients and most of the results are described in the PhD dissertation of the first author of the current paper [1].

Regarding modelling concepts, an extension of the modified Bergman minimal model, the analytical investigation of the high complexity Sorensen-model and model synthesis of a novel molecular-based model are described.

From robust control methods implementation firstly, the minimax method is applied and is described that its limitations can be spanned using Gröbner basis. Secondly, the graphical interpretation of the H_{inf} method under *Mathematica* program is extended with disturbance rejection criteria. Next, an LPV (Linear Parameter Varying) type robust control method is described for the high complexity Sorensen-model; hence it is possible to deal directly with the non-linear model itself. Finally, actual research tasks are summarized related to the previously mentioned topics and further research directions are formulated.

2 New Modeling Concepts for Type I Diabetes

The proposed modeling formalisms cover the analytical investigation of the high complexity Sorensen-model [18] and the extension of the Bergman minimal model [8, 13]. In this way, the proposed approximations are indicating numerical algorithmization for complex optimal control strategies while cover a bigger diabetes population. Additionally, model synthesis results of a novel molecular-based model [19] are summarized.

2.1 Minimal-Model Extension

For the extension of the modified minimal model, an internal insulin control (IIC) device was proposed representing the own insulin control of the human body (Figure 1). Employing parameter estimation, inverse problem solution technique (SOSI – "single output single input") was developed using Chebysev shifted polynomials, and linear identification in time domain based on measured glucose and insulin concentration values was applied. The glucose and insulin input functions have been approximated and the model parameters of IIC were estimated. The IIC part has been identified via dynamical neural network using the proposed SOSI technique. The symbolic and numerical computations were carried out with *Mathematica*. [26].

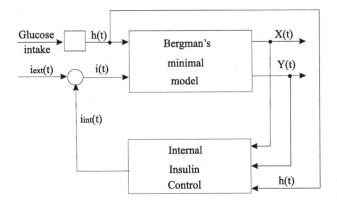

Fig. 1. The block diagram of the extended Bergman model

2.2 LPV Modeling of the Sorensen-Model

In case of the Sorensen-model [18], for an easier handling, inside the physiological boundaries, an LPV (Linear Parameter Varying) modeling formalism was proposed. In this way the model is possible to be reduced to a corresponding degree and consequently to ease the control possibilities and the applicability of the Sorensen-model [27].

Due to the high complexity of the Sorensen-model, it was hard to investigate the global stability of the system (the Lyapunov function is a real function with 19 variables). Therefore, a solution could be to cover the working region with a set of linear systems and in this way to investigate the local stability of them.

Choosing the polytopic points we have restricted to the physiological meanings of the variables. The first point was the normoglycaemic point (glucose concentration 81.1 mg/dL, insulin concentration 26.65 mU/L), while the others were deflections from this point. Firstly, the polytopic region was constructed by 36 points [27]. However, it was shown that the LPV system is stepping out from the physiologically defined region being unable to handle the uncovered region (Figure 2a). Therefore we extend the glucose concentration area of the considered polytope considering other grid points too, while the insulin concentration grid remained the same. With the extended polytope (48 points) it was demonstrated that the LPV model fits well the original one (Figure 2b) [28].

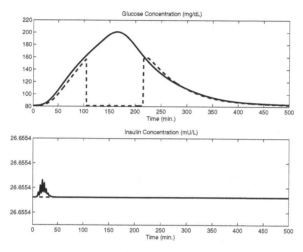

Fig. 2a. The simulation of the nonlinear Sorensen model (continuous) and the 36 points polytope region (dashed)

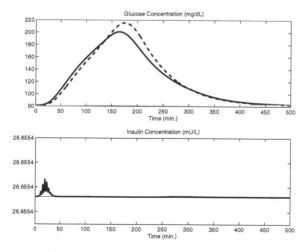

Fig. 2b. The simulation of the nonlinear Sorensen model (solid) and the 48 points polytopic region (dashed)

2.3 Analyzing a Novel Model of the Human Blood Glucose System

A newly appeared biochemical model [19] was also examined to describe the human blood glucose system. As a result of the biochemical point of view the cause-effect relations are more plausible and the processes can be described in a more exact and precise way. Therefore, our investigations went over this model synthesis too [29]. Global control properties were determined by nonlinear analysis (Lie-algebra) followed by steady state linearization. Corner points were defined, but this approach could not ensure proper approximation of the model, hence physiological working points were defined for further LPV (polytopic) modeling. In order to reduce complexity model reduction possibilities were observed with physiological concerns as well as with mathematical ones and the results agreed. Physiological, biochemical and mathematical approaches were applied and conclusions were made by synchronizing the principles of the different fields of study [29].

3   Robust Control Methods for Optimal Insulin Dosage in Case of Type I Diabetic Patients

The proposed robust control methods for insulin dosage were structured on the two considered models: Begman-model [9] and Sorensen-model [18].

3.1 Symbolic-Numerical Implementation of the Minimax Method on the Minimal Bergman-Model

The main idea of the minimax control is that it treats separately the reference input from the unwanted input (disturbance). As a result, the aim of the method is to maximize the cost regarding the disturbance, while the control input is calculated to minimize the maximum cost achievable by the disturbance (by the worst case disturbance). This is a case of the so called "worst-case" design.

Firstly, the minimax method was symbolically implemented under *Mathematica*. It was didactically shown how MATLAB selects its own solution (from the two resulting solutions), and a general formula was determined for the worst-case result in case of the minimal model of Bergman [30].

Secondly, regarding the minimal model it was shown, that the applicability of the minimax method has practical limitations: the optimal control is reached by some kind of negative injection of the glucose in the body, which is physically / physiologically not possible. Therefore, for the modified minimal model of Bergman a solution was proposed, using Gröbner-bases, which spans these limitations and positive definite solution can be found iteratively.

In this way, even if the worst case solution cannot be achieved, it is possible to obtain a better solution than the classical LQ one (Figure 3) [31, 32].

Furthermore, using the μ-synthesis method, parameter uncertainty was taken into account, which supplements the H_{inf} method in guaranteeing the robust performance requirements [33].

Moreover, with suitable parameterization, a quasi-Affine Linear Parameter Varying system-set have been defined and exploiting this result a (nonlinear) controller was designed ensuring quadratic stability [34].

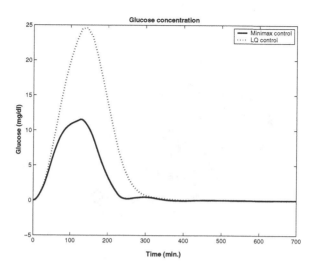

Fig. 3. Blood glucose concentration for classical LQ and "modified" minimax control method

3.2 Extending the H_{inf} Method under Mathematica

The aim of the H_{inf} modern robust control method is to guarantee not only nominal performances (tracking and disturbance rejection), but also robust performances: neglected uncertainties can be taken into account and included in the controller design process. The well-known designing procedure is that implemented under MATLAB where the uncertainties are defined explicitly [35].

A less-known interpretation is the graphical interpretation of the H_{inf} method implemented under *Mathematica* where the robust controller is designed based on a requirement envelope [36].

However, the disturbance rejection criteria formulated for the requirement envelope proved not to be effective enough. As a result, for the disturbance rejection criteria the authors formulated and extended the requirement envelope's criterion-set with an additional criterion. The correctness of this added criterion was demonstrated on the minimal model of Bergman, and it was compared with literature results [32, 37]. It can be seen (Figure 4) that the system remains all the time inside the requirement envelope.

Using the same scenario as [4] it was demonstrated that the obtained controller is robust. [4] showed that setting the p_1 parameter of the Bergman-model zero the system is unable to regulate the glucose level on its own. Figure 5 presents that our case solved this problem, the compensator is able to control the system even in this case [37]. Moreover, it was presented that the constant used in the proposed additional disturbance rejection criteria can be connected with the sensor noise weighting function used under MATLAB [1, 38].

3.3 LPV-Based Robust Control of the Sorensen-Model

Linear Parameter Varying (LPV) systems provide a model paradigm that goes beyond the classical representation of nonlinear and linear systems. Systems with different

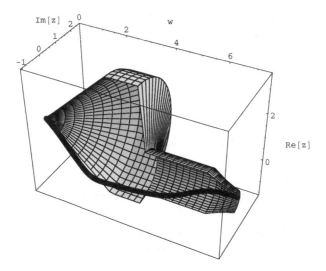

Fig. 4. The numerical values of the optimal transfer function inside the considered constraining envelope

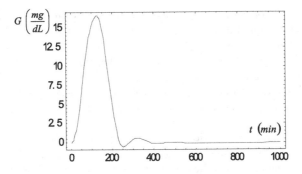

Fig. 5. Controlled dynamics of blood glucose concentration in case of $p_1=0$ of the minimal Bergman-model

parameter variations as non-stationary, nonlinear behavior, dependence on external variables or fast movements between different operating regimes can be handled by LPV framework. Basically, LPV systems can be seen as an extension of linear time-invariant (LTI) systems, where the relations are considered to be linear, but the model parameters are assumed to be functions of a time-varying signal.

Consequently, LPV system can be seen as a representation of nonlinear or time-evolving systems, where the parameter could be an arbitrary time varying, piecewise-continuous and vector valued function denoted by $\rho(t)$, defined on a compact set P. As a result, it can be seen that in the LPV model by choosing parameter variables, the system's nonlinearity can be hidden. In order to evaluate the system, the parameter trajectory is requested to be known either by measurement or by computation.

Regarding the Sorensen-model, using the normoglycaemic insulin input, the high complexity Sorensen-model was parameterized and described with polytopic LTI (Linear Time Invariant) systems (see Section 2.2).

The control of the Sorensen-model in the literature was only realized by linear based methods (H_{inf} or MPC). With the constructed LPV model now it was possible to guarantee the stability of the system in a more generable way. Moreover, it was a challenge for us to create a controller which avoids hypoglycemia as the literature results were not able to do that.

Consequently, with the created LPV model a corresponding controller using induced L_2-norm minimization was designed guaranteeing γ performance level [39].

Furthermore, two additional types of multiplicative uncertainties were additionally included for the system: output (neglected until yet in the literature) and input one. In addition sensor noise was considered, a 10% error for both glucose and insulin measurements (values taken from clinical experience). Finally, for meal disturbances we focused on the worst case carbohydrate intake situation (also not treated until yet in the literature).

The aim of the control design was to minimize the meal disturbance level over the performance output for all possible variation of the parameter within the considered LPV polytope. The obtained results showed that glucose level can be kept inside the normal range, and hypoglycemic episodes can be avoided (Figure 6). Although the simulation on the LPV model presents a bigger hyperglycaemia peak as the nonlinear model one, this is beneficial in our case meaning that the restrictions of the LPV control are much stronger than those of the original nonlinear system.

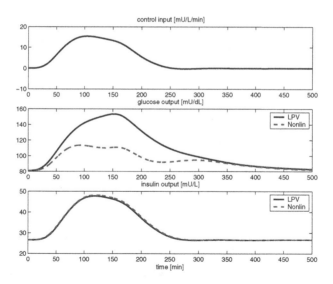

Fig. 6. The LPV-based robust controller (for the case of the considered additional uncertainties) with induced L_2-norm minimization guarantee in case of the original nonlinear Sorensen model (dashed) and the considered polytopic region (solid)

4 Actual and Future Research Tasks

The research results descibed above presents a short summary of the first author's PhD dissertation [1] defended in 2008. Actual research tasks are strictly related to the above presented results [40]:

- Based on the novel molecular-based model [19], robust control design is taken into account: the model was transformed to Type I diabetes, robust controller is designed and simulation on normal/ large meal absorption scenarios investigates the applicability of the applied control strategy on the Liu-model.
- Robust control applicability is tested on a healthy cohort of 10 persons for the Sorensen-model.
- A "model-receipt" is under development to give a useful help for those starting to work in this research field.
- Mixed meal model is under development different starch effects are separated.

Regarding future plans, robust control design is planned for ICU (Intensive Care Unit) cases; the applied robust control methodology is planned to be validated for the considered mathematical models on diabetic patient scenarios using different clinical data.

Model predictive control techniques will be mixed in the future with robust control methods and the controller will be refined to avoid special metabolic scenarios too (e.g. physical activity, nocturnal hypoglycemia, long term hyperglycemia).

Acknowledgments. This work was supported in part by Hungarian National Scientific Research Foundation, Grants No. OTKA T69055, 82066. The authors would like to thank Prof. Béla Paláncz for his great and valuable help in the symbolic programming help under *Mathematica*, for Balázs Kulcsár for his help in LPV modeling as well as to dr. Zsuzsanna Almássy for her valuable comments regarding diabetes.

Abbreviations

BME	Budapest University of Technology and Economics
IIC	Internal Insulin Control
ICU	Intensive Care Unit
LPV	Linear Parameter Varying
SOSI	Single Output Single Input
WHO	World Health Organization

References

[1] Kovács, L.: New Principles and Adequte Control Methods for Insulin Optimization in Case of Type I Diabetes Mellitus, PhD Thesis (in Hungarian) Budapest University of Technology and Economics, Hungary (2008)
[2] Wild, S., Roglic, G., Green, A., Sicree, R., King, H.: Global Prevalence of Diabetes - Estimates for the year 2000 and projections for 2030. Diabetes Care 27(5), 1047–1053 (2004)
[3] Fonyó, A., Ligeti, E.: Physiology (in Hungarian) Medicina, Budapest (2008)
[4] Hernjak, N., Doyle III, F.J.: Glucose Control Design Using Nonlinearity Assessment Techniques. AIChE Journal 51(2), 544–554 (2005)

[5] Parker, R.S., Doyle III, F.J., Ward, J.H., Peppas, N.A.: Robust H_∞ Glucose Control in Diabetes Using a Physiological Model. AIChE Journal 46(12), 2537–2549 (2000)

[6] Ruiz-Velazquez, E., Femat, R., Campos-Delgado, D.U.: Blood Glucose Control for Type I Diabetes Mellitus: A Robust Tracking H_∞ Problem. Elsevier Control Engineering Practice 12, 1179–1195 (2004)

[7] Chee, F., Tyrone, F.: Closed-Loop Control of Blood Glucose. LNCS, vol. 368. Springer, Berlin (2007)

[8] Bergman, B.N., Ider, Y.Z., Bowden, C.R., Cobelli, C.: Quantitive Estimation of Insulin Sensitivity. American Journal of Physiology 236, 667–677 (1979)

[9] Bergman, R.N., Philips, L.S., Cobelli, C.: Physiologic Evaluation of Factors Controlling Glucose Tolerance in Man. Journal of Clinical Investigation 68, 1456–1467 (1981)

[10] Fernandez, M., Acosta, D., Villasana, M., Streja, D.: Enhancing Parameter Precision and the Minimal Modeling Approach in Type I Diabetes. In: Proceedings of 26th IEEE EMBS Annual International Conference, San Francisco, USA, pp. 797–800 (2004)

[11] Morris, H.C., O'Reilly, B., Streja, D.: A New Biphasic Minimal Model. In: Proceedings of 26th IEEE EMBS Annual International Conference, San Francisco, USA, pp. 782–785 (2004)

[12] de Gaetano, A., Arino, O.: Some Considerations on the Mathematical Modeling of the Intra-Venous Glucose Tolerance Test. Journal of Mathematical Biology 40, 136–168 (2000)

[13] Juhász, C.: Medical Application of Adaptive Control, Supporting Insulin-Therapy in case of Diabetes Mellitus. PhD dissertation (in Hungarian), Budapest University of Technology and Economics, Hungary (1997)

[14] Roy, A., Parker, R.S.: Mixed Meal Modeling and Disturbance Rejection in Type I Diabetic Patients. In: Proceedings of the 28th IEEE EMBS Annual International Conference, New York City, USA, pp. 323–326 (2006)

[15] Roy, A., Parker, R.S.: Dynamic Modeling of Free Fatty Acid, Glucose, and Insulin: An Extended "Minimal Model". Diabetes Technology & Therapeutics 8, 617–626 (2006)

[16] Dalla Man, C., Rizza, R.A., Cobelli, C.: Meal Simulation Model of the Glucose-Insulin System. IEEE Transactions on Biomedical Engineering 54(10), 1740–1749 (2007)

[17] Hovorka, R., Shojaee-Moradie, F., Carroll, P.V., Chassin, L.J., Gowrie, I.J., Jackson, N.C., Tudor, R.S., Umpleby, A.M., Jones, R.H.: Partitioning Glucose Distribution/Transport, Disposal, and Endogenous Production during IVGTT. American Journal Physiology Endocrinology Metabolism 282, 992–1007 (2002)

[18] Sorensen, J.T.: A Physiologic Model of Glucose Metabolism is Man and Its use to Design and Assess Improved Insulin Therapies for Diabetes. PhD dissertation, Massachusetts Institute of Technology, USA (1985)

[19] Liu, W., Fusheng, T.: Modeling a Simplified Regulatory System of Blood Glucose at Molecular Levels. Journal of Theoretical Biology 252, 608–620 (2008)

[20] Parker, R.S., Doyle III, F.J., Peppas, N.A.: The Intravenous Route to Blood Glucose Control. A Review of Control Algorithms for Noninvasive Monitoring and Regulation in Type I Diabetic Patients. In: IEEE Engineering in Medicine and Biology, pp. 65–73 (2001)

[21] Chee, F., Fernando, T.L., Savkin, A.V., van Heeden, V.: Expert PID Control System for Blood Glucose Control in Critically Ill Patients. IEEE Transactions on Information Technology in Biomedicine 7(4), 419–425 (2003)

[22] Dazzi, D., Taddei, F., Gavarini, A., Uggeri, E., Negro, R., Pezzarossa, A.: The Control of Blood Glucose in the Critical Diabetic Patient: A Neuro-Fuzzy Method. Journal of Diabetes and Its Complications 15, 80–87 (2001)

[23] Palerm, C.C.: Drug Infusion Control: An Extended Direct Model Reference Adaptive Control Strategy. PhD thesis, Troy, New York (2003)

[24] Hovorka, R.: Management of Diabetes using Adaptive Control. International Journal of Adaptive Control and Signal Processing (2004)

[25] Hovorka, R., Canonico, V., Chassin, L.J., Haueter, U., Massi-Benedetti, M., Orsini Federici, M., Pieber, T.R., Schaller, H.C., Schaupp, L., Vering, T., Wilinska, M.E.: Nonlinear Model Predictive Control of Glucose Concentration in Subjects with Type 1 Diabetes. Physiological Measurement 25, 905–920 (2004)

[26] Kovács, L.: Extension of the Bergman Model – Possible Generalization of the Glucose-Insulin Interaction? Periodica Politechnica Electrical Engineering Budapest 50(1-2), 23–32 (2006)

[27] Kovács, L., Kulcsár, B.: LPV Modeling of Type 1 Diabetes Mellitus. In: Proceedings of the 8th International Symposium of Hungarian Researchers on Computational Intelligence and Informatics, Budapest, Hungary, pp. 163–173 (2007)

[28] Kovács, L., Kulcsár, B.: Robust and Optimal Blood-Glucose Control in Diabetes Using Linear Parameter Varying paradigms. In: Recent Advances in Biomedical Engineering, In-Tech. (in press, 2010)

[29] Kovács, L., György, A., Almássy, Z., Benyó, Z.: Analyzing a Novel Model of Human Blood Glucose System at Molecular Levels. In: Proceedings of the 10th European Control Conference, Budapest, Hungary, pp. 2494–2499 (2009)

[30] Paláncz, B., Kovács, L.: Application of Computer Algebra to Glucose-Insulin Control in H_2/H_∞ Space Using Mathematica. Periodica Politechnica Electrical Engineering Budapest 50(1-2), 33–45 (2006)

[31] Kovács, L., Paláncz, B.: Glucose-Insulin Control of Type1 Diabetic Patients in H_2/H_∞ Space via Computer Algebra. In: Anai, H., Horimoto, K., Kutsia, T. (eds.) AB 2007. LNCS, vol. 4545, pp. 95–109. Springer, Heidelberg (2007)

[32] Kovács, L., Paláncz, B., Borbély, E., Benyó, Z., Benyó, B.: Robust Control Techniques and its Graphical Representation in case of Type I Diabetes using Mathematica. In: Proceedings of the 8th International Symposium on Applied Machine Intelligence and Informatics, Herlany, Slovakia, pp. 71–75 (2010)

[33] Kovács, L., Kulcsár, B., Benyó, Z.: On The Use Of Robust Servo Control In Diabetes Under Intensive Care. In: Proceedings of the 3rd Romanian-Hungarian Joint Symposium on Applied Computational Intelligence, Timisoara, Romania, pp. 236–247 (2006)

[34] Kovács, L., Kulcsár, B., Bokor, J., Benyó, Z.: LPV Fault Detection of Glucose-Insulin System. In: Proceedings of the 14th Mediterranean Conference on Control and Automation, Ancona, Italy, Electronic Publication TLA2-4 (2006)

[35] Zhou, K.: Robust and Optimal Control. Prentice Hall, New Jersey (1996)

[36] Helton, J.W., Merino, O.: Classical Control Using H$_\infty$ Methods: Theory, Optimization and Design. SIAM, Philadelphia (1998)

[37] Kovács, L., Paláncz, B., Benyó, B., Török, L., Benyó, Z.: Robust Blood-Glucose Control using Mathematica. In: Proceedings of the 28th Annual International Conference of the IEEE Engineering in Medicine and Biology Society, New York, USA, pp. 451–454 (2006)

[38] Paláncz, B., Kovács, L., Benyó, B., Benyó, Z.: Robust Blood-Glucose Control of Type I Diabetes Patients under Intensive Care using Mathematica. In: Encyclopaedia of Healthcare Information Systems, Publ. Medical Information Science Reference, pp. 1210–1219 (2008)

[39] Kovács, L., Kulcsár, B., Benyó, B., Benyó, Z.: Induced L$_2$-norm Minimization of Glucose-Insulin System for Type I Diabetic Patients. In: Proceedings of the 7th IFAC Symposium on Modeling and Control in Biomedical Systems Including Biological Systems, Aalborg, Denmark, pp. 55–60 (2009)

[40] Kovács, L., Paláncz, B., Borbély, E., Benyó, Z., Benyó, B.: Robust Control Techniques and its Graphical Representation in case of Type I Diabetes using Mathematica. In: Proceedings of the 8th International Symposium on Applied Machine Intelligence and Informatics, Herlany, Slovakia, pp. 71–75 (2010)

Modern Control Solutions with Applications in Mechatronic Systems

Claudia-Adina Dragoş[1], Stefan Preitl[1], Radu-Emil Precup[1],
Marian Creţiu[1], and János Fodor[2]

[1] Department of Automation and Applied Informatics, "Politehnica" University of Timişoara
Bd. V. Parvan 2, RO-300223 Timişoara, Romania
{claudia.dragos,stefan.preitl,radu.precup}@aut.upt.ro,
cretiu_marian@yahoo.com
[2] Institute of Intelligent Engineering Systems, Óbuda University
Bécsi út 96/b, H-1034 Budapest, Hungary
fodor@uni-obuda.hu

Abstract. The chapter presents a comparison between several modern control solutions for mechatronic systems. A synthesis of the structures included in the modern based-design solutions is offered. The solutions deal with model predictive control, fuzzy control, adaptive control and combined control solutions between different control strategies and structures. Digital simulation and experimental results are presented accepting different modifications of the reference input or disturbances.

Keywords: Control solutions, fuzzy control, magnetic levitation system, mechatronic systems, predictive control.

1 Introduction

Mechatronic systems include mechanical systems, electronic systems, control solutions and information technology. The involvement of microcomputers into the mechatronic systems is useful due to the achieved advantages including the speed control in all conditions and the control of plants with non-measurable or estimated parameters. Electromagnetic actuators are used in mechatronic systems to convert the energy created by electricity to linear motion [1].

Three modern control solutions were developed and analyzed to obtain better performances, PI (PID) referred to also as PI(D) control with possible parameter adaptation [2], fuzzy control [3], model predictive control [4]. It is known [2]–[5] that these methods are based or include the model of the plant (linear or nonlinear) and sometime needs the estimation of parameters and the state variables. The optimal controller design using systems based on Takagi-Sugeno fuzzy models and state variables expressed as orthogonal functions is suggested in [6].

This chapter a synthesis on the development and verification of the performance ensured by the control structures of mechatronic applications in automotive systems.

I.J. Rudas et al. (Eds.): Computational Intelligence in Engineering, SCI 313, pp. 87–102.
springerlink.com © Springer-Verlag Berlin Heidelberg 2010

The presented structures are used for a class of specific applications – electromagnetic actuated clutch – which can provide better performances. They can be implemented easily [7].

The chapter offers comparative conclusions related to the development of the control solutions which are useful for the practitioners. Alternative solutions were analyzed by the authors in previous papers [8], [9].

The chapter is organized as follows. Several low-cost model based control solutions are presented in Section 2. Section 3 gives the mathematical models of two plants, the electromagnetic actuated clutch and the magnetic levitation system. The development of the control structures – PI(D) control, Takagi-Sugeno fuzzy control, model predictive control and state feedback – is outlined in Section 4. Real-time experiments and digital simulations results are presented in Section 5. The conclusions with respect to the performance of the developed control strategies are presented in the last section.

2 Position Control Solutions

The applications presented in this chapter refer to the fixed positioning of the mechanical subsystem. The low-cost developed and analyzed control solutions are: (a) a classical one solution with PI(D) controller with possible parameter adaptation; (b) one solution with Takagi-Sugeno fuzzy controller, and (c) one solution with predictive controller. All solutions were validated by applications concerning: (1) an electromagnetic actuator as part of the vehicular power train system, and (2) a magnetic levitation system (laboratory equipment).

2.1 PI(D) Control Solutions

It is commonly known that PI(D) is the most used controller in industrial applications [10]. The tuning parameters of the controller, i.e. k_r and $T_i (T_d)$, can be adapted also automatically to ensure favourable system behaviour. Furthermore, the integral component ensures the zero control error and the time constants $T_i (T_d)$ can compensate for the large time constant of the plant $\{T_1, T_2\}$, so the system becomes faster. Due to the natural proprieties the PI(D) control solutions offer good performance even in the case of plants with "low" nonlinearities. The adaptation of the controller parameters is advantageous in this case. The PI(D) solutions are frequently used as comparison support for the advanced control solutions. In this chapter the PI(D) controllers were designed based on the Modulus Optimum method [10] using also the pole-zero cancellation technique. The transfer functions (t.f.s) of the controllers were discretized using Tustin's method.

The solution with commutation from one controller to another depending on the operating points is adopted (Fig. 1, presented for two control algorithms referred to as a.r.) if the controllers are developed on the basis of linearized models.

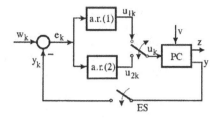

Fig. 1. The commutation from one controller to another one [11]

The bump-less commutation of the controllers needs the modification of the tuning parameters and the reconsidering the past values in the control algorithms:

$$x_{1k\,nec}^{(2)} = [1/(q_1^{(2)} - q_0^{(2)}p_1^{(2)})][q_1^{(1)}x_{1k}^{(1)} + q_0^{(1)}x_{2k}^{(1)}] - [q_0^{(2)}/(q_1^{(2)} - q_0^{(2)}p_1^{(2)})]e_k,$$

$$x_{2,k-1\,nec}^{(2)} = x_{1k\,nec}^{(2)}.$$
(1)

The schematic structure of such a controller is presented in Fig. 2.

2.2 Takagi-Sugeno Fuzzy Controller Solutions

In practical applications, the presence of nonlinearities in the structure of the controlled plant (P) leads to the idea of introducing fuzzy control. It is accepted [12] that Takagi-Sugeno fuzzy controllers (TS-FCs) are flexible to the operating point changes.

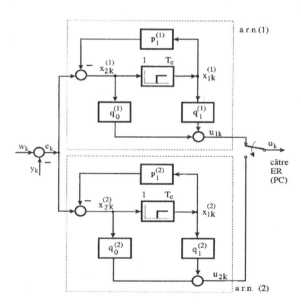

Fig. 2. Detailed block diagram relative to the controller commutation [11]

Therefore they can ensure better performance in the cases of plants with nonlineari-ties. The TS-FC solution can be developed based on PI or PID control solutions [13]. The set of linearized model of the plant used in the TS-FC design can be treated as a linear system with variable parameters. To describe the nonlinear dynamic system, the TS-FC uses *if–then* fuzzy rules expressed as [13]:

$$R_i : \text{if } \{z_1 \text{ is } a_{1i} \text{ and } z_2 \text{ is } a_{2i} \text{ and} \dots \text{ and } z_n \text{ is } a_{ni}\} \text{ then } \{u_k = f(z_1, z_2, \dots, z_n)\} . \qquad (2)$$

2.3 Model Predictive Control Solutions

A class of predictive control systems is built starting with the linear (linearized) model of the plant (model based predictive control, MPC) which provides good re-sults [4], [5]. The control algorithm can be adapted on-line to the current operating points. MPC solutions were used for mechatronic system in several applications due to the presence of a low number of tuning parameters (often only one parameter in in-volved) and to the improved control system performance they offer especially in terms of better robustness.

The modelling of nonlinear plants is sometimes difficult because no standard solu-tions are recommended to represent the nonlinearities. In case of low order plants the MPC advantages consists of the possibility to obtain a reduced order polynomial form for the controller based on t.f.s.

The main idea of the model predictive control is to obtain the next control signal from the actual time moment minimizing the specific objective function. Fig. 3 points out the development steps of an MPC strategy:

- at each time moment, the controlled output (plant's output) is determined in the fu-ture over the prediction horizon; the future values of the plant's output can be pre-dicted from a plant model;
- reference value is planned in advance for the future horizon;
- to calculate the control sequence from the actual time point minimizing a perform-ance criterion that depends on predicted control error;
- a new control signal sequence, different to the previews one, is obtained.

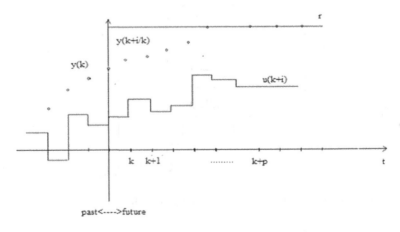

Fig. 3. Model based predictive control operation

For the considered applications two low-cost MPC solutions have been approached: (1) using one-step ahead quadratic objective functions, and (2) using multi-step ahead quadratic objective functions [2]. In both cases, if the plant does not contain integrators the nonzero control error can appear. In order to develop the model predictive controllers, the ARX model of the plant and a first order predictor were used according to (3) and (4), respectively:

$$y(k) = [B(q^{-1}) / A(q^{-1})]u(k-1) + \{C(q^{-1}) / [A(q^{-1})D(q^{-1})]e(k)\}, \tag{3}$$

$$\hat{y}(k+1) = [\hat{B}(q^{-1})\hat{D}(q^{-1}) / T(q^{-1})]u(k) + \{q[T(q^{-1}) - \\ - \hat{A}(q^{-1})\hat{D}(q^{-1})] / T(q^{-1})\}y(k). \tag{4}$$

In the first case, the controller is obtained by minimizing the one-step ahead quadratic objective function based on the control error:

$$J = 0.5[\hat{y}(k+1) - r(k+1)]^2, \tag{5}$$

with the resulted two-degree-or-freedom control algorithm:

$$u(k) = [AT^* / (AR + q^{-1}BS)]r(k+1) - \{[CS / D] / (AR + q^{-1}BS)\}e(k). \tag{6}$$

Multi-step ahead quadratic objective functions with the weighted control are used to solve the problems related to the non-minimum phase zeros:

$$J = \sum_{i=1}^{p} \{[\hat{y}(k+i) - r(k+i)]^2 + \lambda u^2(k+i-1)\}. \tag{7}$$

To obtain the minimum variance, the conditions (minimum prediction horizon, control horizon and the weighting coefficient) differ depending on the plant.

2.4 Adaptive Control Solutions

To control dynamic systems with unknown parameters which are variable in a wide range it is required to do the on-line estimation of these parameters. Two design methods can be adopted [2]: model reference adaptive control and self-tuning control.

The chosen model for adaptive control should be achievable and must to reflect the desired performances. The unknown parameters of the plant and the controller parameters lead to nonzero control error which reflects that the system output is not the same as the reference model output. The control error cancellation is done by an adaptive control law.

3 Applications and Plant Models

The applications used in the paper are related to the applications in mechatronic systems: (1) an electromagnetic actuated clutch, and (2) a magnetic levitation system with two electromagnets (laboratory set). If in the case (2) the bottom electromagnet is neglected and the spring is introduced, then a magnetically actuated mass-spring-damper (MaMSD) system can be obtained. The modelling of an electromagnetic actuator system is based on the MaMSD system's model.

3.1 Mathematical Model of Electromagnetic Actuator

The schematic structure of an electromagnetically actuated MaMSD system is presented in Fig. 4. The system consists of two subsystems: electrical subsystem (system of acting) and mechanical subsystem (acted system). A mass m moves linearly under the effect of three forces: electromagnetic force, F, the force generated by a spring, and one generated by a damper.

The structure given in Fig. 4 enables the first-principle of the system are detailed in [14]. The characteristic variables (position, speed, acceleration, force and the currents) are partly measurable and partly estimated. The nonlinear model is:

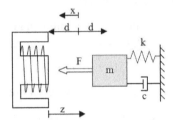

Fig. 4. Schematic structure of magnetically actuated mass-spring-damper system

$$\begin{cases} \dot{x_1} = x_2 \\ \dot{x_2} = -\dfrac{k}{m}x_1 - \dfrac{c}{m}x_2 + \dfrac{k_a}{m(k_b+d-x_1)}x_3^2 \\ \dot{x_3} = \dfrac{R}{2k_a}x_1x_3 - \dfrac{R(k_b+d)}{2k_a}x_3 - \dfrac{1}{2k_a}x_1V + \dfrac{(k_b+d)}{2k_a}V + \dfrac{1}{(k_b+d-x_1)}x_2x_3 \\ y = x_1 \end{cases} \qquad (8)$$

To avoid the detailed manipulation of the nonlinearities in (3.1) and to design the control systems, the nonlinear model (3.1) was linearized around nine operating points from static input-output map, Fig. 5.

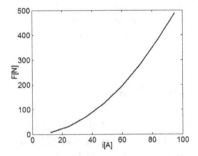

Fig. 5. Static input-output map

The obtained state-space model is:

$$\dot{x} = \underline{A}\,\underline{x} + \underline{b}\,\Delta V,$$

$$\Delta y = \underline{c}^T\,\underline{x},$$

$$\underline{A} = \begin{bmatrix} 0 & 1 & 0 \\ -k/m & -c/m & (2*k_v*x_0)/m \\ 0 & 0 & -R/L_d \end{bmatrix}, \underline{b} = \begin{bmatrix} 0 \\ 0 \\ 1/L_d \end{bmatrix}, \underline{x} = \begin{bmatrix} \Delta x_1 \\ \Delta x_2 \\ \Delta x_3 \end{bmatrix}, \tag{9}$$

$$\underline{c}^T = \begin{bmatrix} 1 & 0 & 0 \end{bmatrix}.$$

Based on this, nine t.f.s were obtained. They are synthesized in Table 1.

Table 1. Transfer functions of plant and consequent in rule base of TS-FC

Operating points	$H_{PC}(s)$	Rule consequents
(1). $x_0 = 0.000043, i_0 = 10, V_0 = 5$	$H_{PC}(s) = \dfrac{0.22}{(1+0.0036s)(1+0.0017s)(1+0.0011s)}$	$u_k = \gamma*(22.82*e_k + 822*\Delta e_k)$
(2). $x_0 = 0.0000624, i_0 = 12, V_0 = 6$	$H_{PC}(s) = \dfrac{0.26}{(1+0.0036s)(1+0.0017s)(1+0.0011s)}$	$u_k = \gamma*(19.16*e_k + 677*\Delta e_k)$
(3). $x_0 = 0.00025, i_0 = 24, V_0 = 12$	$H_{PC}(s) = \dfrac{0.52}{(1+0.0036s)(1+0.0017s)(1+0.0011s)}$	$u_k = \gamma*(9.55*e_k + 338.6*\Delta e_k$
(4). $x_0 = 0.00056, i_0 = 36, V_0 = 18$	$H_{PC}(s) = \dfrac{0.78}{(1+0.0036s)(1+0.0018s)(1+0.0011s)}$	$u_k = \gamma*(6.15*e_k + 218*\Delta e_k)$
(5). $x_0 = 0.001, i_0 = 48, V_0 = 24$	$H_{PC}(s) = \dfrac{1.08}{(1+0.0036s)(1+0.0019s)(1+0.0011s)}$	$u_k = \gamma*(4.3*e_k + 152*\Delta e_k)$
(6). $x_0 = 0.0016, i_0 = 60, V_0 = 30$	$H_{PC}(s) = \dfrac{1.3}{(1+0.0036s)(1+0.0019s)(1+0.0011s)}$	$u_k = \gamma*(3.6*e_k + 126.4*\Delta e_k)$
(7). $x_0 = 0.0022, i_0 = 72, V_0 = 36$	$H_{PC}(s) = \dfrac{1.56}{(1+0.0036s)(1+0.0019s)(1+0.0011s)}$	$u_k = \gamma*(3*e_k + 105.3*\Delta e_k)$
(8). $x_0 = 0.0031, i_0 = 84, V_0 = 42$	$H_{PC}(s) = \dfrac{1.82}{(1+0.0036s)(1+0.002s)(1+0.0011s)}$	$u_k = \gamma*(2.5*e_k + 87.4*\Delta e_k)$
(9). $x_0 = 0.0039, i_0 = 95, V_0 = 47.5$	$H_{PC}(s) = \dfrac{2.05}{(1+0.0036s)(1+0.0021s)(1+0.0011s)}$	$u_k = \gamma*(2.12*e_k + 75*\Delta e_k)$

3.2 Mathematical Model of Magnetic Levitation System

The schematic structure for magnetic levitation system with two electromagnets (MLS2EM), used in many industrial applications, is presented in Fig. 6 [13].

A nonlinear model of the plant can be obtained starting with the following first-principle equations:

$$\begin{cases} \dot{x}_1 = x_2 \\ \dot{x}_2 = -F_{em1}/m + g + F_{em2}/m \\ \dot{x}_3 = [1/f_i(x_1)](k_i u_1 + c_i - x_3) \\ \dot{x}_4 = [1/f_i(x_d - x_1)](k_i u_2 + c_i - x_4) \end{cases}, \tag{10}$$

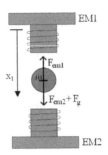

Fig. 6. Schematic structure of magnetic levitation system [13]

$$F_{em1} = x_3^2 (F_{emP1} / F_{emP2}) \exp(-x_1 / F_{emP2})$$
$$F_{em2} = x_4^2 (F_{emP1} / F_{emP2}) \exp[-(x_d - x_1)/ F_{emP2}] \, , \qquad (11)$$
$$f_i(x_1) = (f_{iP1} / f_{iP2}) \exp(-x_1 / f_{iP2})$$

where: x_1 – the sphere position, $x_1 \in [0, 0.016]$; x_2 – the sphere speed; x_3, x_4 – the currents in the top and bottom electromagnets, $x_3, x_4 \in [0.03884, 2.38]$; u_1, u_2 – the control signals for the top and bottom electromagnets, $u_1, u_2 \in [0.00498, 1]$.

The nonlinear model can be linearized around different operating points. Such an example is $x_{10} = 0.007, x_{20} = 0, x_{30} = 0.754, x_{40} = 0.37$. The obtained linearized state-space model and its parameters are:

$$\begin{cases} \Delta \dot{\underline{x}} = \underline{A} \Delta \underline{x} + \underline{b} \, \Delta V \\ \Delta y = \underline{c}^T \Delta \underline{x} \end{cases} \text{ with } \quad A = \begin{bmatrix} 0 & 1 & 0 & 0 \\ a_{21} & 0 & a_{23} & a_{24} \\ a_{31} & 0 & a_{33} & 0 \\ a_{41} & 0 & 0 & a_{44} \end{bmatrix}, \quad \underline{B} = \begin{bmatrix} 0 \\ 0 \\ b_3 \\ b_4 \end{bmatrix}, \quad \underline{c}^T = \begin{bmatrix} 1 & 0 & 0 & 0 \end{bmatrix}. \quad (12)$$

$$a_{2,1} = \frac{x_{30}^2}{m} \frac{F_{emP1}}{F_{emP2}^2} e^{-\frac{x_{10}}{F_{emP2}}} + \frac{x_{40}^2}{m} \frac{F_{emP1}}{F_{emP2}^2} e^{-\frac{x_d - x_{10}}{F_{emP2}}} \, ,$$

$$a_{2,3} = -\frac{2 x_{30}}{m} \frac{F_{emP1}}{F_{emP2}} e^{-\frac{x_{10}}{F_{emP2}}} \, , \quad a_{2,4} = \frac{2 x_{40}}{m} \frac{F_{emP1}}{F_{emP2}} e^{-\frac{x_d - x_{10}}{F_{emP2}}} \, ,$$

$$a_{3,1} = -(k_i u + c_i - x_{30})(x_{10} / f_{iP2}) f_i^{-1}(x_{10}), \quad a_{3,3} = -f_i^{-1}(x_{10}), \qquad (13)$$

$$a_{4,1} = -(k_i u + c_i - x_{40})(x_{10} / f_{iP2}) f_i^{-1}(x_d - x_{10}), \quad a_{4,4} = -f_i^{-1}(x_d - x_{10}),$$

$$b_3 = k_i f_i^{-1}(x_{10}), \quad b_4 = k_i f_i^{-1}(x_d - x_{10}).$$

4   Control Structures

4.1   Control Structures Dedicated to Electromagnetic Actuator

4.1.1   PI(PID) Control Structures

To control the electromagnetic actuated clutch, the modelling of the mass-spring-damper system accepted in literature is used first as the basis for the development of the classical PI(D) control structure. The PI(D) controllers are used due to the simple structure and to good offered performances, even compared with the complex control structures. Based on the mathematical model (MM) (10), the following PID controller was designed using the Modulus Optimum method with pole zero cancellation:

$$H_C(s) = \frac{k_r}{s}(1+T_{r1}s)(1+T_{r2}s)\frac{1}{(1+T_f s)},$$

$$k_r = \frac{1}{2k_p T_3}, T_{r1} = T_1, T_{r2} = T_2, T_f = 0.1 \cdot T_{r2}.$$

(14)

To design a PI controller, the MM (10) was reduced from the third order to the second order. The t.f. of the PI controller is:

$$H_{RG-PI}(s) = \frac{k_r}{s}(1+T_r s) = \frac{k_R}{sT_i}(1+T_i s),$$

$$k_r = \frac{k_R}{T_i} = \frac{1}{2k_{PC}T_\Sigma}, T_r = T_i = T.$$

(15)

The resulting PI and PID controller parameters are detailed in Table 2 for three representative operating points.

4.1.2 Takagi-Sugeno Fuzzy Control Structures

Two control solutions with TS-FC were developed and analysed regarding an electromagnetic actuator (Section 3.1). The controller structures are presented in Fig. 6 as TS-FC with output integration, where: FC – the block fuzzy logic processing, P – controlled plant, w_k – the reference input, $e_k = w_k - y_k$ – the control error, u_k – the control signal, y_k – measured output, x – controlled output (position of the mechanical system) [3].

Table 2. PI and PID controller parameters

Number of oper-ating point	PID controller parameters			PI controller parameters	
	k_r	T_{r1}	T_{r2}	k_r	T_r
(3)	874.13	0.0036	0.0017	343.4	0.0036
(5)	437	0.0036	0.0018	166	0.0036
(9)	221	0.0036	0.0021	76	0.0036

The developed structures differ by the combination of the control rules in the decision table and by the value of an adaptable parameter, γ. This parameter γ is used to introduce additional nonlinearities in order to improve the system's performance. The consequents in the control rules of the TS-FC are shown in Table 1. In the first case, TS-FC(1), $\gamma = 0.00065$ and in the second case, TS-FC(2), $\gamma = 0.001$. These values are set the designer and adapted to plant.

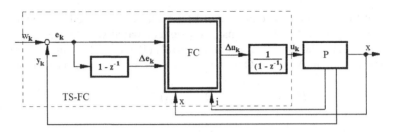

Fig. 7. Block diagram of the control system with Takagi-Sugeno fuzzy controller

The concrete TS-FC structure (with pseudo-PI behaviour) is homogeneous, with four inputs: e_k, Δe_k – the first order increment of the control error, I – the current and x – the position, and one output: V – the control signal. The input variables, i and x, were introduced as additional signals to improve the selective action of the decision table. Three linguistic terms with triangular membership functions (N, ZE, P) are used for each input variable. The decision table of each controller contains 81 rules (*3x3x3x3*). The controllers use the MAX and MIN operators in the inference engine and the weighted average method for defuzification. The appropriate combinations of the 81 rules and the choosing of the parameter value γ can ensure different control system performance.

The quasi-continuous digital PI controller can be obtained from a continual controller, using the Tustin's method [10], with the sampling period T_e. The parameter γ introduces additional nonlinearities to adapt the performance to the application:

$$\Delta u_k^i = \gamma(k_k \Delta e_k + k_i e_k) = \gamma k_k (\Delta e_k + \alpha e_k). \tag{16}$$

where (the superscript i indicates the rule index):

$$K_P = k_R (1 - \frac{T_e}{2T_i}), \, K_I = \frac{k_R T_e}{T_i}, \, \alpha = \frac{K_I}{K_P}. \tag{17}$$

The controller parameters are B_e and $B_{\Delta e}$. B_e is chosen by heuristic rules and $B_{\Delta e}$ is determined from [9]:

$$B_{\Delta e} = \frac{K_P}{K_I} B_e = \alpha B_e. \tag{18}$$

An alternative TS-FC solution based on Takagi-Sugeno fuzzy models uses the optimal linear quadratic controllers design techniques [14].

4.1.3 Model Predictive Control Structures
MPC structures were designed based on the linearized plant model around the nine operating points. In this paper, the simulation results are presented for the controller designed around the average operating point [16]. The ARX model of the linearized plant around the average operating point is:

$$A(q^{-1})y(k) = B(q^{-1})u(k-1),$$

$$A(q^{-1}) = 1 - 1.72q^{-1} + 0.74q^{-2}, \quad B(q^{-1}) = 0.0243 - 0.0032q^{-1}. \tag{19}$$

The controller designed using the one-step ahead quadratic objective function is:

$$u(k) = \frac{0.0243}{0.0243(0.0243 - 0.0032q^{-1}) + 0.001} r(k+1) - \frac{0.0243(1.72 - 0.74q^{-1})}{0.0243(0.0243 - 0.0032q^{-1}) + 0.001} y(k). \tag{20}$$

The standard form of the predictive control structure can be obtained based on (4.6) and (4.7), and the parameters, $\lambda = 0.001$ and $c = 1$:

$$R = \frac{B + \lambda}{b_0 + \lambda} = 1 + 0.1265q^{-1}, \quad T^* = \frac{b_0 c}{b_0 + \lambda} = 0.9605,$$

$$S = \frac{qb_0(1 - A)}{b_0 + \lambda} = 1.6513 - 0.7107q^{-1}. \tag{21}$$

and for the multi-step ahead quadratic objective function the controller is:

$$\begin{bmatrix} u(k) \\ u(k+1) \\ u(k+2) \end{bmatrix} = \begin{bmatrix} 0.00032 & 0.00053 & 0.00023 \\ 0.0098 & 0.00149 & 0.00053 \\ 0.00059 & 0.00093 & 0.00032 \end{bmatrix} \begin{bmatrix} r(k+1) \\ r(k+2) \\ r(k+3) \end{bmatrix} - \begin{bmatrix} -26.8596 & 17.1013 & 0.0702 \\ -39.9832 & 24.876 & 0.1197 \\ -45.479 & 28.798 & 0.1183 \end{bmatrix} \begin{bmatrix} y(k) \\ y(k+1) \\ u(k-1) \end{bmatrix}. \tag{22}$$

4.2 Control Structures Dedicated to Magnetic Levitation System

Two control structures were designed for the levitation and the stabilization of the magnetic ball [17]: a state feedback control structure and a PID control structure. The control signal is applied to the top electromagnet.

First, a state feedback control structure was designed for the plant, Fig. 8.

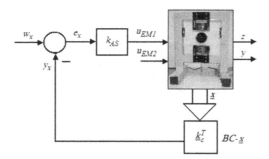

Fig. 8. State feedback control structure for magnetic levitation system

The pole placement method was applied to design the block $\underline{k}_c^T$. So, the poles $p_1^* = -0.25, p_2^* = -240$ were chosen and the following parameters were obtained: $\underline{k}_c^T = [40 \quad 5]$.

To ensure the performance requirements (zero steady-state control error, the phase margin of 60^0, small settling time), a cascade control structure was designed, with a state feedback control structure for the inner loop and a PID controller for the external loop.

The t.f. obtained for the inner loop is:

$$H(s) = \underline{c}^T (sI - \underline{A}_x)\underline{b}_u = \frac{-0.11}{(1+0.2s)(0.000023s^2 + 0.0034s + 1)}. \tag{23}$$

To design the external loop, a PID controller extended with a first order filter was developed using the pole-zero cancellation [10]:

$$H_C(s) = \frac{k_r}{s}(1 + 2\zeta_r T_r s + T_r^2 s^2)\frac{1}{(1+sT_f)}, \quad \zeta_r < 1. \tag{24}$$

4.3   Simulation and Experimental Results

The designed control structures were tested by simulation in Matlab&Simulink and by real-time experiments. Part of these results is presented as follows.

4.3.1   Simulation Results for Electromagnetic Actuator

The control structures behaviour was checked for all nine operating points. For each structure two simulation scenarios have been used: (1) system response with respect to the step modifications of the reference input and (2) system response with respect to the rectangular modifications of the input [18]. In this section, the simulation results are presented only for the control structures designed around an average operating point and with respect to the rectangular reference input of the CS designed for MaMSD-s.

The simulation results for the control structure with PI(D) controllers designed for the electromagnetic actuated clutch are presented in the Fig. 9(a) and Fig. 9(b).

The simulation results for the control structure with fuzzy controllers designed for the electromagnetic actuated clutch are presented in Fig. 10(a) for $\gamma = 0.00065$ and Fig. 10(b) for $\gamma = 0.001$.

The simulation results for the predictive control structure which use multi-step ahead quadratic objective function are presented in the Fig. 11(a). The results for the control system with controller designed to minimize the one-step ahead quadratic objective function are illustrated in Fig. 11(b).

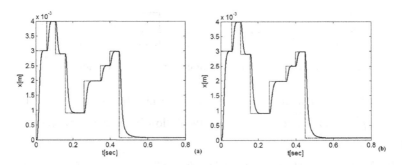

Fig. 9. Simulation results for the control system with PI(D) controller

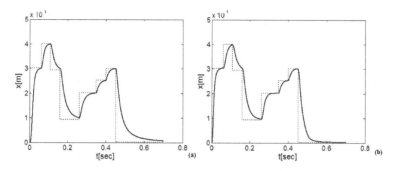

Fig. 10. Simulation results for the control system with Takagi-Sugeno fuzzy controllers

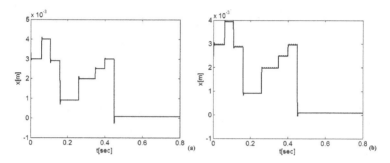

Fig. 11. Simulation results for the control system with predictive controllers

In the simulation results, the output of the predictive control system presents a small overshoot ($\sigma_1 = 2\%$) in comparison with the second case, where the system's output presents a higher overshoot ($\sigma_1 = 4.8\%$). It is highlighted that the settling time is reduced, $t_r = 0.004$, and in the second case the settling time is slightly improved.

4.3.2 Experimental Results for the Magnetic Levitation System

The block diagram illustrated in Fig. 8 was used to verify the state feedback control structure (used for stabilization) and the PID controller for the magnetic levitation system in real-time experiments.

The experimental results related to the both control structure behaviour are detailed in Figs. 12 and 13: sphere position (a), sphere speed (b), current in top and bottom electromagnets (c), and control signal for both electromagnets (d).

The results show the real-time behaviour of the control structure designed for the MLS2EM. The oscillations at the beginning of transience response (around the value of the reference input) are due to the complex conjugated poles and the nonlinearities of the plant. The reference input is tracked.

The results for the current in both electromagnets present oscillations at the beginning. All these results do not modify the speed, which remains zero.

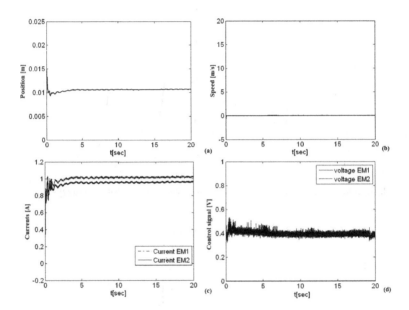

Fig. 12. Real-time experimental results for the state feedback control structure: (a) sphere position; (b) speed; (c) currents; (d) control signals

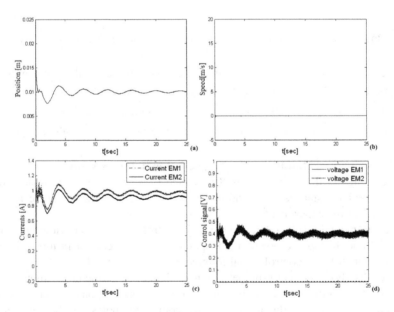

Fig. 13. Real-time experimental results for cascade control structure: (a) sphere position; (b) speed; (c) currents; (d) control signals

5 Conclusions

The paper presents control systems for two applications: an electromagnetic actuator clutch and a magnetic levitation system. For the first application, four control solutions were designed and for the second a state feedback control structure and a conventional PID control loop were designed.

In the case of MaMSD modelling, the nonlinear model has been linearized around a several operating points (nine operating points) and then the linearized models have been obtained. These models have been used to design the control structure with PI(D) controller, the Takagi-Sugeno fuzzy control structure, the model predictive control systems.

The simulation and experimental results have proved that all solutions are viable. Similar results were obtained also for the second category of test signals.

Future research will be dedicated to other controller models. Real-time experimental tests are necessary in all applications [19]–[25] beyond the automotive mechatronic systems.

Acknowledgments. This paper was supported by the CNMP & CNCSIS of Romania, "Politehnica" University of Timisoara, Romania, Óbuda University, Budapest, Hungary, and University of Ljubljana, Slovenia, from the Hungarian-Romanian and Slovenian-Romanian Intergovernmental Science & Technology Cooperation Programs. This work was partially supported by the strategic grant POSDRU 6/1.5/S/13 (2008) of the Ministry of Labour, Family and Social Protection, Romania, cofinanced by the European Social Fund – Investing in People.

References

[1] Isermann, R.: Mechatronic Systems: Fundamentals. Springer, Heidelberg (2005)
[2] Škrjanc, I., Blažič, S., Matko, D.: Model-reference Fuzzy Adaptive Control as a Framework for Nonlinear System Control. J. Intell. Robot Syst. 36(3), 331–347 (2003)
[3] Precup, R.-E., Preitl, S.: Fuzzy Controllers. Editura Orizonturi Universitare Publishers, Timisoara (1999)
[4] Camacho, E.F., Bordons, C.: Model Predictive Control, 2nd edn. Springer, Heidelberg (2004)
[5] Brosilow, C., Joseph, B.: Techniques of Model-based Predictive Control. Prentice Hall PTR, Upper Sanddle River (2002)
[6] Ho, W.H., Chou, J.H.: Design of Optimal Controllers for Takagi-Sugeno Fuzzy-Model-based Systems. IEEE Trans. Syst. Man Cybern. A 37(3), 329–339 (2007)
[7] Van der Heijden, A.C., Serrarens, A.F.A., Camlibel, M.K., Nijmejer, H.: Hybrid optimal control of dry clutch engagement. Int. J. Control 80(11), 1717–1728 (2007)
[8] Dragos, C.A., Preitl, S., Precup, R.-E., et al.: Modern Control Solutions for Mechatronic Servosystems. Comparative Case Studies. In: Proceedings of 10th International Symposium of Hungarian Researchers on Computational Intelligence and Informatics (CINTI 2009), Budapest, Hungary, November 12-14, pp. 69–82 (2009)
[9] Dragos, C.A., Preitl, S., Radac, M.B., Precup, R.-E.: Nonlinear and Linearized Models and Low Cost Control Solution for an Electromagnetic Actuator. In: Proceedings of 5th International Symposium on Applied Computational Intelligence and Informatics (SACI 2009), Timisoara, Romania, pp. 89–94 (2009)

[10] Åström, K.J., Hägglund, T.: PID Controllers, Theory, Design and Tuning. In: Instrument Society of America, Research Triangle Park, NC, USA (1995)

[11] Preitl, S., Precup, R.-E., Preitl, Z.: Control Algorithms and Structures, vol. 1, 2. Editura Orizonturi Universitare Publishers, Timisoara (2009) (in Romanian)

[12] Tanaka, K., Wang, H.O.: Fuzzy Control Systems Design and Analysis: A Linear Matrix Inequality Approach. John Wiley & Sons, New York (2001)

[13] Coppier, H., Chadli, M., Bruey, S., et al.: Implementation of a Fuzzy Logic Control for a Silo's Level Regulation in Stone Quarries. In: Preprints of 3rd IFAC Workshop on Advanced Fuzzy and Neural Control (AFNC 2007), Valenciennes, France, p. 6 (2007) CD-ROM

[14] Blažič, S., Škrjanc, I.: Design and Stability Analysis of Fuzzy Model-based Predictive Control – A Case Study. J. Intell. Robot Syst. 49(3), 279–292 (2007)

[15] Kiencke, U., Nielsen, L.: Automotive Control Systems for Engine, Driveline and Vehicle, 2nd edn. Springer, New York (2005)

[16] Cairano, S., Bemporad, A., Kolmanovsky, V., et al.: Model Predictive Control of Magnetically Actuated Mass Spring Dampers for Automotive Applications. Int. J. Control 80(11), 1701–1716 (2007)

[17] Inteco Ltd., Magnetic Levitation System 2EM (MLS2EM), User's Manual (laboratory set), Krakow, Poland (2008)

[18] Research staff, Real-Time Informatics Technologies for Embedded-System-Control of Power-Train in Automotive Design and Applications (SICONA). Research Report of a CNMP Research Grant, "Gh. Asachi" Technical University of Iasi, Iasi, Romania (2009) (in Romanian)

[19] Horváth, L., Rudas, I.J.: Modeling and Problem Solving Methods for Engineers. Academic Press, Elsevier (2004)

[20] Baranyi, P.: TP Model Transformation as a Way to LMI-based Controller Design. IEEE Trans. Ind. Electron. 51(2), 387–400 (2004)

[21] Škrjanc, I., Blažič, S., Agamennoni, O.E.: Interval Fuzzy Modeling Applied to Wiener Models with Uncertainties. IEEE Trans. Syst. Man Cybern. B 35, 1092–1095 (2005)

[22] Johanyák, Z.C., Kovács, S.: Sparse Fuzzy System Generation by Rule Base Extension. In: Proceedings of 11th International Conference on Intelligent Engineering Systems (INES 2007), Budapest, Hungary, pp. 99–104 (2007)

[23] Vaščák, J.: Navigation of Mobile Robots Using Potential Fields and Computational Intelligence Means. Acta Polytechnica Hungarica 4(1), 63–74 (2007)

[24] Baranyi, P., Korondi, P., Tanaka, K.: Parallel-distributed Compensation-based Stabilization of a 3-DOF RC Helicopter: a Tensor Product Transformation-based Approach. J. Adv. Comput. Intell. Inform. 13(1), 25–34 (2009)

[25] Ahn, H.K., Anh, H.P.H.: Inverse Double NARX Fuzzy Modeling for System Identification. IEEE/ASME Trans. Mechatronics 15(1), 136–148 (2010)

Adaptive Tackling of the Swinging Problem for a 2 DOF Crane – Payload System

József K. Tar[1], Imre J. Rudas[2], János F. Bitó[2], José A. Tenreiro Machado[3], and Krzysztof R. Kozłowski[4]

[1] Telematics and Communications Informatics Knowledge Centre
[2] Institute of Intelligent Engineering Systems, John von Neumann Faculty of Informatics, Óbuda University, Bécsi út 96/B, H-1034 Budapest, Hungary
tar.jozsef@nik.uni-obuda.hu, {rudas,bito}@uni-obuda.hu
[3] Department of Electrotechnical Engineering,
Institute of Engineering of Porto,
Rua Dr. Antonio Bernardino de Almeida, 4200-072 Porto, Portugal
jtm@isep.ipp.pt
[4] Chair of Control and Systems Engineering, Computing Science & Management,
Poznan University of Technology, Piotrowo 3a, 60-965 Poznan, Poland
krzysztof.kozlowski@put.poznan.pl

Abstract. The control of a crane carrying its payload by an elastic string corresponds to a task in which precise, indirect control of a subsystem dynamically coupled to a directly controllable subsystem is needed. This task is interesting since the coupled degree of freedom has little damping and it is apt to keep swinging accordingly. The traditional approaches apply the input shaping technology to assist the human operator responsible for the manipulation task. In the present paper a novel adaptive approach applying fixed point transformations based iterations having local basin of attraction is proposed to simultaneously tackle the problems originating from the imprecise dynamic model available for the system to be controlled and the swinging problem, too. The most important phenomenological properties of this approach are also discussed. The control considers the 4th time-derivative of the trajectory of the payload. The operation of the proposed control is illustrated via simulation results.

Keywords: Adaptive Control; Fixed Point Transformations; Cauchy Sequences; Iterative Learning; Local Basin of Attraction.

1 Introduction

Any payload carried by some crane normally is connected to the directly controllable engine via an elastic string that has very little damping, therefore it is apt to have long-lasting swinging. Precise positioning of swinging bodies traditionally is solved by the so-called "input shaping approach" that goes back to the nineties of the past century. The main idea of input shaping is generation command signals that can

I.J. Rudas et al. (Eds.): Computational Intelligence in Engineering, SCI 313, pp. 103–114.
springerlink.com

efficiently reduce payload oscillations by slightly modifying the operator's command by convolving it with a series of impulses [1], [2]. This technique can cancel out the system's own motion-induced oscillations. It was successfully used to reduce transient and residual oscillation in various systems, e.g. in coordinate measuring machines [3], and even recently in various cranes [4], [5], [6], [7]. The positive effect of this technique has been shown in the reduction of task completion time and obstacle collisions in a number of crane operator performance studies. It is a present trend to further improve operator performance by assisting crane operators in the estimation of the crane's stopping location. From mathematical point of view this technique is strictly related to linear systems and linear approximation of nonlinear ones as well as to linear control solutions (e.g. [8]).

An alternative approach to this problem may be the simultaneous tackling of the imprecision of the available dynamic model of the system to be controlled and the swinging problem. The most sophisticated adaptive control elaborated for robots is the Slotine-Li controller [9] that tries to learn certain parameters of the dynamic model using Lyapunov's 2nd Method. It has the main deficiency that it is unable to compensate the effects of lasting unknown external perturbations [10], and is unable to identify the parameters of strongly nonlinear phenomena as friction for which sophisticated techniques have to be applied (e.g. [11]). Furthermore, due to insisting on the use of the Lyapunov function technique the order of the ordinary differential equations to be handled by this method is limited to 2. For getting rid of the formal restrictions that normally originate from the use of Lyapunov functions alternative possibilities were considered for developing adaptive controllers. In the modern literature various different control approaches can be found. Due to its way of thinking that partly is similar to our approach the idea of "situational control" using models only in typical regimes of operation [12] applied in the control of a small turbo jet engine [13], "anytime control" [14] and other progressive modeling efforts [15] can be mentioned.

In our particular approach published e.g. in [16] the mathematical model of the system to be controlled was considered as a mapping between its *desired* and *realized responses* in which the *desired response* was calculated on purely kinematical basis, and the appropriate excitation to obtain this response was computed by the use of a partial and approximate dynamic model of the system, while the *realized response* was measured. It was shown that in this approach the "*response*" of the system may be arbitrary order derivative of the state variables, it can be even a fractional order one. Its robust variant was successfully applied even for a strongly nonlinear system as e.g. the Van der Pol Oscillator [17]. The essence of this method is obtaining a convergent iteration using contractive mapping in Banach spaces, and to some extent it is akin to other iterative approaches as e.g. iterative tuning techniques (e.g. [18]). The same idea was successfully applied for reduction of the swinging of two mass-points connected by an elastic string by directly acting with control force only one of the mass-points [19]. In this preliminary approach the directly accelerated mass-point corresponded to a very simplified model of a crane. In the present paper this simple model replaced by a 2 Degree of Freedom (DOF) crane and the applicability of the same adaptive approach is demonstrated via simulations. The details of the control approach proposed were considered in connection with various physical systems, e.g. in [20]. In this paper it will not be detailed again. In the next section the control problem will be mathematically defined, and following that, simulation results will be obtained. The paper will be closed by conclusions and the other usual components necessary.

2 Mathematical Description of the Control Task

In the model applied the crane consists of a chassis of mass M being able to move back and forth along a straight in the horizontal direction. The height of this body is y_0. At the top of this body a rotational axis is placed that is also horizontal and perpendicular to the direction of the horizontal motion of the body. Around this axis a beam of length L and mass m can be rotated by angle q_1 (measured clockwisely from the vertical axis). It is assumed that the beam has even mass-distribution. To the end of the beam a string of zero force length L_0 and stiffness k is attached. To the other end of the string a point-like payload of mass m_A is attached. The crane body's motion is also controlled along the horizontal line, the appropriate "generalized coordinate" is q_2. In this approach 2D motion is considered. (It is worth noting that this fact does not mean any significant simplification for the mathematical problems to be tackled.) The Euler-Lagrange equations of motion of this sub-system are as follows:

$$
\begin{bmatrix} \dfrac{mL^2}{4} & \dfrac{mL\cos q_1}{2} \\ \dfrac{mL\cos q_1}{2} & (m+M) \end{bmatrix}\begin{bmatrix} \ddot{q}_1 \\ \ddot{q}_2 \end{bmatrix} + \begin{bmatrix} -\dfrac{mgL}{2}\sin q_1 \\ -\dfrac{mL}{2}\sin q_1\dot{q}_1^2 \end{bmatrix} = \begin{bmatrix} Q_1 \\ Q_2 \end{bmatrix}
\tag{1}
$$

in which g denotes the gravitational acceleration (it acts in the vertical direction), and Q_i (i=1,2) denote the appropriate components of the generalized forces action on the crane, i.e. the sum of the generalized forces of the carne's drives and the contribution of the contact force that is needed for carrying the payload connected to the crane by the string. If the contact force acting on the payload is denoted by $\mathbf{F}^{Contact}$, its reaction force gives the following contribution to the generalized forces of the crane:

$$
\begin{bmatrix} Q_1^{Contact} \\ Q_2^{Contact} \end{bmatrix} = -\begin{bmatrix} F_{1=x}^{Contact}L\cos q_1 - F_{2=y}^{Contact}L\sin q_1 \\ F_{1=x}^{Contact} \end{bmatrix}.
\tag{2}
$$

The motion of the payload is described by the equation

$$
\ddot{\mathbf{x}} = \frac{\mu_A}{m_A}\dot{\mathbf{x}} + \frac{1}{m_A}\mathbf{F}_{cont} + \mathbf{g}_{grav}
$$

$$
\mathbf{F}^{Contact} = \frac{\mathbf{y}-\mathbf{x}}{\|\mathbf{y}-\mathbf{x}\|}k\left(\|\mathbf{y}-\mathbf{x}\| - L_0\right) \equiv \frac{\mathbf{y}-\mathbf{x}}{s(\mathbf{y},\mathbf{x})}k\left(s(\mathbf{y},\mathbf{x}) - L_0\right)
\tag{3}
$$

in which $\mathbf{y}$ denotes the vector of the Cartesian coordinates of the end-point of the crane's beam, $\mathbf{x}$ is the Cartesian position of the payload, and s denotes the distance between them. Parameter μ_A denotes the viscous damping coefficient of the payload slowly moving in the air (in this case laminar motion of the air is assumed), and $\mathbf{g}_{grav}$ denotes the vector of the gravitational acceleration. In the simulations the 2nd order equations (1) and (3) have to be numerically integrated. However, due to the laws of

Classical Mechanics, directly we can manipulate only $\ddot{\mathbf{y}}$ by the control force and torque components $\mathbf{Q}$ directly determining $\ddot{\mathbf{q}}$ as

$$\underbrace{\begin{bmatrix} L\cos q_1 & 1 \\ -L\sin q_1 & 0 \end{bmatrix}}_{\mathbf{J}(\mathbf{q})}\begin{bmatrix} \ddot{q}_1 \\ \ddot{q}_2 \end{bmatrix}+\underbrace{\begin{bmatrix} -L\sin q_1\dot{q}_1 & 0 \\ -L\cos q_1\dot{q}_1 & 0 \end{bmatrix}}_{\mathbf{j}(\mathbf{q},\dot{\mathbf{q}})}\begin{bmatrix} \dot{q}_1 \\ \dot{q}_2 \end{bmatrix}=\begin{bmatrix} \ddot{y}_1 \\ \ddot{y}_2 \end{bmatrix}. \tag{4}$$

According to (3) $\ddot{\mathbf{y}}$ occurs only in $\dot{\mathbf{F}}^{Conatct}$ that is directly related to the 4th time-derivative of $\mathbf{x}$. Therefore the control has to be elaborated for a 4th order system. A plausible possibility for that is the introduction of a "nominal trajectory" $\mathbf{x}^N(t)$ for the payload that can be designed by the crane-driver. By introducing a $\Lambda>0$ positive number the following tracking error relaxation can be prescribed on purely kinematical basis: $0=\left(\dfrac{d}{dt}+\Lambda\right)^4\left(\mathbf{x}^N-\mathbf{x}\right)$ from which a *desired 4th time-derivative for* $\mathbf{x}$ can be prescribed. Equation (4) can be used for relating $\ddot{\mathbf{q}}^{Des}$ to $\ddot{\mathbf{y}}^{Des}$ that is mathematically connected to the *desired acceleration of the crane's end-point* as

$$\ddot{\mathbf{x}}+\left[\frac{k(s-L_0)}{m_A s}\mathbf{I}+\frac{L_0 k}{m_A s^3}(\mathbf{y}-\mathbf{x})(\mathbf{y}-\mathbf{x})^T\right]^{-1}\times$$

$$\times\left[\mathbf{x}^{(4)Des}+\frac{\mu_A}{m_A}\mathbf{x}^{(3)}-\frac{2L_0\dot{s}k}{m_A s^2}(\dot{\mathbf{y}}-\dot{\mathbf{x}})-(\mathbf{y}-\mathbf{x})\frac{L_0 k}{m_A}\frac{(\dot{\mathbf{x}}-\dot{\mathbf{y}})^T(\dot{\mathbf{x}}-\dot{\mathbf{y}})-3\dot{s}^2}{s^3}\right]= \tag{5}$$

$$=\ddot{\mathbf{y}}^{Des}.$$

It is assumed that $\ddot{\mathbf{x}}$ can directly be measured by cheap, small acceleration sensors attached to the end-knot of the string carrying the payload, with respect to a system of coordinates rigidly attached to the sensors. As the initial position and initial zero velocity of the payload with respect to the approximately inertial "workshop system of reference" is known, via integration and coordinate transformations during a short trip $\dot{\mathbf{x}}$ and $\mathbf{x}$ can be considered as measurable, known values. (In this paper there is no enough room to detail the relationships between the direct sensorial data and the acceleration with respect to the inertial frame. It can briefly be noted that the swinging payload is rotated with respect to this frame.) Since (5) contains the value of $\ddot{\mathbf{x}}$ directly and it occurs in $\mathbf{x}^{(4)Des}$, too, its estimated value has to be used for the control. In similar manner the estimated value of the realized $\mathbf{x}^{(4)}$ also will be used by the controller. For model-based estimation for this purpose it is enough to know the lower order derivatives since

$$\dddot{\mathbf{x}}\equiv\mathbf{x}^{(3)}=-\frac{\mu_A}{m_A}\ddot{\mathbf{x}}+\frac{1}{m_A}\dot{\mathbf{F}}^{Contact},\dot{\mathbf{F}}_{cont}=\frac{\dot{\mathbf{y}}-\dot{\mathbf{x}}}{s}k(s-L_0)+L_0 k\dot{s}\frac{\mathbf{y}-\mathbf{x}}{s^2}$$

$$\mathbf{x}^{(4)}=\left[\frac{k(s-L_0)}{m_A s}\mathbf{I}+\frac{L_0 k}{m_A s^3}(\mathbf{y}-\mathbf{x})(\mathbf{y}-\mathbf{x})^T\right](\ddot{\mathbf{y}}-\ddot{\mathbf{x}})-\frac{\mu_A}{m_A}\mathbf{x}^{(3)}+\frac{2L_0\dot{s}k}{m_A s^2}(\dot{\mathbf{y}}-\dot{\mathbf{x}})+ \tag{6}$$

$$+(\mathbf{y}-\mathbf{x})\frac{L_0 k}{m_A}\frac{(\dot{\mathbf{x}}-\dot{\mathbf{y}})^T(\dot{\mathbf{x}}-\dot{\mathbf{y}})-3\dot{s}^2}{s^3}$$

Therefore we can use only the directly measurable coordinate-derivatives and instead of the exact model parameters m_A, μ_A, L_0, k their available estimations $\hat{m}_A, \hat{\mu}_A, \hat{L}_0, \hat{k}$ can be applied in the estimation of the realized $\ddot{x}$ and $x^{(3)}$ derivatives.

Regarding the adaptive control suggested it can be briefly noted that several control tasks can be formulated by using the concepts of the appropriate "excitation" $\mathbf{Q}$ of the controlled system to which it is expected to respond by some prescribed or "desired response" $\mathbf{r}^d$. The appropriate excitation can be computed by the use of some inverse dynamic model $\mathbf{Q} = \varphi(\mathbf{r}^d)$. Since normally this inverse model is neither complete nor exact, the actual response determined by the system's dynamics, ψ, results in a *realized response* $\mathbf{r}^r$ that differs from the desired one: $\mathbf{r}^r = \phi(\varphi(\mathbf{r}^d)) := \mathbf{f}(\mathbf{r}^d)$. It is worth noting that these functions may contain various hidden parameters that partly correspond to the dynamic model of the system, and partly pertain to unknown external dynamic forces acting on it. Due to phenomenological reasons the controller can manipulate or "deform" the input value from $\mathbf{r}^d$ so that $\mathbf{r}^d = \mathbf{f}(\mathbf{r}_*^d)$. Other possibility is the manipulation of the output of the rough model as $\mathbf{r}^d = \phi(\varphi_*(\mathbf{r}^d))$. In the sequel it will be shown that for *SISO* systems the appropriate deformation can be defined as some *Parametric Fixed Point Transformation*. The latest version elaborated for SISO systems was the function

$$G(r; r^d) = (r + K)[1 + B \tanh(A[f(r) - r^d])] - K \tag{7}$$

with the following properties: if $f(r_*) = r^d$ then $G(r_*, r^d) = r_*$, $G(-K, r^d) = -K$, and

$$G' = (r + K)\frac{BAf'(r)}{\cosh^2(A[f(r) - r^d])} + [1 + B \tanh(A[f(r) - r^d])] \tag{8}$$

that can be made contractive in the vicinity of r_* by properly setting the parameters A, B, and K, in which case the iterative sequence $r_{n+1} = G(r_n, r^d) \to r_*$ as $n \to \infty$. The saturated nonlinear behavior of the *tanh* function plays very important role in (7). The generalization of (7) for *Multiple Input – Multiple Output (MIMO)* systems may be done in different manners. A possibility that is used in this paper is giving the definition of the *response error* and its *direction* in the nth control step as $\mathbf{h}_n = \mathbf{f}(\mathbf{r}_n) - \mathbf{r}^d$, $\mathbf{e}_n := \mathbf{h}_n / \|\mathbf{h}_n\|$, and applying the following transformation:

$$\text{if } \|\mathbf{h}_n\| > \varepsilon \text{ then } \mathbf{x}_{n+1} = (1 + \tilde{B})\mathbf{x}_n + \tilde{B}K\mathbf{e} \text{ else } \mathbf{x}_{n+1} = \mathbf{x}_n, \ \tilde{B} := B\sigma(A\|\mathbf{h}_n\|) \tag{9}$$

in which ε is a small positive threshold value for the response error. If the response error is quite small, the system already attained the fixed point and no any manipulation is needed with the unit vector the computation of which would be singular. In the case of this implementation we have four control parameters, ε, A, B, and K, and a single sigmoid function $\sigma()$. This realization applies correction in the direction of the response error, and normally leads to more precise tracking than the more complicated one using separate control parameters for various directions. In the sequel this realization will be applied for the swinging reduction problem. The command signal given to the model-based controller will be referred to as "*required*" signal. In the non-adaptive case the *"required"* and *"desired"* values are equal to each other, while in the adaptive case they differ from each other according to the adaptive law coded in (9). In the sequel simulation results will be given to illustrate the applicability of this adaptive approach for automatic and precise tracking of the path of lifting determined by the crane driver in advance.

3 Simulation Results

In the simulations the more or less "exactly available" dynamic parameters of the crane itself can be separated from the less exactly known ones of the payload and the elastic cable. It can be assumed that the *stiffness of the cable "k"* itself depends on the zero force length of the cable. If it is assumed that at certain movements the cable is partly is coiled up on a spindle that has strong friction to keep the already wounded up part at fixed position we can assume that k can vary during the lifting process. In the simulations we consider a constant rotational position of the spindle to which not exactly known stiffness k pertains. The lifting considered happens only via the horizontal movement of the crane as a whole and by tilting its beam.

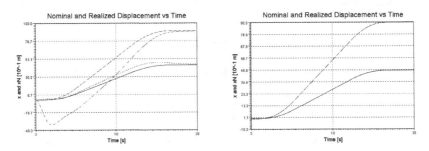

Fig. 1. The operation of the non-adaptive (LHS) and the adaptive (RHS) controllers: the nominal displacement of the payload (x^N_1: solid, x^N_2: dashed), and the realized one ending up in the vicinity of the nominal motion (x_1: dotted, x_2: dash-dot)

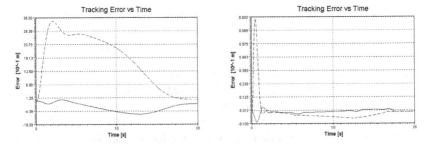

Fig. 2. The operation of the non-adaptive (LHS) and the adaptive (RHS) controllers: the trajectory tracking error of the payload (for x_1: solid, for x_2: dashed)

In the simulations the following numerical values were considered: $m_A = 50\,kg$, $\mu_A = 0.01\,Ns/m$, $L_0 = 3\,m$, $k = 500\,N/m$ for the payload and the cable, and $m = 10\,kg$, $M = 400\,kg$, $L = 12\,m$, $g = 9.81\,m/s^2$. The appropriate approximations in use were as $\hat{m}_A = 0.8m_A$, $\hat{\mu}_A = 0.02\,Ns/m$, $\hat{L}_0 = 2.4\,m$, $\hat{k} = 1.1k$ and $\hat{m} = 1.1m$, $\hat{M} = 0.9M$, $\hat{L} = L$, $\hat{g} = g$.

For the control $\Lambda = 7/s$, $A = 2 \times 10^{-6}$, $B = 1$, $K = -32000$ were used. The time-resolution of the numerical integration was $\delta t = 10^{-3}\,s$. For calculating the 3rd and 4th derivatives the model-based estimations were used. Figures 1 and 2 display the displacement and the tracking error of the payload, respectively, and Fig. 3 describes the variation of the crane's generalized co-ordinates. As it is well revealed by Figs. 1-3 the application of the adaptive deformation considerably improved the precision of trajectory tracking. The angle of tilt of the crane decreased while its body was translated in the q_2 direction. Accordingly, the negative torque Q_1 that was needed for lifting the payload continuously decreased in its absolute value (the solid line in Fig. 4).

It is worth noting that the nominal trajectory consisted of constant 3rd derivative, constant acceleration, and constant velocity segments, therefore the 4th time-derivative of the nominal motion was constantly zero. The 4th time-derivatives appear in the controlled motion only due to feeding back the tracking error in the control. Accordingly in Fig. 4 in the Q_2 component (apart from the initial "learning stage") only short accelerating and decelerating segments can be revealed.

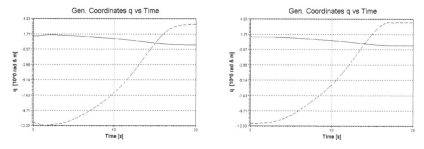

Fig. 3. The operation of the non-adaptive (LHS) and the adaptive (RHS) controllers: the displacement of the generalized co-ordinates of the crane (for q_1 in *rad:* solid, for q_2 in *m:* dashed)

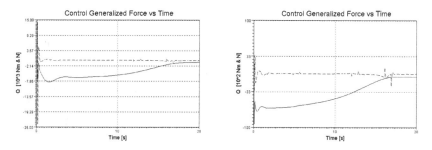

Fig. 4. The operation of the non-adaptive (LHS) and the adaptive (RHS) controllers: the generalized force components of the crane (for Q_1 in *Nm:* solid, for Q_2 in *N:* dashed)

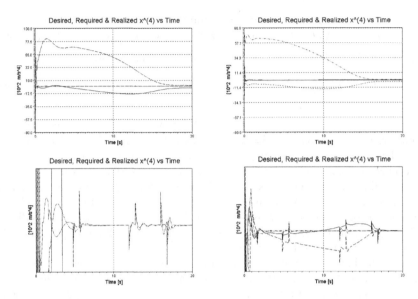

Fig. 5. The operation of the non-adaptive (LHS) and the adaptive (RHS) controllers: in the non-adaptive case the rough-model based "desired" 4th derivatives, in the adaptive case the deformed, "required" values are put into the approximate dynamic model ($x_1^{(4)d}$: solid, $x_2^{(4)d}$: dashed, $x_1^{(4)}$: long dash – dot – dot, $x_2^{(4)}$: long dash – dot; (in the adaptive case the "simulated" and "desired" curves run together, the 2nd row contains zoomed excerpts of the appropriate graphs)

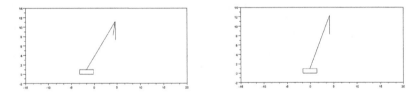

Fig. 6. The operation of the non-adaptive (LHS) and the adaptive (RHS) controllers: isometric view of the crane and the payload at time $t=13.766\ s$ (the end of the crane's beam is connected to the nominal and the simulated positions of the payload)

In Fig. 5 the quite considerable extent of "deformation" in the input value of the approximate dynamic model caused by the adaptive approach can be well identified.

Figure 6 well reveals of the significant extent of swinging in the non-adaptive case and that it considerably can be reduced by the proposed adaptive law.

It is interesting to investigate the significance of using the rough-model based approximations for observing $\mathbf{x}^{(3)}$ and $\mathbf{x}^{(4)}$. For this purpose the counterpart of the above simulations was run in which the estimation was replaced by the exactly calculated values (Fig. 7).

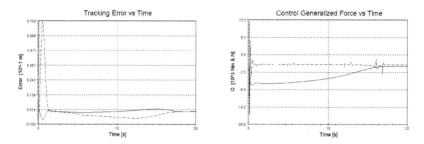

Fig. 7. The operation of the adaptive controller with the exact values of $\mathbf{x}^{(3)}$ and $\mathbf{x}^{(4)}$: the trajectory tracking error and the generalized force components exerted by the drives of the crane

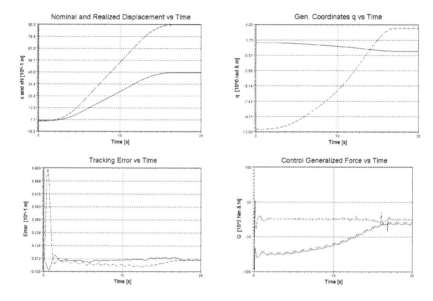

Fig. 8. The operation of the adaptive controller under external disturbances added to the generalized forces $\mathbf{Q}$

Comparing the so obtained values with the appropriate adaptive ones in Figs. 2 and 4 reveal that the use of the approximate values in the control results in small differences in the tracking accuracy and in the exerted generalized forces.

It is interesting to see the effect of disturbances acting on the driven axles of the crane on the adaptive control. For this purpose 3rd order polynomials were used to simulate the additive noises. Figure 8 reveals that quite considerable noise is quite well compensated by the adaptive controller.

It is also important to note that setting the adaptive control parameters is not a very difficult task. Via simulations proper order of magnitude easily can be found for these parameters. It can be expected that the appropriate values of parameters A, B, and K concerns the width of the basin of attraction of the convergence and its speed.

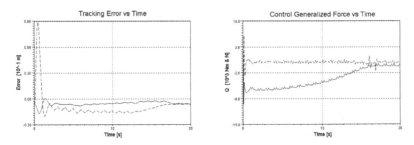

Fig. 9. The counterpart of Fig. 8 with modified adaptive control parameters $A=3\times10^{-6}$, $B=1$, $K=-16000$

To illustrate the robustness of the method for setting the adaptive control parameter a counterpart of Fig. 8 was created as Fig. 9. No significant differences can be observed in the controlled motion under the same noise influence.

4   Concluding Remarks

In this paper a robust fixed point transformations based adaptive approach to the problem of swinging payloads of cranes was presented. In this approach only 2nd order time-derivatives have to be numerically integrated though the control was formulated by the use of the 3rd and 4th time-derivatives of the displacement of the mass carried by the crane.

The system can design a lifting path consisting of segments of constant 3rd, 2nd, and 1st time-derivatives of the payload's displacement. The controller's task is to automatically execute this motion.

It was shown and illustrated via simulations that the simple adaptive approach suggested can efficiently compensate the effects of the model uncertainties of the crane and the payload as well as that of the unknown external disturbances. It was shown that it is satisfactory to measure only the 2nd time-derivatives of the payload via using simple and cheap acceleration sensors, and the approximate model values can be used for the estimation of the necessary 3rd and 4th time-derivatives without noise-sensitive numerical derivation.

In the simulations a 2 DOF crane model was applied and the motion under consideration was assumed to happen in a 2D plane. However, from the structure of the equations used it became clear that the extension of the results into a three dimensional trajectory does not require significant increase in the complexity of the problem: instead of 2D vectors simply 3D vectors have to be used in the same formulae, and the present crane model has to be replaced by a 3 DOF construction.

In the future the simulations can be improved by modeling the variation of the stiffness of the carne's cable as it is wound on a spindle, and 3D perturbations have to be introduced and studied. Furthermore, studying motions in which the designed trajectory simultaneous winding of the spindle and motion of the end point of the crane's beam would be expedient, too.

Acknowledgments. The authors gratefully acknowledge the support by the *National Office for Research and Technology* (*NKTH*) using the resources of the *Research and Technology Innovation Fund* within the projects *No. OTKA K063405, OTKA CNK-78168*. We also express our thanks for the support our research obtained from the *Bilateral Science & Technology Research Projects No. PT-12/2007* and *No. PL-14/2008*.

References

[1] Singer, N.C., Seering, W.P.: Preshaping Command Inputs to Reduce System Vibration. Journal of Dynamic Systems, Measurement, and Control 112, 76–82 (1990)

[2] Singhose, W.E., Porter, L.J., Tuttle, T.D., Singer, N.C.: Vibration Reduction Using Multi-Hump Input Shapers. J. of Dynamic Systems, Measurement, and Control 119, 320–326 (1997)

[3] Singhose, W., Singer, N., Seering, W.: Improving Repeatability of Coordinate Measuring Machines with Shaped Command Signals. Precision Engineering 18, 138–146 (1996)

[4] Khalid, A., Huey, J., Singhose, W., Lawrence, J., Frakes, D.: Human Operator Performance Testing Using an Input-shaped Bridge Crane. Journal of Dynamic Systems, Measurement and Control 128(4), 835–841 (2006)

[5] Kim, D., Singhose, W.: Manipulation with Tower Cranes Exhibiting Double-Pendulum Oscillations. In: Proc. of the 2007 IEEE International Conference on Robotics and Automation, Rome, Italy, April 10-14 (2007)

[6] Sorensen, K.L., Singhose, W.E., Dickerson, S.: A Controller Enabling Precise Positioning and Sway Reduction in Bridge and Gantry Cranes. Control Engineering Practice 15(7), 825–837 (2007)

[7] Singhose, W., Kim, D., Kenison, M.: Input Shaping Control of Double-Pendulum Bridge Crane Oscillations. Journal of Dynamic Systems, Measurement, and Control 130(3), 1–7 (2008)

[8] Huey, J.R.: The Intelligent Combination of Input Shaping and Pid Feedback Control. Dissertation, Georgia Institute of Technology (2006)

[9] Slotine, Jean-Jacques, E., Li, W.: Applied Nonlinear Control. Prentice Hall International, Inc., Englewood Cliffs (1991)

[10] Tar, J.K., Rudas, I.J., Hermann, G., Bitó, J.F., Tenreiro Machado, J.A.: On the Robustness of the Slotine-Li and the FPT/SVD-based Adaptive Controllers. WSEAS Transactions on Systems and Control 3(9), 686–700 (2008)

[11] Márton, L., Lantos, B.: Identification and Model-based Compensation of Stribeck Friction. Acta Polytechnica Hungarica 3(3), 45–58 (2006)

[12] Andoga, R., Madarász, L.: Situational Control Methology in Control of Aircraft Turbocompressor Engines. In: Proc. of the 3rd International Conference on Computional Cybernetics (ICCC 2005), Mauritius, April 13-16, pp. 345–350 (2005)

[13] Andoga, R., Főző, L., Madarász, L.: Digital Electronic Control of a Small Turbojet Engine MPM 20. Acta Polytechnica Hungarica 4(4), 83–96 (2007)

[14] Andoga, R., Főző, L., Madarász, L.: Use of Anytime Control Algorithms in the Area of Small Turbojet Engines. In: Proc. of the 6th IEEE International Conference on Computational Cybernetics (ICCC 2008), Stará Lesná, Slovakia, Slovakia, November 27-29, pp. 33–36 (2008)

[15] Madarász, L., Andoga, R., Főző, L.: Turbojet Engines – Progressive Methods of Modeling. Acta Mechanica Slovaca 13(3), 64–74 (2009)

[16] Tar, J.K., Rudas, I.J., Kozłowski, K.R.: Fixed Point Transformations-based Approach in Adaptive Control of Smooth Systems. In: Thoma, M., Morari, M. (eds.) Robot Motion and Control 2007. Lecture Notes in Control and Information Sciences, vol. 360 (2007); Kozłowski, K.R. (ed.) pp. 157–166. Springer, London (2007)

[17] Tar, J.K., Bitó, J.F., Rudas, I.J., Kozłowski, K.R., Tenreiro Machado, J.A.: Possible Adaptive Control by Tangent Hyperbolic Fixed Point Transformations Used for Controlling the Φ^6-Type Van der Pol Oscillator. In: Proc. of the 6th IEEE International Conference on Computational Cybernetics (ICCC 2008), Stará Lesná, Slovakia, November 27-29, pp. 15–20 (2008)

[18] Preitl, S., Precup, R.-E., Fodor, J., Bede, B.: Iterative Feedback Tuning in Fuzzy Control Systems. Theory and Applications, Acta Polytechnica Hungarica 3(3), 81–96 (2006)

[19] Tar, J.K., Rudas, I.J., Bitó, J.F., Tenreiro Machado, J.A., Kozłowski, K.R.: A Higher Order Adaptive Approach to Tackle the Swinging Problem. In: Proc. of the 10th International Symposium of Hungarian Researchers on Computational Intelligence and Informatics (CINTI 2009), Budapest, November 12-14 (2009)

[20] Tar, J.K., Bitó, J.F.: Adaptive Control Using Fixed Point Transformations for Nonlinear Integer and Fractional Order Dynamic Systems. In: Fodor, J., Kacprzyk, J. (eds.) Proc. of the Budapest Tech Jubilee Conference, Aspects of Soft Computing, Intelligent Robotics and Control, Budapest. SCI, vol. 241, pp. 253–267. Springer, Heidelberg (2009)

Robots as In-Betweeners

Jana Horáková[1] and Jozef Kelemen[2]

[1] Faculty of Arts, Masaryk University, Brno, Czech Republic
horakova@phil.muni.cz
[2] VŠM College of Management, Bratislava, Slovakia, and
Institute of Computer Science, Silesian University, Opava, Czech Republic
jkelemen@vsm.sk, kelemen@fpf.slu.cz

Abstract. The contribution sketches the emergence of the present day robotic art as the result and reflection of activities in fields of ancient mythology, the development of science, technology, art creativity and science fiction (both in the literature and in cinematography) in the first half of the previous century, and the convergence of art and scientific and technical development mainly during the second half of the 20th Century.

Keywords: robot, robotics, robotic art, science-fiction, cybernetics, engineering, art, cyberart, post-humanism, Khepera, R. U. R., Metropolis, Trial.

1 Introduction

The intellectual, ethical and aesthetical state of a society might be in certain extent understood on the base of relation of this society to some ideas living in it. One of such idea in the European culture[1] is the idea of artificially created more or less autonomous human-like artifacts. In this contribution we try to sketch the developmental line leading from the early concepts of human-like autonomous creatures (the Adam, golems, homunculi, mechanical androids from the past up to the robots and cyborgs of the present time) towards some interesting abstractions, the concept of the robot, for instance, and a new system aesthetics[2], up to the relatively common acceptance of autonomy in the work of art.

The development in above mentioned directions leads towards results, which deeply influenced the development of the culture of all of the Western type civilizations. Moreover, this line crossed in a new branch of the artistic creativity. The technology of robot design and engineering touch concepts dealing with emergence of the society and the culture built for and by human beings as well as by autonomous machines.

---

[1] However, the concept is, at least from the 18th Century, present also in the Japanese tradition; for more details see e.g. [18].

[2] For more about system aesthetics see [4]. Jack Burnham in [5] traces the developmental line in artistic creativity from static, inert objects through kinetic installations towards autonomous interacting open systems.

I.J. Rudas et al. (Eds.): Computational Intelligence in Engineering, SCI 313, pp. 115–127.
springerlink.com   © Springer-Verlag Berlin Heidelberg 2010

In order to contribute to better understanding of legal nature of our society attributed by all of the mentioned activities[3], our main goal in this contribution is to sketch the image of robots as some kind of *in-betweeners* staying somewhere between fictions and facts, engineering and arts, and science and dreams.

2 The First Robot

It is commonly known that the word *robot* appeared first in the play *R. U. R.* (Rossum's Universal Robots) – see e.g. [6] – by the Czech writer and journalist Karel Čapek (1890-1938). He wrote the *R. U. R.* during his and his brother's Josef (1887-1945) vacation in the house of their parents in the spa Trenčianske Teplice, in former Czechoslovakia (now in Slovakia) during the summer 1920. The official premier of the *R. U. R.* was in the Prague National Theatre in January 25, 1921 under the direction of Vojta Novák. Costumes have been designed by Josef Čapek, the stage for the performance was designed by Bedřich Feuerstein.

Čapek opened two among the most appealing topics of the 20th Century intellectual discourse by his play:

1) He replaced the old (eternal) human dream about construction of *human-like machines* to the modern age, called the *age of machine*, by robot characters.

2) He reacted in his play to the automation of the mass-production, which brought about questions emerging from intensive *human-machine interaction*[4], in different ways: e.g. (a) by a theme of crowds of robots produced by assembly lines of R. U. R. factory, (b) by a background of the scenes going on processes of dehumanization of humankind (man becomes machine-like), and humanization of artificial beings (through increasing machines human-like abilities), and (c) by situations of confusion cased by inability to distinguish between man and machine because of their similar behavior caused by generally shared adoration of machine-qualities of human body and mind in the age of machines.[5]

However, the artificial humanoid beings introduced by Čapek might be understood also as his humanistic reaction to the trendy concepts dominating the modernistic view of human beings in the first third of the 20th Century – the concept of a "new man" – which appeared most significantly in the numerous manifestos of the futurism. In connection with that Čapek's robots can be considered also as reflection of social and political situation of Europe immediately after the end of the World War I, and thus as a metaphor of workers dehumanized by the hard stereotypical work, and consequently as both an abused social class that reacts on its situation by revolts, and as a dehumanized crowd dangerously unpredictable in its actions.

Very soon after the first night of the *R. U. R.*, robots became to be understood as mechanical creatures not determined to be forever part of the drama-fiction, but sooner or later constructed in their realistic form, and replaced from the realm of

---

[3] This contribution is based on [17], and represents some further development of our ideas.
[4] More about continuous development from mechanization through automation to interactive technologies see in [19].
[5] More about the development and the transformation of the concept of robots we presented in [15].

imagery to the realm of facts – and implemented to become parts of our reality. But Čapek never accepted this position and never agreed with the interpretation, that his robots are some kind of a „tin and cogwheels" mechanisms. In the first act of his play, is his conviction declared by with words of the character Harry Domin: *And then, Miss Glory, old Rossum wrote among his chemical formulae: "Nature has found only one process by which to organize living matter. There is, however, another process, simpler, more moldable and faster, which nature has not hit upon at all. It is this other process, by means of which the development of life could proceed, that I have discovered this very day." Imagine, Miss Glory, that he wrote these lofty words about some phlegm of a colloidal jelly that not even a dog would eat. Imagine him sitting over a test tube and thinking how the whole tree of life would grow out of it, starting with some species of worm and ending – ending with man himself. Man made from a different matter than we are. Miss Glory, that was a tremendous moment* [7, p. 6].

So, no any "tin and cogwheels", no any engineering, but some kind of science, and some strange kind of colloidal jelly "not even a dog would eat" is the starting point for robots, according to Čapek's opinion. He requires in author's notes at the beginning of the play – see e.g. [6, p. 6] – that robots appearing in the *Prologue* of the *R. U. R.* should be dressed as humans, but on the other hand *...their* (the robots) *movements and pronunciation is brief, and faces are without emotional expressions* (ibid.). He expressed his position also in the Prague newspaper *Lidové noviny* (June 9, 1935) some years after the Prague first night: "*...robots are not mechanisms. They have not been made up from tin and cogwheels. They have been built up not for glory of the mechanical engineering. Having the author in his mind some admire of the human mind, it was not the admiration of technology, but that of the science. I am terrified by the responsibility for the idea that machines may replace humans in the future, and that in their cogwheels may emerge something like life, love or revolt*". So, Čapek recognized *robots* as a metaphor of a *simplified man* (more precisely, a simplified worker) not as a *sophisticated machine*. However, the author is never owner of his works and ideas. Real robots were soon built, even though autonomy and complexity of their behavior was rather imitated by the first real robots, than they would really possessed expected qualities of humanoid machines physically and intellectually transcending human.

The best way to characterize the first real robots of 20s and 30s of the 20th Century is to label them as *curiosities of the age of machines*. One among these very first real robots was made by Westinghouse Corp. in the USA in 1927, and was called *Mr. Televox*, according to the device (televox) controlled remotely by human voice, which *Mr. Televox* advertised in its productions. *Mr. Televox* was in fact the machine televox + humanoid limbs and head made from cardboard. His British robot-colleague, *Eric the Robot*, was a mechanical puppet built from steal with a support by captain W. H. Richards only one year later, in 1928, and was first time presented during Model Engineers' Society exhibition in London. *Eric* resembled a middle age knight look (in full outfit from steel) and it has written *R. U. R.* on his chest to not allow his viewers to hesitate about his lodestars.

3 The First Humanoid in Cinemas

The first robots – more or less close to the original concept of Čapek – appeared soon also in the cinematography, We can detect e.g. certain similarities with *R. U. R.* drama topics in very different works of art of the "machine age" period. We can see certain analogies in imagination of Charlie Chaplin, who in his famous movie *Modern Times* (1936) presented in his typical grotesque manner the assembly line "as a DJ", who governs the rhythm of workers movements, as well as social unrest of the time.

Even more fruitful is to compare *R. U. R.* with another example from the film production of the time of its first night. We can see many analogical themes with the *R. U. R.* play in a classical expressionistic sci-fi movie *Metropolis*. In the period of 1925-1926, the German film director Fritz Lang (1890-1976) has been working on the completion of this film. Financial and technical expenses had been unparalleled in history. This film is one of the most famous and influential silent films and sci-fi films to that date. Thea von Harbou wrote the screenplay after her own novel of the same title published in 1926. Cameramen were Karl Feund and Günter Rittau. Universum-Film AG in Berlin produced the film. Aenne Willkomm was the costume designer. The premiere was in Berlin, October 1, 1927 (the premier of the cut version was in New York City, May 3, 1927).

As well as this of the *R. U. R.* also the *Metropolis* story is moved from presence to imaginary realms of future and utopias: *Metropolis* – according to the note on the DVD containing the restored version[6] – takes place in the year 2026, when the population is divided between crowd of workers who must work hard and live in the dark underground and factory owners who enjoy a futuristic city of splendor above.

This separation between owners and worker and thus lack of communication between them seems to be shared opinion about social problems of the time. Visually is this separation described very similarly: Čapek, Lang (and Chaplin at least in certain extent, too) present the factory directors isolated in their offices and only remotely communicating with the employees. There are windows, telephones, kind of cameras and other devices to be able to govern and control from distance.

Similar to that one in *R. U. R.* is also the central conflict in *Metropolis*. It is a conflict between the classes of the "owners" and "workers" which ends with revolution and destruction. In both cases a revolution of the class of "factory workers" against the "factory-owners" is a culminating moment of the plot, and in both cases the solution of the problems is founded in the spiritual level. (Even though Lang is more optimistic than Čapek, because Lang bank on the young generation – the son of the factory owner – who serves as a "heard" or "mediator" between these two conflicting sides in *Metropolis*, described as "head and hands". Čapek, on a contrary, doesn't see solution in the level of human activities and he gives us, inspired with the bergsonian philosophy, only a hope into the vital power of nature and life).

The further point to compare is the robot characters, which appears in the *R. U. R.*, and in the *Metropolis*. While, Čapek's robots are beings which escape its precise definition in favor to metaphor of mechanized man of crowd and a product of factory

---

[6] *Metropolis* – restored authorized version with the original 1927 orchestral score (124 min.), licensed by Transit Films on behalf of the Friedrich-Wilhelm-Murnau-Stiftung, Wiesbaden, copyright 2002 by Kino International Corporation, New York, USA.

production, Lang works with robot character as an artificial creature with body from steel and "cogs and wheels" inside in more tradition way. We can say that Lang works with robot character in more traditional way. The Hel, creature made to resemble living woman Maria looks like she, but "inside" is her opposite. Maria-woman embodies Christian values of non-violence, patience and belief. But her double, the robot Maria cases revolution of workers, chaos and destruction. Lang presents the robot as an embodiment of evil, and as an instigator of the dark side of human character. Hel's costume, for instance, is clearly influenced by the "tin and cogwheels" idea, which replaced in robots "outlook" very quickly the original, Čapek's "organic" one. In fact, the "tin and cogwheels" robot-look has survived up to our days also thank to technology used in case of present day robots.

4   Robot as a Result of Science and Technology

First of all, it seems to be appropriate to clarify, the meaning of the word robot as we will use it. According to [22, p. 2], a robot is an autonomous system which exists in the physical world, can sense its environment, and can act on it to achieve some goals. Accepting this definition, all of the above mentioned robots and mechanical (wo)men, moreover, all human beings, and all animals are from this technical point of view robots. However, it seems to be useful in certain contexts to distinguish between all human beings and generally all animals, and the robots! So, in order to separate human beings and animals from what we intuitively understand as robots, it seems to us as useful to start the Matarić's definition as follows: A robot is by human beings constructed autonomous system which....

The history of robots, if accept our modification of Matarić's definition, started hundred years ago, when the first mechanical systems capable to react to its environments in order to prolong their functionality, are robots. An example of such old robotic concepts (having its roots several centuries before, somewhere in the Far-East, probably in Japan or China) is the autonomous mechanism (having the camouflage of a mechanical ladybug as in the Fig. 1, in today mechanical toy market).[7] The example of present sophisticated electromechanical robots for laboratory experiments is the Khepera robot (designed preferably for experimentation in the field of sociorobotics; cf. [23]) by Switzerland-based K-Team in the same figure.

The first attempt to make a machine that would imitate some aspects of the behavior of living creatures, e.g. the familiar test of animal intelligence in finding the way out of a maze, started from first third of the 20th Century. According [10, p. 110] Thomas Ross in the USA made a machine, which successfully imitated this experiment. R. A. Wallace built another creature of the same species, also in the USA in 1952. Claude Shannon has also devised a maze-learning creature, a sort of electromechanical mouse that fidgets its way out of confinement. In 1950 Alan Turing published his influential paper, which started researches in Artificial Intelligence, a

---

[7] More information on the rationality of the behavior of this type of machines, on their "intelligence" provides [21].

Fig. 1. The self-maintaining mechanical ladybug (a mechanical toy) which does not fall from the table top, and the experimental robotic platform Khepera by K-Team, both from the end of the 20th Century (photo: courtesy of J. Kelemen)

conceptual base of advanced robotics up to now, and – last but not least – Isaac Asimov named the field of scientific and engineering interests connected with robots by the word *robotics*. The word appeared for the first time in 1941, in the sentence: ... *advances in robotics these days were tremendous* – cf. e.g. [1, p. 31] – which represent everlasting fascination, which robots and activities connected with the effort to build them invoke in human.

William Grey Walter experimented with different electro-mechanical creatures, which contain also a sort of artificial neurons for decision making and learning from own experience. In [10] he provides also some detailed (electro-)technical descriptions on some of these creatures.[8] In [2] a sequence of machines with subsequently growing "psychic" capacities is proposed by Valentino Braitenberg.

Generally speaking, the robotics today has as its principal goal the development of as much as *necessary* autonomous systems, developed precisely according the needs of the research and the industry. This is the principal societal demand reflected also in financial supports of the robotic research. The development of humanoids as similar as possible to the real human beings (without any relation to their applications) remains to be a nice dream (however, time to time supported by grants for advanced laboratory research). In this meaning the old dreams meet the today more or less autonomous machine; Fig. 2 documents one (very symbolic) meeting of this kind.

---

[8] These above mentioned purely mechanical or electromechanical creatures (including ythe above mentioned mechanical ladybug) are real robots in the sense of Matarić's definition. In other side, creatures like the famous Honda's *Asimo* seem to be more *teleoperators* as real robots according to the same definition.

Fig. 2. The robot *Asimo* placing a bunch of flowers to the bust of Karel Čapek in August 2003 in the Prague National Museum (photo: courtesy of Lidové novin, used with permission)

5   Robot and Cyborg as Metaphors, and the Post-Human

Way of Euro-American society from the modern age to the post-modern era goes in parallel with transition of general (philosophical) thinking of the West from humanism to post-humanism. This change is clearly illustrated by replacement the metaphor of robot which dominated the beginning of the 20th Century by the metaphor of cyborg, referring to the 21st Century. (This change we have analyzed earlier, e.g. in [13]).

The *robot metaphor* is a product of a culminating phase of modern age (the machine age) identifying humanism with intellectual abilities (that enable human kind to overrule and subjugate its environment by means of instruments/machines that we construct). Robot is at the same time the figure of discourse defending differences between men and machines in favor of distance between the two concepts. This is a condition of both mirroring each other (man and robot/machine) on one hand, and fearing from the *other* (robot) on the other hand.

The *cyborg metaphor* refers to an other way of thinking joined with a post-industrial, information society, which is self-conscious about the dependence of the humankind on information and communication technologies. The cyborg functions also as a figure of the rejection of the (modern humanistic) anthropocentric views and critiques of its failures, and became an expression of the post-human condition that is usually characterized by blurring and questioning some of the borders between traditional dichotomy of the concepts by which we grasp our world. Thus, the cyborg metaphor forces our thinking on the relationship between human and machine beyond categories of irreconcilable oppositions[9].

The appearance of *cybernetics paradigm* during the 50s and 60s of the past Century that includes changes in understanding of human-machine relationship is closely connected with a fundamental transformation of machines themselves. This transformation is characterized by transition from factory work at assembly lines of the industrial societies towards the massive use of the information and communication technologies in post-industrial societies. Typical shape of contemporary machine is not anymore "mega-machine" set up from smaller machines and working according the principle of organized set of cog-wheels, but as a system which is possible to describe only in categories of each other overlapping, conditioning, influencing and each other permeating information (sub-)systems – in categories of an information net. So, we can conclude that the development of informatics and communication science together with the massive and mass use of highly sophisticated information and communication technologies markedly contributed to the spread of the development of the cybernetic paradigm and to its further modifications, and to the emergence of the cyborg metaphor for explaining the place of the human in his social and cultural context.

6   The In-Betweeners

The cybernetic paradigm has influenced not only the ways of the philosophical and sociological thinking on human-machine relationship but has spread over the field of art, more precisely over the domain of different kinds of so-called *living arts* (events, happenings, performances). It was the conceptual apparatus of cybernetics that systematically articulated relationships and processes (by terms as loop, feedback, close or open circuit). These terms were able to describe aesthetical happenings/experiences of artists and their audience, artworks, spectators and environment, so exactly these categories that dominated to the field of art in 1960ties. We provide some examples of that.

Nam June Paik in his manifesto entitled *Cybernated Art* wrote: *Cybernetics, the science of pure relationship itself, has its origin in karma. Marshall McLuhan´s famous phrase "medium is a message" was formulated by Norbert Wiener in 1948 as "The signal, where the message is sent, plays equally important role as the signal, where the message is not sent* [24, p. 229]. The manifesto clearly refers to categories that were the most evaluated on cybernetics by artists of 50-ies and 60-ies of the 20th Century.

---

[9] A famous text which redefined the understanding of the concept of the cyborg is e.g. [11].

Owing to the atmosphere in the field of the art after the World War II, in which we can trace down an accent on process, system, environment and participation of spectators, it was possible that cybernetics became certain theoretical model of a second half of the 20th Century art. This remarkable influence of cybernetics on field of art was mediated by aesthetical context that corresponded with scientific theories that appear in 1940ies and was possible thanks to the complementarily of cybernetics with central tendencies of the experimental art of the 20th Century.

As we have already mentioned above, it is possible to characterize this trend as a process of general process of convergence of science, technology, and art. An expression of this trend is artists' increasing interest in categories of scientific and engineering knowledge. However, this doesn't mean that this convergence or blurring of boundaries between spheres of art, science and technology necessarily have to mean an increasing just the type of domination of technology in our society with mostly negative connotations and fears. It can equally refer to an opposite tendency – to the extension of sphere of artistic creativity that in the late modernity phase has overflowed from ghettos of galleries to the streets, to the artists' as well as audience's private spaces and in thanks to the widespread of the cybernetics paradigm closely connected with information technologies find another space of inspiration in laboratories of scientists and engineers. This tendency has dominated to the art of the second half of the 20th Century as an expression of heading beyond boundaries of mimeses and re-presentation of the world in an artistic production. These processes are part of the very important shift in understanding of artist´s relationship to his/her work and creative process as such, that has certain analogies in technical discoveries e.g. in the fields of Artificial Intelligence (AI) and Artificial Life (AL)[10], and in constructions of autonomous robotic systems indicating certain elements of intelligence. Reading [26] we can make a picture of wide use of the newest scientific knowledge and technological inventions in a contemporary media art (or field called art and technology) as well as on the base of the conceptual heterogeneity of artistic creativity that realizes artistic potentials of new technologies.

It is not an accident that many artists working in 1960s in the field of conceptual or minimal art have become pioneers of so called cybernetics art. In cybernetics art generally we can usually meet with artistic presentation of the concept of machine (organic-artificial relationship) in a shape of hybrid penetration of biological and technological and blurring boundaries between them.

A good example of this approach to the biological-technological relationship, even not from the field of artistic but genuinely scientific creativity, is an experiment executed and described by British professor of cybernetics Kevin Warwick [25]. During this experiment/event Mr. and Mrs. Warwick communicated in a close circuit (between his and her nervous system) through connection of both nervous systems to the computer.

In the context of the cyberart we can demonstrate the above mentioned tendency also pointing to the well-known work of Australian performer Stelarc. He focuses on evolution and adaptation of human on highly technological environments reminding cyberpunk fiction in his performances/experiments. For example, he experiments/performs with prosthetic technologies as in case of a third hand or with

---

[10] We have connected AL with robots appearing in the Čapek's *R. U. R.* in [16].

telepresence technologies by which he connects his body through the www, and in this way he enables remote participants of his performances to stimulate motions of his muscles leading into unintentional gestures and motions of his body.

An inclination to blurring and to mudding boundaries between organic and digital systems implies efforts of artists to make experience of human-machine interaction as ´natural´ for human as possible, and an interface through witch we communicate with machines as invisible (transparent) as possible. This trend leads towards different shapes of digital environments (virtual realities) into which we can immerse and which react "spontaneously" (in real time) on even the smallest input/stimulation from side of users. In [3] a specification of this aesthetic strategy is formulated by J. D. Bolter and R. Grusin as a tendency towards *immediacy*. The authors argue that there exists another, complementary tendency in design of mediated experiences focused to make the formal and material qualities of the used media and technology a part of the experience, equal to the "message". They call this second aesthetic strategy as *hypermediacy*.

Robotic art as postulated e.g. in [20], can be understood through the prism of this second tendency, using the strategy of hypermediacy, from its origins in abound 1960s of the 20th Century. This aesthetic strategy is in the case of the robotic art joined with the concept of (human-machine) *interactivity*, which artists share with scientists and engineers working in robotics and in AI and AL, the scientific branches developed from preconditions established by cybernetics and informatics. They have been inspired directly by some of the concrete outcomes of scientists and technicians efforts to create kind of system, which behavior would simulate human behavior (in case of AI) or behavior of living organisms (in the case of AL).

The trend towards hypermediacy and elaboration on a concept of human-machine interactivity are well reflected in robotic art since very first works – e.g. in the case of *Robot K-456* (1964) by Nam June Paik, but notably in the case of Edward Ihnatowicz´s first autonomous robotic construct, *The Senster* (1969-1970), that showed certain marks of independent behavior contrasting with traditional associations connected with robot.

We can say that the concept of the robot (with it's long and contradictory cultural history) together with new technologies that enable artists to create unusual interactive communication scripts in a physical or virtual worlds, possibly telematic, spaces is in a context of contemporary art connected with a new aesthetical dimension that prefers modeling of behavior (an artist creates not only the form but also actions and reactions of robotic system according to inner or outer stimulations) over creation of static objects. Preference of behavior over form and system over object is understood as a general and characteristic feature of robotic and cybernetic art.

The just sketched cybernetic or computational approaches to the human beings lies in a heart of post-human thinking which evolution and metamorphoses in different cultural contexts (from literature to informatics) maps K. N. Hayles in [12]. She connects the acceptance of the post-humanism with a general spread of concept of cyborg as an expression or an image that characterizes the contemporary state of humanity. However, Hayles does not consider a cyborg as a human with added technological prostheses, repeatedly constructed from sci-fi imagery. She recognizes the shift from humanism to *post-humanism* as a process that have come on a conceptual level when we accepted definition of the human (under general influence of concepts of cybernetics and informatics discourses) as an information processing system with qualities similar to other kinds of such systems especially to intelligent computers and advanced robots as a sufficient description. In correspondence with the last opinions in the fields of AI and AL, Hayles accents the role of an "embodied cognition" that makes nonsense to think about functioning of any "processing system" without considering also the body of the system. *The human mind without human body is not human mind* she emphasizes [12, p. 246]. This sentence can be understood in two ways: On the one hand as a defense of humanity, and on the other hand, when we relate it to "postmodern machines"-computers, as an argument for acceptance of their "otherness", their irreducibility into analogies with functions of minds and bodies of their creators. The "otherness" of machines, their emancipation from our binary structured imagination, is one among the central themes of contemporary *robotic art.*

As an interesting and inspiring contribution to the understanding of human-machine relationship in a context of the post-humanism or the era of cyborg culture we refer to the robotic artwork of two Canadian artists – Luis-Philippe Demers and Bill Vorn. They create different types of noisy, violent, and uneasy "robotic ecosystems", as they call their installations (see Fig. 3, for instance, a snapshot from their robotic performance *The Trial*, inspired by Franz Kafka's famous novel). Their robots are not humanoid but purposely technomorphic, these robots do not act as a man, only perform their autonomous "machinity" – they become to be the *in-betweeners.*

Fig. 3. The stage with two robotic arms in the robotic performance *The Trial* (1999) by Bill Vorn and Luis-Philippe Demers (photo: courtesy of J.-Ph. Demers, used with permission)

Demers and Vorn call their robotic environments as "theatre of affect" to stress the emotional level on which their works communicate with viewers, in a frame of theatre situation known as a "suspension of disbelieve" [8]. Their robotic installations are constructed to reflect our experience of life in always more and more technological societies. However, they don't see this as a human treating. They believe that machines are (natural) part of our life and evolution [25, p. 124]. They describe their concept of human-machine relationship by comparison: *We understand machines as entities different from us insofar as we differ ourselves from a nature* [9]. So, we can say that they postpone the duty to formulate an own individual answer to everyone according to his/her individual perception of, understanding, and approach to this problem.

7 Conclusions

At the beginning of this contribution we promised to discuss the developmental line leading from the early dreams about the human-like autonomous creatures through literal and cinematographic fictions dealing with the concept of robots and similar artificially created human-like more or less autonomous beings, and from the scientific researches and technological engineering activities on the field of informatics, artificial intelligence, and robotics during the 20th Century towards new developmental lines of the art. We conclude, that this development led not only to the new forms of artistic expressions of the human being about himself, but also to the new situation, characterized very precisely in [9], according the opinion of which we start to ... *understand machines as entities different from us insofar as we differ ourselves from a nature*. So, we conclude, that somewhere between the dreams in the past, the present dreams, and between the engineering and scientific activities of the past and the present remains at east a small but important place for the in-betweeners – the creation which express the artistic reflection of the whole of that context.

Acknowledgments. The research of Jana Horáková is sponsored by the Grant Agency of the Czech Republic grant No. LC544. Jozef Kelemen's research is partially supported by Gratex International Corp., Bratislava, Slovakia.

References

[1] Asimov, I.: I, Robot. Bantam Books, New York (1991)
[2] Braitenberg, V.: Vehicles – Experiments in Synthetic Psychology. The MIT Press, Cambridge (1984)
[3] Bolter, J.D., Grusin, R.: Remediation–Understanding New Media. The MIT Press, Cambridge (1999)
[4] Burnham, J.: System Aestehtics. Artforum, (September 1968a),
     http://www.volweb.cz/horvitz/burnham/systems-esthetics.html
[5] Burnham, J.: Beyond Modern Sculpture: The Effects of Science and Technology on the Sculpture of This Century. Penguin Press, London (1968b)
[6] Čapek, K.: R. U. R. Aventinum, Praha (1923)
[7] Čapek, J.K.: R. U. R. and The Insect Play. Oxford University Press, Oxford (1961)

[8] Demers, L.-P., Horáková, J.: Anthropocentrism and the Staging of Robots. In: Adams, R., et al. (eds.) Transdisciplinary Digital Art, pp. 434–450. Springer, Berlin (2008)

[9] Demers, L.-P., Vorn, B.: Real Artificial Life as an Immersive Medium. In: Convergence – the 5th Biennial Symposium of Arts and Technology, pp. 190–203. Connecticut College, Connecticut (1995)

[10] Grey Walter, W.: The Living Brain. Penguin Books, Harmondsworth (1961)

[11] Haraway, D.: A Cyborg Manifesto – Science, Technology, and Socialist Feminism in the Late Twentieth Century. In: Simians, Cyborgs and Women – The Reinvention of Nature, pp. 149–181. Routledge, New York (1991)

[12] Hayles, K.N.: How We Became Posthuman, vol. Ill. The University of Chicago Press, Chicago (1999)

[13] Horáková, J.: Staging Robots – Cyborg Culture as a Context of Robots Emancipation. In: Trappl, R. (ed.) Cybernetics and System Research, pp. 312–317. Austrian Society for Cybernetics Studies, Vienna (2006)

[14] Horáková, J., Kelemen, J.: Robots – some Cultural Roots. In: Proc. 4th International Symposium on Computational Intelligence, pp. 39–50. Budapest Polytechnic, Budapest (2003)

[15] Horáková, J., Kelemen, J.: The Robot Story – Why Robots were Born and How They Grew Up. In: Husbands, P., et al. (eds.) The Mechanical Mind in History, pp. 283–306. The MIT Pess, Cambridge (2008)

[16] Horáková, J., Kelemen, J.: Artificial Living Beings and Robots – One Root, Variety of Influences. Artificial Life and Robotics 13, 555–560 (2009)

[17] Horáková, J., Kelemen, J.: Robots between Fiction and Facts. In: Proc. 10th International Conference on Computational Intelligence and Informatics. CINTI 2009, pp. 21–39. Budapest Tech, Budapest (2009)

[18] Hornyak, T.N.: Loving the Machine – The Art and Science of Japanese Robots. Kodanasha International, Tokyo (2006)

[19] Huhtamo, E.: From Cybernation to Interaction – a Contribution to an Archeology of Interactivity. In: Lunenfeld, P. (ed.) The Digital Dialectic – New Essays on New Media. The MIT Press, Cambridge (2001)

[20] Kac, E.: Origin and Development of Robotic Art. Art Journal 56, 60–67 (1997)

[21] Kelemen, J.: A Note on Achieving Low-Level Rationality from Pure Reactivity. Journal of Experimental and Theoretical Artificial Intelligence 8, 121–127 (1996)

[22] Matarić, M.J.: The Robotics Primer. The MIT Press, Cambridge (2007)

[23] Murphy, R.R.: Introduction to AI Robotics. The MIT Press, Cambridge (2000)

[24] Paik, N.J.: From Manifestos. In: Wardrip-Fruin, N., Montford, N. (eds.) The New Media Reader, p. 229. The MIT Press, Cambridge (2003)

[25] Warwick, K.: I, Cyborg, vol. Ill. University of Illinois Press, Chicago (2002)

[26] Whitelaw, M.: Metacreation – Art and Artificial Life. The MIT Press, Cambridge (2004)

Comparative Investigation of Various Evolutionary and Memetic Algorithms

Krisztián Balázs[1], János Botzheim[2], and László T. Kóczy[1,3]

[1] Department of Telecommunications and Media Informatics,
Budapest University of Technology and Economics, Hungary
[2] Department of Automation, Széchenyi István University, Győr, Hungary
[3] Institute of Informatics, Electrical and Mechanical Engineering, Faculty of Engineering
Sciences, Széchenyi István University, Győr, Hungary
{balazs,koczy}@tmit.bme.hu, {botzheim,koczy}@sze.hu

Abstract. Optimization methods known from the literature include gradient techniques and evolutionary algorithms. The main idea of gradient methods is to calculate the gradient of the objective function at the actual point and then to step towards better values according to this value. Evolutionary algorithms imitate a simplified abstract model of evolution observed in nature. Memetic algorithms traditionally combine evolutionary and gradient techniques to exploit the advantages of both methods. Our current research aims to discover the properties, especially the efficiency (i.e. the speed of convergence) of particular evolutionary and memetic algorithms. For this purpose the techniques are compared on several numerical optimization benchmark functions and on machine learning problems.

Keywords: evolutionary algorithms, memetic algorithms, fuzzy rule-based learning.

1 Introduction

The scope of engineering applications based on soft computing methods is continuously expanding in the field of complex problems, because of their favorable properties. Evolutionary computation (and evolutionary-based, e.g. memetic) methods form a huge part of these techniques. However, both theory and application practice still contain many unsolved questions, hence researching the theory and applicability of these methods is obviously an important and actual task. As the results of the investigations on applicability mean some kind of labeling for the involved methods, there are two outcomes of these investigations. One is the fact that they result in practical knowledge for industrial users concerning which techniques offer better possibilities and which ones are worth to be selected for integration into their respective products. The other one is feedback to the researchers regarding to in which direction they should continue their work.

Our work aims to investigate evolutionary based algorithms, which are numerical optimization techniques, so their efficiency can be characterized by the speed of

I.J. Rudas et al. (Eds.): Computational Intelligence in Engineering, SCI 313, pp. 129–140.
springerlink.com

convergence to the global optimum. Since so far there have not been invented any methods to obtain this property exactly, it can be figured out mostly by simulation. Therefore, this investigation is based on simulation carried out by using a modular software system implemented in C language, introduced in [1] and discussed deeper in [2].

It contains two larger units: a machine learning frame and a main optimization module. Thus, the system is able to deal with both optimization and machine learning problems.

The learning frame implements fuzzy rule-based learning with two inference methods, one using dense whereas the other one spare rule bases. The former method is called Mamdani-inference [3] and the latter one is the stabilized KH-interpolation technique [4], [5].

The optimization main module contains various sub-modules, each one implementing an optimization method, such as steepest descent [6] and Levenberg-Marquardt [7], [8] from the family of gradient based techniques, genetic algorithm [9] and bacterial evolutionary algorithm [10] both being evolutionary methods, furthermore particle swarm optimization technique [11], which is a type of swarm intelligence method. Obviously, memetic techniques [12] are also available in the software as the combination of the previous algorithm types.

The methods have been compared by their respective performance on various optimization benchmark functions (that are typically used in the literature to 'evaluate' global optimization algorithms) and on machine learning problems.

Although, many results have been published comparing particular evolutionary and memetic techniques (see e.g. [10], [13]), these discussions considered only a few methods and mainly focused on the convergence of the algorithms in terms of number of generations. However, different techniques have very differing computational demands. This difference is sometimes two or three orders of magnitude in computational time. Therefore the question arises: what is the relation of these methods compared to each other in terms of time? This has been set as the main question of this research.

Actually, our work is far from being complete, because we definitely have not implemented and compared all optimization and inference methods that can be found in the literature. This paper first of all tries to give a concept how such comparative investigations can be carried out.

A former state of this work can be found in [14].

The next section gives a brief overview of the algorithms and techniques used. After that, the benchmark functions and machine learning problems applied in the simulations will be described shortly. The simulation results and the observed behavior will be discussed in the fourth section. Finally, we summarize our work and draw some conclusions.

2   Overview of the Algorithms and Techniques Used

In order to carry out this investigation, it is necessary to overview two related theoretical topics and to point at the connection between them. One of these is numerical optimization and the other one is supervised machine learning.

The following subsections aim to give a brief overview of some important points of these theoretical aspects, which will be referred to later repeatedly in the paper.

2.1 Numerical Optimization

Numerical optimization [6] is a process, where the (global) optimum of an objective function $f_{obj}(\mathbf{p})$ is being searched for by choosing the proper variable (or parameter) vector $\mathbf{p}$. The optimum can be the maximum or the minimum of the objective function depending on the formulation of the problem.

There are several deterministic techniques as well as stochastic algorithms for optimization. Some of them will be presented below; these are the ones that were investigated in our work.

2.1.1 Gradient Methods

A family of iterative deterministic techniques is called gradient methods. The main idea of these methods is to calculate the gradient of the objective function at the actual point and to step towards better (greater if the maximum and smaller if the minimum is being searched) values using it by modifying $\mathbf{p}$. In case of advanced algorithms additional information about the objective function may also be applied during the iterations. For example, steepest descent (SD) [6] and Levenberg-Marquardt [7], [8] algorithms are members of this family of optimization techniques.

After a proper amount of iterations, as a result of the gradient steps, the algorithms find the nearest local minimum quite accurately. However, these techniques are very sensible to the location of the starting point. In order to find the global optimum, the starting point must be located close enough to it, in the sense that no local optima separate these two points.

2.1.2 Evolutionary Computation Methods

A family of iterative stochastic techniques is called evolutionary algorithms. These methods, like the genetic algorithm (GA) [9] or the bacterial evolutionary algorithm (BEA) [10], imitate the abstract model of the evolution observed in the nature. Their aim is to change the individuals in the population by the evolutionary operators to obtain better and better ones. The goodness of an individual can be measured by its 'fitness'. If an individual represents a solution for a given problem, the algorithms try to find the optimal solution for the problem. Thus, in numerical optimization the individuals are potentially optimal parameter vectors and the fitness function is a transformation of the objective function. If an evolutionary algorithm uses an elitist strategy, it means that the best ever individual will always survive and appear in the next generation. As a result, at the end of the algorithm the best individual will hold the (quasi-) optimal values for $\mathbf{p}$, i.e. the best individual will represent the (quasi-) optimal parameter vector.

2.1.3 Swarm Intelligence Techniques

Another type of iterative methods is called swarm intelligence techniques. These algorithms, like the particle swarm optimization technique (PSO) [11], are inspired by social behavior observed in nature, e.g. bird flocking, fish schooling. In these methods a number of individuals try to find better and better places by exploring their environment led by their own experiences and the experiences of the whole community. Since these methods are also based on processes of the nature, like GA or BEA, and there is also a type of evolution in them ('social evolution'), they can be categorized amongst evolutionary algorithms.

Similarly, like it was mentioned above, these techniques can also be applied as numerical optimization methods, if the individuals represent parameter vectors.

2.1.4 Memetic Algorithms

Evolutionary computation techniques explore the whole objective function, because of their characteristic, so they find the global optimum, but they approach it slowly, while gradient algorithms find only the nearest local optimum, however, they converge to it faster.

Avoiding the disadvantages of the two different technique types, evolutionary algorithms (including swarm intelligence techniques) and gradient-based methods may be combined (e.g. [12], [13]), for example, if in each iteration for each individual some gradient steps are applied. Expectedly, this way the advantages of both gradient and evolutionary techniques can be exploited: the local optima can be found quite accurately on the whole objective function, i.e. the global optimum can be approximated well.

There are several results in the literature confirming this expectation in the following aspect. Usually, the more difficult the applied gradient step is, the higher convergence speed the algorithm has in terms of number of generations. It must be emphasized, that most often these results discuss the convergence speed in terms of number of generations. However, the more difficult an algorithm is, the greater computational demand it has, i.e. each iteration takes longer.

Therefore the question arises: how does the speed of the convergence change in terms of time if the gradient technique applied in the method is changed?

Apparently, this is a very important question of applicability, because in real world applications time as a resource is a very important and expensive factor, but the number of generations the algorithm executes does not really matter.

This is the reason why the efficiency in terms of time was chosen to be investigated in this paper.

2.1.5 Variable Individual Length

Although, it often holds that a parameter vector is better than another one if it produces a better objective function value than the other one, sometimes there is another objective of the optimization, namely to minimize the length of $\mathbf{p}$, i.e. to minimize the number of the parameters. (This is a kind of multi-objective optimization.) In evolutionary algorithms this aim can be achieved by allowing the modification of the length of the individuals within the evolutionary operators applied during the optimization process [15] and using a penalty factor in the fitness function that makes the fitness value lower, if the length of the individual is higher and makes it greater, if the length is lower. In this case a shorter individual may have better fitness than a longer one, even if it produces a worse objective function value.

2.1.6 Algorithms Considered in this Paper

Our work considers nine algorithms with fixed individual lengths:

- Genetic algorithm, GA (without gradient steps)
- Genetic steepest descent, GSD (GA using SD steps)
- Genetic memetic algorithm, GMA (GA using LM steps)
- Bacterial evolutionary algorithm, BEA (without gradient steps)
- Bacterial steepest descent, BSD (BEA using SD steps)
- Bacterial memetic algorithm, BMA (BEA using LM steps)
- Particle swarm optimization, PSO (without gradient steps)
- Particle steepest descent, PSD (PSO using SD steps)
- Particle memetic algorithm, PMA (PSO using LM steps)

furthermore three algorithms with variable individual lengths:

- Extension of the bacterial evolutionary algorithm, BEAv (without gradient steps, with variable individual lengths)
- Extension of the bacterial steepest descent, BSDv (BEA using SD steps, with variable individual lengths)
- Extension of the bacterial memetic algorithm, BMAv (BEA using LM steps, with variable individual lengths)

2.2 Supervised Machine Learning

Supervised machine learning [16] means a process where parameters of a 'model' are being adjusted so that its behavior becomes similar to the behavior of the 'system', which is to be modeled. Since the behavior can be characterized by input-output pairs, the aim of the learning process can be formulated so that the modeling system should give similar outputs for the input as the original system does.

The model can be for example a simple function (e.g. a polynomial function), where the parameters are the coefficients, or it can be a neural network, where the parameters are the weights, or it can be a fuzzy rule base together with an inference engine [17]. In this case the parameters can be the characteristic points of the membership functions of the rules in the rule base. In our work we applied a fuzzy rule base combined with an inference engine using both Mamdani-inference [3] and stabilized KH-interpolation techniques [4], [5].

If a function $\phi(\mathbf{x})$ denotes the system and $f(\mathbf{x},\mathbf{p})$ denotes the model, where $\mathbf{x} \in \mathbf{X}$ is the input vector and $\mathbf{p}$ is the adjustable parameter vector, the previous requirement can be expressed as follows:

$$\forall \mathbf{x} \in \mathbf{X} : \phi(\mathbf{x}) \approx f(\mathbf{x},\mathbf{p})$$

In a supervised case the learning happens using a set of training samples (input-output pairs). If the number of samples is m, the input in the ith sample is $\mathbf{x}_i$, the desired output is $d_i = \phi(\mathbf{x}_i)$ and the output of the model is $y_i = f(\mathbf{x}_i,\mathbf{p})$, the following formula can be used:

$$\forall i \in [1,m] : d_i \approx y_i$$

The error (ε) shows how similar the modeling system to the system to model is. It is the function of the parameter vector, so it can be denoted by $\varepsilon(\tilde{\mathbf{p}})$. Let us use a widely applied definition for the error, the Mean of Squared Errors (MSE):

$$\varepsilon(\mathbf{p}) = \frac{\sum_{i=1}^{m}(d_i - y_i)^2}{m}$$

Obviously, the task is to minimize this $\varepsilon(\tilde{\mathbf{p}})$ function. It can be done by numerical optimization algorithms.

This way machine learning problems can be traced back to optimization problems, furthermore they can be applied to discover the efficiency of evolutionary and memetic algorithms.

3   Benchmark Functions and Machine Learning Problems

The optimization benchmark functions and machine learning problems, which were used during our investigations, will be described in this section.

In case of the benchmark functions the minimum value, in case of the learning problems the minimum error is searched.

3.1   Benchmark Functions

In our investigations five benchmark functions were applied: Ackley's [18], Keane's [19], Rastrigin's [20], Rosenbrock's [21] and Schwefel's function [22]. These functions are widely used in the literature to evaluate global optimization methods, like evolutionary techniques. They are generic functions according to dimensionality, i.e. the number of dimensions of these functions can be set to an arbitrary positive integer value. In our simulations this value was set to 30, because it is a typical value in the literature for these functions.

Ackley's, Keane's, Rastrigin's and Schwefel's functions are multimodal, i.e. they have more than one local optima (actually they have a number of local optimum).

Rastrigin's function is separable as well as Schwefel's function. This means that the minimization along each dimensions result the minimum.

For example Ackley's benchmark function is given as follows (k denotes the number of dimensions):

$$f_{Ack}(\mathbf{x}) = 20 + e - 20e^{\left(-0.2\sqrt{\frac{1}{k}\sum_{i=1}^{k}x_i^2}\right)} - e^{\left(\frac{1}{k}\sum_{i=1}^{k}\cos(2\pi x_i)\right)}, \quad \forall x_i \in [-30,30]$$

3.2   Machine Learning Problems

In our investigations three machine learning problems were applied: the one dimensional pH [13], the two dimensional ICT [13] and a six dimensional problem that was also used by Nawa and Furuhashi to evaluate the performance of bacterial evolutionary algorithm [10].

For example this six dimensional function is defined as follows:

$$f_{6\dim} = x_1 + \sqrt{x_2} + x_3 x_4 + 2e^{2(x_5-x_6)},$$

$$x_1, x_2 \in [1,5], \quad x_3 \in [0,4], \quad x_4 \in [0,0.6], \quad x_5 \in [0,1], \quad x_6 \in [0,1.2]$$

4 Results and Observed Behavior

Simulation runs were carried out in order to compare the efficiency of the various numerical optimization methods. All possible combinations of the mentioned inference methods, optimization algorithms, optimization benchmark functions and machine learning problems were considered in case of fix length optimization techniques, i.e. in case of methods working with fixed parameter vector lengths.

Variable length methods, i.e. techniques coping with variable parameter vector lengths, were applied only on machine learning problems, because the optimization benchmark functions had fixed dimensionality, whereas in the learning problems the size of the rule base determined by the length of the parameter vector could modify. Since bacterial based algorithms performed better than genetic and particle swarm based ones (see later), for variable length simulations the modifications of bacterial based techniques (BEAv, BSDv, BMAv) were applied.

In the simulations the parameters had the following values, because after a number of test runs these values seemed to be suitable. The number of rules in the rule base was 4 in case of fix rule base size and the maximum number of rules was 15 in case of variable rule base size. The number of individuals in a generation was 14 in genetic and bacterial algorithms, whereas it was 80 in particle swarm methods. In case of genetic techniques the selection rate was 0.3 and the mutation rate was 0.1, in case of bacterial techniques the number of clones was 5 and 4 gene transfers were carried out in each generation. The genetic methods applied elitist strategy. In case of memetic algorithms 8 iteration long gradient steps were applied. The gradient vector and the Jacobian matrix computing functions were not given, hence pseudo-gradients and pseudo-Jacobians were computed where steepest descent or Levenberg-Marquardt gradient steps were used.

The numbers of training samples were between 100 and 200 in the learning processes.

During the runs the fitness values of the best individuals were monitored in terms of time. In case of variable length runs for learning problems the fitness values were calculated based on the MSE values (measured on the training samples) as follows:

$$F = \frac{10}{MSE+1}(1-0.02(N_{rules}-1)) = \frac{10m}{\sum\limits_{i=1}^{m}(d_i-y_i)^2 + m}(1-0.02(N_{rules}-1)),$$

where N_{rules} is the number of rules in the rule base and $(1-0.02(N_{rules}-1))$ is the penalty factor (mentioned in Section 2.1.5), which was defined according to the experiences we gathered in the test runs. This factor serves the goal that the optimization process to produce a smaller resultant rule base.

In case of fix length runs for learning problems the fitness was calculated similarly, but it did not contain the penalty factor.

For the optimization benchmark functions the penalty factor also did not appear and the difference between the value at the actual point and the minimum value was substituted for MSE in the previous equation.

In case of all algorithms for all benchmark functions and learning problems 10 runs were carried out. Then we took the mean of the obtained values. These means were presented in figures to get a better overview. The horizontal axes show the elapsed computation time in seconds and the vertical axes show the fitness values of the best individuals at the current time.

On the figures (see later) dashed lines show the result of the pure evolutionary algorithms (GA, BEA and PSO), dotted lines denote the techniques using steepest descent gradient steps and solid lines present the graphs of methods using Levenberg-Marquard technique.

The results of the runs and their short explanations follow in the next subsections. Every simulation will not appear though, because their great number does not allow it, rather some example results will be presented in the next three subsections. The results for the other optimization benchmark functions and learning problems are mostly, but not always similar, therefore these examples present only a behavior that were observed most often.

In subsection 4.4 conclusions will be drawn about the behavior of the methods considering all of the simulations.

4.1 Results for Ackley's Benchmark Function

This example presents the performance of the compared techniques on Ackley's benchmark function.

Figure 1 shows the fitness values of the best individuals in terms of time.

As it can be observed, bacterial algorithms gave better results than genetic techniques. Actually bacterial methods found the global optimum (the 10 fitness value indicates this fact). At the beginning BEA and BSD were better than BMA, but after an adequate time BMA was not worse than any other algorithm.

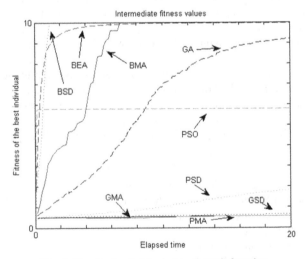

Fig. 1. Results for Ackley's benchmark function

Bacterial techniques were better than the corresponding genetic and particle swarm methods (i.e. BEA was better than GA and PSO, BSD was better than GSD and PSD, etc.).

4.2 Results in Case of Applying Mamdani-Inference-Based Learning together with Fix Length Optimization Techniques for the ICT Problem

This example presents the performance of the compared techniques on the two dimensional ICT problem, when Mamdani-inference based learning was applied.

Like in the previous case, there is an adequate time again from when BMA was not worse than any other technique; however at the beginning BEA was better.

At the end GMA gave the second best result. Thus it was better than any other genetic algorithms (see Figure 2).

The methods using steepest descent gradient steps were the worst among both bacterial and genetic algorithms.

Again, bacterial techniques were always better than the corresponding genetic and particle swarm methods.

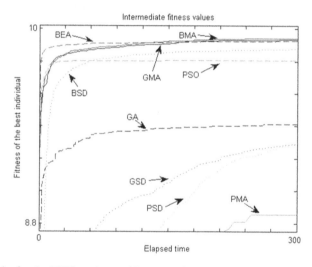

Fig. 2. Results for the ICT learning problem applying Mamdani-inference-based learning

4.3 Results in Case of Applying Stabilized KH-Interpolation-Based Learning together with Variable Length Optimization Techniques for the Six Dimensional Problem

This example presents the performance of the compared variable length techniques on the six dimensional learning problem, when stabilized KH-interpolation based learning was applied.

In this case from the beginning BMAv gave far the best fitness values (see Figure 3).

The method using steepest descent gradient steps was the worst among the variable length algorithms.

4.4 Observed Behavior

Based on the simulation results we observed the following behaviors:

- Generally, bacterial techniques seemed to be better than the corresponding genetic and particle swarm methods.
- Except Rosenbrock's function, in case of optimization benchmark functions PSO seemed to be the worst among techniques using no gradient steps, however for learning problems it always performed better than GA and sometimes relatively close to BEA.
- Generally, among particle swarm methods PSO gave better results than the algorithms using gradient steps.
- Generally, PMA seemed to be the worst technique using Levenberg-Marquardt gradient steps.
- Generally, for learning problems GMA had the highest convergence speed among genetic algorithms.
- Usually (but not always), after a sufficient time, BMA was not worse than any other fix length algorithm and BMAv was not worse than any other variable length technique. The more difficult the problem is, the better the advantage of the bacterial memetic techniques appears.

Considering the fix length methods, it might be said that BMA advances 'slowly but surely' to the optimum. 'Slowly', because in most of the cases at the beginning it did not have the highest convergence speed. 'Surely', because during the simulations it did not lose so much from its efficiency than the other techniques.

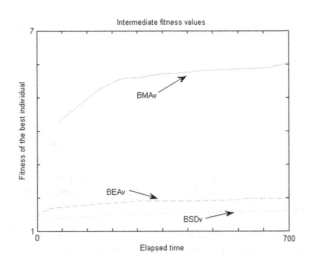

Fig. 3. Results for the six dimensional learning problem applying stabilized KH-interpolation based learning

5 Conclusions

In our work various evolutionary and memetic algorithms have been compared on general optimization benchmark functions and on machine learning problems.

After we carried out simulation runs and investigated the obtained results, we drew some conclusions about the behavior of the various techniques compared to each other. They can be summed up shortly as follows.

Generally, bacterial techniques seemed to be better than the corresponding genetic and particle swarm methods. For optimization benchmark functions (except Rosenbrock's function) PSO was outperformed by both other techniques using no gradient steps, however for learning problems it performed better than GA and sometimes relatively close to BEA. Usually, in case of genetic and bacterial methods algorithms applying LM technique seemed to be better for learning problems than methods not using gradient steps or using SD, however among particle swarm techniques the algorithm applying no gradient steps seemed to be the best.

To reinforce these tendencies, as a continuation of this work, carrying out more simulations is necessary. Further research may aim to compare other global optimization algorithms that can be found in the literature.

Acknowledgments. This paper was supported by the National Scientific Research Fund Grant OTKA K75711, a Széchenyi István University Main Research Direction Grant and the Social Renewal Operation Programme TÁMOP–4.2.2 08/1–2008–0021.

References

[1] Balázs, K., Kóczy, L.T., Botzheim, J.: Comparison of Fuzzy Rule-based Learning and Inference Systems. In: Proceedings of the 9th International Symposium of Hungarian Researchers on Computational Intelligence and Informatics, CINTI 2008, Budapest, Hungary, pp. 61–75 (2008)

[2] Balázs, K.: Comparative Analysis of Fuzzy Rule-based Learning and Inference Systems, Master's Thesis, Department of Telecommunications and Media Informatics, Budapest University of Technology and Economics, Budapest, Hungary, 97 p (2009)

[3] Mamdani, E.H.: Application of Fuzzy Algorithms for Control of Simple Dynamic Plant. IEEE Proc. 121(12), 1585–1588 (1974)

[4] Kóczy, L.T., Hirota, K.: Approximate Reasoning by Linear Rule Interpolation and General Approximation. Internat. J. Approx. Reason. 9, 197–225 (1993)

[5] Tikk, D., Joó, I., Kóczy, L.T., Várlaki, P., Moser, B., Gedeon, T.D.: Stability of Interpolative Fuzzy KH-Controllers. Fuzzy Sets and Systems 125, 105–119 (2002)

[6] Snyman, J.A.: Practical Mathematical Optimization: An Introduction to Basic Optimization Theory and Classical and New Gradient-based Algorithms. Springer, New York (2005)

[7] Levenberg, K.: A Method for the Solution of Certain Non-Linear Problems in Least Squares. Quart. Appl. Math. 2(2), 164–168 (1944)

[8] Marquardt, D.: An Algorithm for Least-Squares Estimation of Nonlinear Parameters. J. Soc. Indust. Appl. Math. 11(2), 431–441 (1963)

[9] Holland, J.H.: Adaption in Natural and Artificial Systems. The MIT Press, Cambridge (1992)

[10] Nawa, N.E., Furuhashi, T.: Fuzzy System Parameters Discovery by Bacterial Evolutionary Algorithm. IEEE Transactions on Fuzzy Systems 7(5), 608–616 (1999)

[11] Kennedy, J., Eberhart, R.: Particle Swarm Optimization. In: Proceedings of the IEEE International Conference on Neural Networks (ICNN 1995), Perth, WA, Australia, vol. 4, pp. 1942–1948 (1995)

[12] Moscato, P.: On Evolution, Search, Optimization, Genetic Algorithms and Martial Arts: Towards Memetic Algorithms, Technical Report Caltech Concurrent Computation Program, Report. 826, California Institute of Technology, Pasadena, California, USA (1989)

[13] Botzheim, J., Cabrita, C., Kóczy, L.T., Ruano, A.E.: Fuzzy Rule Extraction by Bacterial Memetic Algorithms. In: Proceedings of the 11th World Congress of International Fuzzy Systems Association, IFSA 2005, Beijing, China, pp. 1563–1568 (2005)

[14] Balázs, K., Botzheim, J., Kóczy, L.T.: Comparative Analysis of Various Evolutionary and Memetic Algorithms. In: Proceedings of the 10th International Symposium of Hungarian Researchers on Computational Intelligence and Informatics, CINTI 2009, Budapest, Hungary, pp. 193–205 (2009)

[15] Drobics, M., Botzheim, J., Kóczy, L.T.: Increasing Diagnostic Accuracy by Meta Optimization of Fuzzy Rule Bases. In: IEEE International Conference on Fuzzy Systems, FUZZ-IEEE 2007, London, UK, pp. 271–275 (2007)

[16] Alpaydin, E.: Introduction to Machine Learning, 445 p. MIT Press, Cambridge (2004)

[17] Driankov, D., Hellendoorn, H., Reinfrank, M.: An Introduction to Fuzzy Control, 316 p. Springer, New York

[18] Ackley, D.: An Empirical Study of Bit Vector Function Optimization. Genetic Algorithms and Simulated Annealing, 170–215 (1987)

[19] Keane, A.: Experiences with Optimizers in Structural Design. In: Parmee, I.C. (ed.) Proceedings of the 1st Conf. on Adaptive Computing in Engineering Design and Control, University of Plymouth, pp. 14–27. University of Plymouth, UK (1994)

[20] Rastrigin, L.A.: Extremal Control Systems. Theoretical Foundations of Engineering Cybernetics Series. Moscow, Russian (1974)

[21] Rosenbrock, H.H.: An Automatic Method for Finding the Greatest or Least Value of a Function. Computer Journal (3), 175–184 (1960)

[22] Schwefel, H.P.: Numerical Optimization of Computer Models. John Wiley & Sons, Chichester (1981); english translation of Numerische Optimierung von Computermodellen mittels der Evolutionsstrategie (1977)

A Novel Approach to Solve Multiple Traveling Salesmen Problem by Genetic Algorithm

András Király and János Abonyi

Department of Process Engineering, University of Pannonia
P.O. Box 158, H-8200 Veszprém, Hungary
`kiralya@fmt.uni-pannon.hu`

Abstract. The multiple Traveling Salesman Problem (mTSP) is a complex combinatorial optimization problem, which is a generalization of the well-known Traveling Salesman Problem (TSP), where one or more salesmen can be used in the solution. The optimization task can be described as follows: given a fleet of vehicles, a common depot and several requests by the customers, find the set of routes with overall minimum route cost which service all the demands. Because of the fact that TSP is already a complex, namely an NP-complete problem, heuristic optimization algorithms, like genetic algorithms (GAs) need to take into account. The extension of classical GA tools for mTSP is not a trivial problem, it requires special, interpretable encoding to ensure efficiency. The aim of this paper is to review how genetic algorithms can be applied to solve these problems and propose a novel, easily interpretable representation based GA.

Keywords: mTSP, VRP, genetic algorithm, multi-chromosome, optimization.

1 Introduction

In logistics, the main goal is to get the right materials to the right place at the right time, while optimizing some performance measure, like the minimization of total operating cost, and satisfying a given set of constraints (e.g. time and capacity constraints). In logistics, several types of problems could come up; one of the most remarkable is the set of route planning problems. One of the most studied route planning problem is the Vehicle Routing Problem (VRP), which is a complex combinatorial optimization problem that can be described as follows: given a fleet of vehicles with uniform capacity, a common depot, and several requests by the customers, find the set of routes with overall minimum route cost which service all the demands. The complexity of the search space and the number of decision variables makes this problem notoriously difficult.

The relaxation of VRP is the multiple traveling salesman problem (mTSP) [3], which is a generalization of the well-known traveling salesman problem (TSP) [10], where one or more salesman can be used in the solution. Because of the fact that TSP

I.J. Rudas et al. (Eds.): Computational Intelligence in Engineering, SCI 313, pp. 141–151.

belongs to the class of NP-complete problems, it is obvious that mTSP is an NP-hard problem thus it's solution require heuristic approach.

In this paper tools developed for a modified mTSP related to the optimization of one to many distribution systems will be studied and a novel genetic algorithm based solution will be proposed.

In the case of mTSP, a set of nodes (locations or cities) are given, and all of the cities must be visited exactly once by the salesmen who all start and end at the single depot node. The number of cities is denoted by n and the number of salesman by m. The goal is to find tours for all salesmen, such that the total travelling cost (the cost of visiting all nodes) is minimized. The cost metric can be defined in terms of distance, time, etc. Some possible variations of the problem are as follows:

- *Multiple depots:* If there exist multiple depots with a number of salesmen located at each, a salesman can return to any depot with the restriction that the initial number of salesmen at each depot remains the same after all the travel.
- *Number of salesmen:* The number of salesmen in the problem can be a fixed number or a bounded variable.
- *Fixed charges*: If the number of salesmen is a bounded variable, usually the usage of each salesman in the solution has an associated fixed cost. In this case the minimization of this bounded variable may be involved in the optimization.
- *Time windows:* Certain cities must be visited in specific time periods, named as time windows. This extension of mTSP is referred to as multiple Traveling Salesman Problem with Time Windows (mTSPTW).
- *Other restrictions:* These additional restrictions can consist of the maximum or minimum distance or travelling duration a salesman travels, or other special constraints.

mTSP is more capable to model real life applications than TSP, since it handles more than one salesmen. An overview of application areas can be found in [3] and in [10]. In the paper, an mTSPTW problem will be optimized with a novel approach, where the number of salesmen is an upper bounded variable, and there exist additional constraints, like the maximum travelling distance of each salesman.

Usually, mTSP is formulated by integer programming formulations. One variation is presented in equations (1.1)-(1.7). The mTSP problem is defined on a graph $G = (V,A)$, where V is the set of n nodes (vertices) and A is the of arcs (edges). Let $\mathbf{C} = (c_{ij})$ be a cost (distance) matrix associated with A. The matrix $\mathbf{C}$ is *symmetric* if $c_{ij} = c_{ji}, \forall (i, j) \in A$ and *asymmetric* otherwise. Here $x_{ij} \in \{0,1\}$ is a binary variable used to represent that an arch is used on the tour and c_m represents the cost of the involvement of one salesman in the solution. Further mathematical representations can be found in [3].

$$\min \sum_{i=1}^{n} \sum_{j=1}^{n} c_{ij} x_{ij} + m c_m \qquad (1.1)$$

so that

$$\sum_{j=2}^{n} x_{1j} = m \tag{1.2}$$

$$\sum_{j=2}^{n} x_{j1} = m \tag{1.3}$$

$$\sum_{i=1}^{n} x_{ij} = 1, \quad j = 2, \hbar, n \tag{1.4}$$

$$\sum_{j=1}^{n} x_{ij} = 1, \quad i = 2, \hbar, n \tag{1.5}$$

$$+ \text{ subtour elimination constratins} \tag{1.6}$$

$$x_{ij} \in \{0,1\}, \forall (i, j) \in A \tag{1.7}$$

2 Literature Review

In the last two decades the traveling salesman problem received quite big attention, and various approaches have proposed to solve the problem, e.g. branch-and-bound [7], cutting planes [17], neural network [4] or tabu search [9]. Some of these methods are exact algorithms, while others are near-optimal or approximate algorithms. The exact algorithms use integer linear programming approaches with additional constraints.

The mTSP is much less studied like TSP. [3] gives a comprehensive review of the known approaches. There are several exact algorithms of the mTSP with relaxation of some constraints of the problem, like [15], and the solution in [1] is based on Branch-and-Bound algorithm.

Due to the combinatorial complexity of mTSP, it is necessary to apply some heuristic in the solution, especially in real-sized applications. One of the first heuristic approach were published by Russell [23] and another procedure is given by Potvin et al. [20]. The algorithm of Hsu et al. [12] presented a Neural Network-based solution.

More recently, genetic algorithms (GAs) are successfully implemented to solve TSP [8]. Potvin presents a survey of GA approaches for the general TSP [21].

2.1 Application of Genetic Algorithms to Solve mTSP

Lately GAs are used for the solution of mTSP too. The first result can be bound to Zhang et al. [25]. Most of the work on solving mTSPs using GAs has focused on the vehicle scheduling problem (VSP) ([16, 18]). VSP typically includes additional constraints like the capacity of a vehicle (it also determines the number of cities each vehicle can visit), or time windows for the duration of loadings. Recent application can be found in [4], where GAs were developed for hot rolling scheduling. It converts the mTSP into a single TSP and apply a modified GA to solve the problem.

A new approach of chromosome representation, the so-called *two-part chromosome technique* can be found in [5] which reduces the size of the search space by the elimination of redundant solutions. According to the referred paper, this representation is the most effective one so far.

There are several representations of mTSP, like *one chromosome technique* [20], the *two chromosome technique* [16, 18] and the latest *two-part chromosome technique*. Each of the previous approaches has used only a single chromosome to represent the whole problem, although salesmen are physically separated from each other. The novel approach presented in the next chapter use multiple chromosomes to model the tours.

3   The Proposed GA-Based Approach to Solve the mTSP

GAs are relatively new global stochastic search algorithms which based on evolutionary biology- and computer science principles [11]. Due to the effective optimization capabilities of GAs [2], it makes these technique suitable solving TSP and mTSP problems.

3.1   The Novel Genetic Representation for mTSP

As mentioned in the previous chapter, every GA-based approach for solving the mTSP has used single chromosome for representation so far. The new approach presented here is a so-called multi-chromosome technique, which separates the salesmen from each other thus may present a more effective approach.

This approach is used in notoriously difficult problems to decompose complex solution into simpler components. It was used in mixed integer problem [19], a usage of routing problem optimization can be seen in [18] and a lately solution of a symbolic regression problem in [6]. This section discusses the usage of multi-chromosomal genetic programming in the optimization of mTSP.

Fig. 1 illustrates the new chromosome representation for mTSP with 15 locations ($n=15$) and with 4 salesperson ($m=4$). The figure above illustrates a single individual of the population. Each individual represents a single solution of the problem. The first chromosome represents the first salesman itself so each gene denotes a city (depot is not presented here, it is the first and the last station of each salesman). This encoding is so-called permutation encoding. It can be seen in the example that salesperson 1 visits 4 cities: city 2,5,14 and 6, respectively. In the same way, chromosome 2 represents salesperson 2 and so on. This representation is much similar to the characteristic of the problem, because salesmen are separated from each other "physically".

3.2   Special Genetic Operators

Because of our new representation, implementation of new genetic operators became necessary, like mutation operators. There are two sets of mutation operators, the so-called *In-route mutations* and the *Cross-route mutations*. Only some example of the

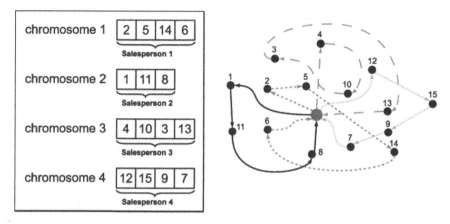

Fig. 1. Example of the multi-chromosome representation for a 15 city mTSP with 4 salesmen

Fig. 2. In-route mutation – gene sequence inversion

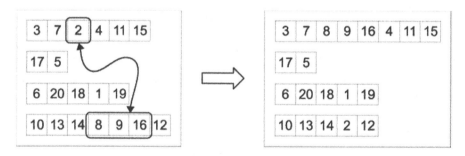

Fig. 3. Cross-route mutation – gene sequence transposition

newly created operators are given in this section. Further information with several examples about the novel operators can be found in [13].

In-route mutation operators work inside one chromosome. An example is illustrated on Fig. 2. The operator chooses a random subsection of a chromosome and inverts the order of the genes inside it.

Cross-route mutation operates on multiple chromosomes. If we think about the distinct chromosomes as individuals, this method could be similar to the regular crossover operator. Fig. 3 illustrates the method when randomly chosen subparts of two chromosomes are transposed. If the length of one of the chosen subsections is equal to zero, the operator could transform into an interpolation.

3.3 Genetic Algorithm

Every genetic algorithm starts with an initial solution set consists of randomly created chromosomes. This is called population. The individuals in the new population are generated from the previous population's individuals by the predetermined genetic operators. The algorithm finishes if the stop criteria is satisfied.

Obviously for a specific problem it is a much more complex task, we need to define the encoding, the specific operators and selection method. The encoding is the so-called permutation encoding (see previous section). Detailed description of the related operators can be found in [14] and an example can be seen in the previous section.

3.3.1 Fitness Function

The fitness function assigns a numeric value to each individual in the population. This value define some kind of goodness, thus it determines the ranking of the individuals. The fitness function is always problem dependent.

In this case the fitness value is the total cost of the transportation, i.e. the total length of each round trip. The fitness function calculates the total length for each chromosome, and summarizes these values for each individual. This sum is the fitness value of a solution. Obviously it is a minimization problem, thus the smallest value is the best.

3.3.2 Selection

Individuals are selected according to their fitness. The better the chromosomes are, the more chances to be selected they have. The selected individuals can be presented in the new population without any changes (usually with the best fitness), or can be selected to be a parent for a crossover. We use the so-called tournament selection because of its efficiency.

In the course of tournament selection, a few (tournament size, min. 2) individuals are selected from the population randomly. The winner of the tournament is the individual with the best fitness value. Some of the first participants in the ranking are selected into the new population (directly or as a parent).

3.4 Complexity Analysis

Using the multi-chromosome technique for the mTSP reduces the size of the overall search space of the problem. Let the length of the first chromosome be k_1, let the length of the second be k_2 and so on. Of course $\sum_{i=1}^{m} k_i = n$. Determining the genes of the first chromosome is equal to the problem of obtaining an ordered subset of k_1 element from a set of n elements. There are $\dfrac{n!}{(n-k_1)!}$ distinct assignment. This number is $\dfrac{(n-k_1)!}{(n-k_1-k_2)!}$ for the second chromosome, and so on. Thus, the total search space of the problem can be formulated as equation (3.1).

$$\frac{n!}{(n-k_1)} * \frac{(n-k_1)!}{(n-k_1-k_2)!} * \hbar * \frac{(n-k_1-\hbar-k_{m-1})!}{(n-k_1-\hbar-k_m)!} = \frac{n!}{(n-n)!} = n! \quad (3.1)$$

It is necessary to determine the length of each chromosome too. It can be represented as a positive vector of the lengths $(k_1, k_2, ..., k_m)$ that must sum to n. There are $\binom{n-1}{m-1}$ distinct positive integer-valued vectors that satisfy this requirement [22].

Thus, the solution space of the new representation is $n!\binom{n-1}{m-1}$. It is equal with the solution space in [5], but this approach is more similar to the characteristic of the mTSP, so it can be more problem-specific therefore more effective.

4 Implementation Issues

To analyze the new representation, a novel genetic algorithm using this approach was developed in MATLAB. This novel approach was compared with the most effective one so far (the two-part chromosome) which is available on MATLAB Central[1]. The novel algorithm can optimize the traditional mTSP problems, furthermore, it is capable to handle the additional constraints and time windows (see Sect. 1).

It requires two input sets, like the coordinates of the cities and the distance table which contains the travelling distances between any pair of cities. Naturally, the determination of the constraints, time windows and the parameters of the genetic algorithms are also necessary.

The fitness function simply summarizes the overall route lengths for each salesman inside an individual. The selection is tournament selection, where tournament size i.e. the number of individuals who compete for survival is 8. Therefore population size must be divisible by 8. The winner of the tournament is the member with the smallest fitness, this individual is selected for new individual creation, and this member will get into the new population without any modification.

The penalty of the too long routes (over the defined constraint) instead of a proportionally large fitness value assignment is implemented by a split operator, which separates the route into smaller routes, which do not exceed the constraints (but the number of salesmen is incremented). Because there exists a constraint for the number of the salesmen, the algorithm involves the minimization of this amount, hence this penalty has a remarkable effect in the optimization process.

Further information about the implemented algorithm can be found in [14].

5 Illustrative Example

Although the algorithm was tested with a big number of problems, only an illustrative result is presented here. As it was mentioned earlier, the algorithm has implemented

---

[1] http://www.mathworks.com/matlabcentral/

in MATLAB, tiny refinements in constraints are in progress. The exmaple represents a whole process of a real problem's solution. The initial input is given in a Google Maps map, and the final output is a route system defined by a Google Maps map also.

The first step is the determination of the distance matrix. The input data is given by a map as it can see on Fig. 4 and a portion of the resulted distance table is shown on Table 1. It contains 25 locations (with the depot). The task is to determine the optimal routes for these locations with the following constraints: the maximum number of salesmen is 5 and the maximum travelling distance of each salesman is 450 km.

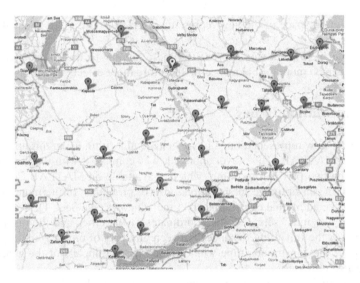

Fig. 4. The map of the example application (initial input)

Table 1. Example distance table - kilometers

Kilometers	Adony	Celldömölk	Kapuvár
Adony	0	169.81	147.53
Celldömölk	169.41	0	44.42
Kapuvár	146.56	44.43	0

After distance table determination, the optimizer algorithm can be executed to determine the optimal routes using the novel representation. The GA ran with a population size 320 and it did 200 iterations. The result of the optimization is shown on Fig. 5. It resulted that 4 salesman is enough to satisfy the constraints. After the optimization, we can visualize the results on a Google Maps map, as it is shown on Fig. 6. The length of the routes are 364 km, 424 km, 398 km and 149 km respectively, i.e. they satisfy the constraints, thus the algorithm provided a feasible solution of the problem.

In every case, the running time was between 1 and 2 minutes. The genetic algorithm has made 200 iterations, because experiences have shown that this number is sufficient for the optimization.

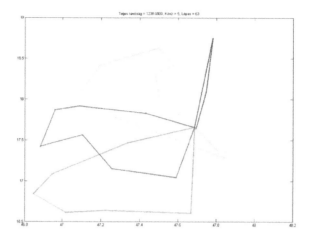

Fig. 5. The result of the optimization by MATLAB

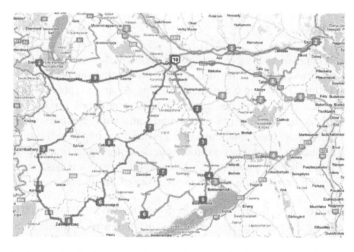

Fig. 6. Result of the optimization on a Google Maps map for 25 locations with at most 5 salesmen and at most 450 km tour length per salesman

Obviously the algorithm is highly sensitive for the number of iterations. The running time is directly proportional to the iteration number, but the resulted best solution can't get better after a specific time. If the constraints become tighter, the duration time will increase slightly. With 500 maximal tour lengths, it is about 90 seconds, and with 450 it is about 110 seconds. The maximal tour length (or equivalently the maximal duration per tour) has a big effect of the number of salesman needed. The tighter the constraints are, the bigger the number of salesman we need. However narrower restrictions forth more square round trips. Furthermore, the resulted optima can depend on the initial population. On Fig. 7 it can be seen that the algorithm can find a near optimal solution in more than 80% of the cases. The effectiveness of the

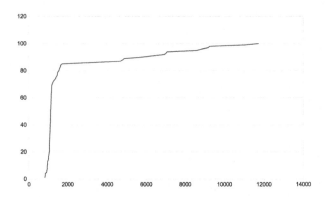

Fig. 7. Results of the optimization from different initial values

calculation can be enhanced by applying additional heuristics. Obviously these results can be further improved by executing more iteration also.

6   Conclusions

In this paper a detailed overview was given about the application of genetic algorithms in vehicle routing problems. It has been shown that the problem is closely related to the multiple Traveling Salesman Problem. A novel representation based genetic algorithm has been developed to the specific one depot version of mTSPTW. The main benefit is the transparency of the representation that allows the effective incorporation of heuristics and constrains and allows easy implementation. Some heuristics can be applied to improve the effectiveness of the algorithm, like the appropriate choice of the initial population. After some final touches, the supporting MATLAB code will be also available at the website of the authors.

Acknowledgments. The financial support from the TAMOP-4.2.2-08/1/2008-0018 (Élhetőbb környezet, egészségesebb ember - Bioinnováció és zöldtechnológiák kutatása a Pannon Egyetemen, MK/2) project is gratefully acknowledged.

References

[1] Ali, A.I., Kennington, J.L.: The Asymmetric m-Traveling Salesmen Problem: a Duality-based Branch-and-Bound Algorithm. Discrete Applied Mathematics 13, 259–276 (1986)

[2] Back, T.: Evolutionary Algorithms in Theory and Practice: Evolution Strategies, Evolutionary Programming, Genetic Algorithms. Oxford University Press, Oxford (1996)

[3] Bektas, T.: The Multiple Traveling Salesman Problem: an Overview of Formulations and Solution Procedures. Omega 34, 209–219 (2006)

[4] Bhide, S., John, N., Kabuka, M.R.: A Boolean Neural Network Approach for the Traveling Salesman Problem. IEEE Transactions on Computers 42(10), 1271 (1993)

[5] Carter, A.E., Ragsdale, C.T.: A New Approach to Solving the Multiple Traveling Salesperson Problem Using Genetic Algorithms. European Journal of Operational Research 175, 246–257 (2006)

 [6] Cavill, R., Smith, S., Tyrrell, A.: Multi-Chromosomal Genetic Programming. In: Proceedings of the 2005 conference on Genetic and evolutionary computation, pp. 1753–1759. ACM, New York (2005)
 [7] Finke, G., Claus, A., Gunn, E.: A Two-Commodity Network Flow Approach to the Traveling Salesman Problem. Congressus Numerantium 41, 167–178 (1984)
 [8] Gen, M., Cheng, R.: Genetic Algorithms and Engineering Design. Wiley Interscience, Hoboken (1997)
 [9] Glover, F.: Artificial Intelligence, Heuristic Frameworks and Tabu Search. Managerial and Decision Economics 11(5) (1990)
[10] Gutin, G., Punnen, A.P.: The Traveling Salesman Problem and Its Variations. In: Combinatorial Optimization, Kluwer Academic Publishers, Dordrecht (2002)
[11] Holland, J.H.: Adaptation in Natural and Artificial Systems. The University of Michigan Press, Ann Arbor (1975)
[12] Hsu, C.Y., Tsai, M.H., Chen, W.M.: A Study of Feature-mapped Approach to the Multiple Travelling Salesmen Problem. In: IEEE International Symposium on Circuits and Systems, vol. 3, pp. 1589–1592 (1991)
[13] Király, A., Abonyi, J.: Optimization of Multiple Traveling Salesmen Problem by a Novel Representation-based Genetic Algorithm. In: 10th International Symposium of Hungarian Researchers on Computational Intelligence and Informatics, Budapest, Hungary, pp. 315–326 (2009)
[14] Király, A., Abonyi, J.: Optimization of Multiple Traveling Salesmen Problem by a Novel Representation-based Genetic Algorithm. In: Koeppen, M., Schaefer, G., Abraham, A., Nolle, L. (eds.) Intelligent Computational Optimization in Engineering: Techniques & Applications, Advances in Intelligent and Soft Computing. Springer, Heidelberg (excepted to 2010)
[15] Laporte, G., Nobert, Y.: A Cutting Planes Algorithm for the m-Salesmen Problem. Journal of the Operational Research Society 31, 1017–1023 (1980)
[16] Malmborg, C.J.: A Genetic Algorithm for Service Level-based Vehicle Scheduling. European Journal of Operational Research 93(1), 121–134 (1996)
[17] Miliotis, P.: Using Cutting Planes to Solve the Symmetric Travelling Salesman Problem. Mathematical Programming 15(1), 177–188 (1978)
[18] Park, Y.B.: A Hybrid Genetic Algorithm for the Vehicle Scheduling Problem with Due Times and Time Deadlines. International Journal of Productions Economics 73(2), 175–188 (2001)
[19] Pierrot, H.J., Hinterding, R.: Using Multi-Chromosomes to Solve a Simple Mixed Integer Problem. In: Canadian AI 1997. LNCS, vol. 1342, pp. 137–146. Springer, Heidelberg (1997)
[20] Potvin, J., Lapalme, G., Rousseau, J.: A Generalized k-opt Exchange Procedure for the mtsp. INFOR 21, 474–481 (1989)
[21] Potvin, J.Y.: Genetic Algorithms for the Traveling Salesman Problem. Annals of Operations Research 63(3), 337–370 (1996)
[22] Ross, S.M.: Introduction to Probability Models. Macmillian, New York (1984)
[23] Russell, R.A.: An Effective Heuristic for the m-tour Traveling Salesman Problem with some Side Conditions. Operations Research 25(3), 517–524 (1977)
[24] Tanga, L., Liu, J., Rongc, A., Yanga, Z.: A Multiple Traveling Salesman Problem Model for Hot Rolling Scheduling in Shangai Baoshan Iron & Steel Complex. European Journal of Operational Research 124, 267–282 (2000)
[25] Zhang, T., Gruver, W., Smith, M.: Team Scheduling by Genetic Search. In: Proceedings of the Second International Conference on Intelligent Processing and Manufacturing of Materials, vol. 2, pp. 839–844 (1999)

Some Examples of Computing the Possibilistic Correlation Coefficient from Joint Possibility Distributions

Robert Fullér[1], József Mezei[2], and Péter Várlaki[3]

[1]IAMSR, Åbo Akademi University
Joukahaisenkatu 3-5 A, FIN-20520 Åbo, Finland
robert.fuller@abo.fi

[2]Turku Centre for Computer Science
Joukahaisenkatu 3-5 B, FIN-20520 Turku, Finland
jmezei@abo.fi

[3]Budapest University of Technology and Economics
Bertalan L. u. 2, H-1111 Budapest, Hungary
and
Széchenyi István University
Egyetem tér 1, H-9026 Győr, Hungary
varlaki@kme.bme.hu

Abstract. In this paper we will show some examples for computing the possibilistic correlation coefficient between marginal distributions of a joint possibility distribution. First we consider joint possibility distributions, $(1-x-y)$, $(1-x^2-y^2)$, $(1-\sqrt{x}-\sqrt{y})$ and $(1-x^2-y)$ on the set $\{(x,y) \in \mathbb{R}^2 |\ x \geq 0, y \geq 0, x+y \leq 1\}$, then we will show (i) how the possibilistic correlation coefficient of two linear marginal possibility distributions changes from zero to $-1/2$, and from $-1/2$ to $-3/5$ by taking out bigger and bigger parts from the level sets of a their joint possibility distribution; (ii) how to compute the autocorrelation coefficient of fuzzy time series with linear fuzzy data.

1   Introduction

A fuzzy number A is a fuzzy set in $\mathbb{R}$ with a normal, fuzzy convex and continuous membership function of bounded support. The family of fuzzy numbers is denoted by $\mathcal{F}$. Fuzzy numbers can be considered as possibility distributions. A fuzzy set C in $\mathbb{R}^2$ is said to be a joint possibility distribution of fuzzy numbers $A, B \in \mathcal{F}$, if it satisfies the relationships $\max\{x \mid C(x;\ y)\} = B(y)$ and $\max\{y \mid C(x;\ y)\} = A(y)$ for all $x, y \in \mathbb{R}$. Furthermore, A and B are called the marginal possibility distributions of

I.J. Rudas et al. (Eds.): Computational Intelligence in Engineering, SCI 313, pp. 153–169.
springerlink.com   © Springer-Verlag Berlin Heidelberg 2010

C. Let $A \in \mathscr{F}$ be fuzzy number with a γ-level set denoted by $[A]^\gamma = [a_1(\gamma), a_2(\gamma)]$, $\gamma \in [0,1]$ and let U_γ denote a uniform probability distribution on $[A]^\gamma$, $\gamma \in [0,1]$.

In possibility theory we can use the principle of *expected value* of functions on fuzzy sets to define variance, covariance and correlation of possibility distributions. Namely, we equip each level set of a possibility distribution (represented by a fuzzy number) with a uniform probability distribution, then apply their standard probabilistic calculation, and then define measures on possibility distributions by integrating these weighted probabilistic notions over the set of all membership grades. These weights (or importances) can be given by weighting functions. A function $f: [0;1] \rightarrow \mathbb{R}$ is said to be a weighting function if f is non-negative, monotone increasing and satisfies the following normalization condition $\int_0^1 f(\gamma)d\gamma = 1$. Different weighting functions can give different (case-dependent) importances to level-sets of possibility distributions.

In 2004 Fullér and Majlender [2] introduced the notion of covariance between marginal distributions of a joint possibility distribution C as the expected value of their interactivity function on C. That is, the f-weighted measure of interactivity between $A \in \mathscr{F}$ and $B \in \mathscr{F}$ (with respect to their joint distribution C) is defined by their measure of possibilistic covariance [2], as

$$\mathrm{Cov}_f(A,B) = \int_0^1 \mathrm{cov}(X_\gamma, Y_\gamma) f(\gamma) d\gamma,$$

where X_γ and Y_γ are random variables whose joint distribution is uniform on $[C]^\gamma$ for all $\gamma \in [0,1]$, and $\mathrm{cov}(X_\gamma, Y_\gamma)$ denotes their probabilistic covariance. They interpreted this covariance as a measure of interactivity between marginal distributions. They also showed that non-interactivity entails zero covariance, however, zero covariance does not always imply non-interactivity. The measure of interactivity is positive (negative) if the expected value of the interactivity relation on C is positive (negative). It is easy to see that the possibilistic covariance is an absolute measure in the sense that it can take any value from the real line. To have a relative measure of interactivity between marginal distributions we have introduced the normalized covariance in 2010 (see [3]).

Definition 1.1 ([3]) *The f-weighted normalized measure of interactivity between $A \in \mathscr{F}$ and $B \in \mathscr{F}$ (with respect to their joint distribution C) is defined by*

$$\rho_f(A,B) = \int_0^1 \rho(X_\gamma, Y_\gamma) f(\gamma) d\gamma, \tag{1}$$

where

$$\rho(X_\gamma, Y_\gamma) = \frac{\mathrm{cov}(X_\gamma, Y_\gamma)}{\sqrt{\mathrm{var}(X_\gamma)}\sqrt{\mathrm{var}(Y_\gamma)}}$$

Following the terminology of Carlsson, Fullér and Majlender [1] we will call this improved measure of interactivity as the f-weighted possibilistic correlation ratio.

In other words, the f-weighted possibilistic correlation coefficient is nothing else, but the f-weighted average of the probabilistic correlation coefficients $\rho(X_\gamma, Y_\gamma)$ for all $\gamma \in [0,1]$.

2   Some Illustrations of Possibilistic Correlation

2.1 Joint Distribution: (1-x-y)

Consider the case, when $A(x) = B(x) = (1\text{-}x)\cdot\chi_{[0,1]}(x)$, for $x \in \mathbb{R}$, that is $[A]^\gamma = [B]^\gamma = [0, 1\text{-}\gamma]$, for $\gamma \in [0, 1]$. Suppose that their joint possibility distribution is given by $F(x,y) = (1\text{-}x\text{-}y)\cdot\chi_T(x,y)$, where $T=\{(x,y) \in \mathbb{R}^2 |\ x\geq0, y\geq0, x+y\leq1\}$. Then we have $[F]^\gamma = \{(x,y) \in \mathbb{R}^2 |\ x\geq0, y\geq0, x+y\leq1\text{-}\gamma\}$.

This situation is depicted on Fig. 1, where we have shifted the fuzzy sets to get a better view of the situation. In this case the f-weighted possibilistic correlation of A and B is computed as (see [3] for details),

$$\rho_f(A,B) = \int_0^1 -\frac{1}{2} f(\gamma) d\gamma = -\frac{1}{2}.$$

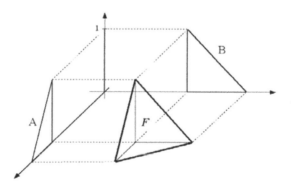

Fig. 1. Illustration of joint possibility distribution F

Consider now the case when $A(x) = B(x) = x\cdot\chi_{[0,1]}(x)$, for $x \in \mathbb{R}$, that is $[A]^\gamma = [B]^\gamma = [\gamma, 1]$, for $\gamma \in [0, 1]$. Suppose that their joint possibility distribution is given by $W(x,y) = \max\{x+y\text{-}1,0\}$. Then we get $\rho_f(A,B)=-1/2$. We note here that W is nothing else but the Lukasiewitz t-norm, or in the statistical literature, W is generally referred to as the lower Fréchet-Hoeffding bound for copulas.

2.2 Joint Distribution: $(1-x^2-y^2)$

Consider the case, when $A(x) = B(x) = (1-x^2) \cdot \chi_{[0,1]}(x)$, for $x \in \mathbb{R}$, that is $[A]^\gamma = [B]^\gamma = [0, \sqrt{1-\gamma}]$, for $\gamma \in [0, 1]$. Suppose that their joint possibility distribution is given by: $C(x,y) = (1-x^2-y^2) \cdot \chi_T (x,y)$, where $T = \{(x,y) \in \mathbb{R}^2 |\ x \geq 0, y \geq 0, x^2+y^2 \leq 1\}$. A γ-level set of C is computed by $[C]^\gamma = \{(x,y) \in \mathbb{R}^2|\ x \geq 0, y \geq 0, x^2+y^2 \leq 1-\gamma\}$.

The density function of a uniform distribution on $[C]^\gamma$ can be written as

$$f(x,y) = \begin{cases} \dfrac{1}{\int_{[C]^\gamma} dxdy}, & \text{if } (x,y) \in [C]^\gamma \\ 0 & \text{otherwise} \end{cases} = \begin{cases} \dfrac{4}{(1-\gamma)\pi}, & \text{if } (x,y) \in [C]^\gamma \\ 0 & \text{otherwise} \end{cases}$$

The marginal functions are obtained as

$$f_1(x) = \begin{cases} \dfrac{4\sqrt{1-\gamma-x^2}}{(1-\gamma)\pi}, & \text{if } 0 \leq x \leq 1-\gamma \\ 0 & \text{otherwise} \end{cases}$$

$$f_2(y) = \begin{cases} \dfrac{4\sqrt{1-\gamma-y^2}}{(1-\gamma)\pi}, & \text{if } 0 \leq y \leq 1-\gamma \\ 0 & \text{otherwise} \end{cases}$$

We can calculate the probabilistic expected values of the random variables X_γ and Y_γ, whose joint distribution is uniform on $[C]^\gamma$ for all $\gamma \in [0, 1]$:

$$M(X_\gamma) = \frac{4}{(1-\gamma)\pi} \int_0^{\sqrt{1-\gamma}} x\sqrt{1-\gamma-x^2}dx = \frac{4\sqrt{1-\gamma}}{3\pi}$$

$$M(Y_\gamma) = \frac{4}{(1-\gamma)\pi} \int_0^{\sqrt{1-\gamma}} y\sqrt{1-\gamma-y^2}dx = \frac{4\sqrt{1-\gamma}}{3\pi}.$$

We calculate the variations of X_γ and Y_γ with the formula $var(X) = M(X^2) - M(X)^2$:

$$M(X_\gamma^2) = \frac{4}{(1-\gamma)\pi} \int_0^{\sqrt{1-\gamma}} x^2\sqrt{1-\gamma-x^2}dx = \frac{1-\gamma}{4}$$

$$var(X_\gamma) = M(X_\gamma^2) - M(X_\gamma)^2 = \frac{1-\gamma}{4} - \frac{16(1-\gamma)}{9\pi^2} = \frac{(1-\gamma)(9\pi^2-64)}{36\pi^2}.$$

And similarly we obtain

$$var(Y_\gamma) = \frac{(1-\gamma)(9\pi^2-64)}{36\pi^2}.$$

Using that

$$\mathrm{cov}(X_\gamma, Y_\gamma) = M(X_\gamma Y_\gamma) - M(X_\gamma)M(Y_\gamma) = \frac{(1-\gamma)(9\pi - 32)}{18\pi^2},$$

we can calculate the probabilistic correlation of the random variables:

$$\rho(X_\gamma, Y_\gamma) = \frac{\mathrm{cov}(X_\gamma, Y_\gamma)}{\sqrt{\mathrm{var}(X_\gamma)}\sqrt{\mathrm{var}(Y_\gamma)}} = \frac{2(9\pi - 32)}{(9\pi^2 - 64)} \approx -0.302.$$

And finally the f-weighted possibilistic correlation of A and B:

$$\rho_f(A,B) = \int_0^1 \frac{2(9\pi - 32)}{(9\pi^2 - 64)} f(\gamma)d\gamma = \frac{2(9\pi - 32)}{(9\pi^2 - 64)}.$$

2.3 Joint Distribution: $(1 - \sqrt{x} - \sqrt{y})$

Consider the case, when $A(x) = B(x) = (1 - \sqrt{x})\cdot\chi_{[0,1]}(x)$, for $x \in \mathbb{R}$, that is $[A]^\gamma = [B]^\gamma = [0, (1-\gamma)^2]$, for $\gamma \in [0, 1]$. Suppose that their joint possibility distribution is given by: $C(x,y) = (1 - \sqrt{x} - \sqrt{y})\cdot\chi_T(x,y)$, where

$$T = \{(x,y) \in \mathbb{R}^2 | x \geq 0, y \geq 0, \sqrt{x} + \sqrt{y} \leq 1\}.$$

A γ-level set of C is computed by $[C]^\gamma = \{(x,y) \in \mathbb{R}^2 | x \geq 0, y \geq 0, \sqrt{x} + \sqrt{y} \leq 1 - \gamma\}$.
The density function of a uniform distribution on $[C]^\gamma$ can be written as

$$f(x,y) = \begin{cases} \dfrac{1}{\int_{[C]^\gamma} dxdy}, & \text{if } (x,y) \in [C]^\gamma \\ 0 & \text{otherwise} \end{cases} = \begin{cases} \dfrac{6}{(1-\gamma)^4}, & \text{if } (x,y) \in [C]^\gamma \\ 0 & \text{otherwise} \end{cases}$$

The marginal functions are obtained as

$$f_1(x) = \begin{cases} \dfrac{6(1-\gamma-\sqrt{x})^2}{(1-\gamma)^4}, & \text{if } 0 \leq x \leq (1-\gamma)^2 \\ 0 & \text{otherwise} \end{cases}$$

$$f_2(y) = \begin{cases} \dfrac{6(1-\gamma-\sqrt{y})^2}{(1-\gamma)^4}, & \text{if } 0 \leq y \leq (1-\gamma)^2 \\ 0 & \text{otherwise} \end{cases}$$

We can calculate the probabilistic expected values of the random variables X_γ and Y_γ, whose joint distribution is uniform on $[C]^\gamma$ for all $\gamma \in [0, 1]$:

$$M(X_\gamma) = \frac{6}{(1-\gamma)^4} \int_0^{(1-\gamma)^2} x(1-\gamma-\sqrt{x})^2 dx = \frac{(1-\gamma)^2}{5}$$

$$M(Y_\gamma) = \frac{6}{(1-\gamma)^4} \int_0^{(1-\gamma)^2} y(1-\gamma-\sqrt{y})^2 dy = \frac{(1-\gamma)^2}{5}$$

The variations of X_γ and Y_γ with the formula $var(X) = M(X^2) - M(X)^2$:

$$M(X_\gamma^2) = \frac{6}{(1-\gamma)^4} \int_0^{(1-\gamma)^2} x^2(1-\gamma-\sqrt{x})^2 dx = \frac{(1-\gamma)^4}{14}$$

$$var(X_\gamma) = M(X_\gamma^2) - M(X_\gamma)^2 = \frac{(1-\gamma)^4}{14} - \frac{(1-\gamma)^4}{25} = \frac{9(1-\gamma)^4}{350}.$$

And similarly we obtain

$$var(Y_\gamma) = \frac{9(1-\gamma)^4}{350}.$$

Using that

$$cov(X_\gamma, Y_\gamma) = M(X_\gamma Y_\gamma) - M(X_\gamma)M(Y_\gamma) = -\frac{13(1-\gamma)^4}{700},$$

we can calculate the probabilistic correlation of the random variables:

$$\rho(X_\gamma, Y_\gamma) = \frac{cov(X_\gamma, Y_\gamma)}{\sqrt{var(X_\gamma)}\sqrt{var(Y_\gamma)}} = -\frac{13}{18} \approx -0.722.$$

And finally the f-weighted possibilistic correlation of A and B:

$$\rho_f(A, B) = -\int_0^1 \frac{13}{18} f(\gamma) d\gamma = -\frac{13}{18}.$$

2.4 Joint Distribution: (1-x²-y)

Consider the case, when $A(x) = (1-x^2)\cdot\chi_{[0,1]}(x)$, $B(x) = (1-x)\cdot\chi_{[0,1]}(x)$, for $x \in \mathbb{R}$, that is $[A]^\gamma = [0, \sqrt{1-\gamma}\,]$, $[B]^\gamma = [0, 1-\gamma]$, for $\gamma \in [0, 1]$. Suppose that their joint possibility distribution is given by: $C(x,y) = (1-x^2-y)\cdot\chi_T(x,y)$, where

$$T = \{(x,y) \in \mathbb{R}^2 | x \geq 0, y \geq 0, x^2 + y \leq 1\}.$$

A γ-level set of C is computed by $[C]^\gamma = \{(x,y) \in \mathbb{R}^2 | x \geq 0, y \geq 0,\ x^2 + y \leq 1-\gamma\}$. The density function of a uniform distribution on $[C]^\gamma$ can be written as

$$f(x,y) = \begin{cases} \dfrac{1}{\int_{[C]^\gamma} dxdy}, & \text{if } (x,y) \in [C]^\gamma \\ 0 & \text{otherwise} \end{cases} = \begin{cases} \dfrac{3}{2(1-\gamma)^{\frac{3}{2}}}, & \text{if } (x,y) \in [C]^\gamma \\ 0 & \text{otherwise} \end{cases}$$

The marginal functions are obtained as

$$f_1(x) = \begin{cases} \dfrac{3(1-\gamma-x^2)}{2(1-\gamma)^{\frac{3}{2}}}, & \text{if } 0 \leq x \leq \sqrt{1-\gamma} \\ 0 & \text{otherwise} \end{cases}$$

$$f_2(y) = \begin{cases} \dfrac{3\sqrt{1-\gamma-y}}{2(1-\gamma)^{\frac{3}{2}}}, & \text{if } 0 \leq y \leq 1-\gamma \\ 0 & \text{otherwise} \end{cases}$$

We can calculate the probabilistic expected values of the random variables X_γ and Y_γ, whose joint distribution is uniform on $[C]^\gamma$ for all $\gamma \in [0, 1]$:

$$M(X_\gamma) = \frac{3}{2(1-\gamma)^{\frac{3}{2}}} \int_0^{\sqrt{1-\gamma}} x(1-\gamma-x^2)dx = \frac{3\sqrt{1-\gamma}}{8}$$

$$M(Y_\gamma) = \frac{3}{2(1-\gamma)^{\frac{3}{2}}} \int_0^{\sqrt{1-\gamma}} y\sqrt{1-\gamma-y}\,dy = \frac{2(1-\gamma)}{5}.$$

We calculate the variations of X_γ and Y_γ, with the formula $var(X) = M(X^2) - M(X)^2$:

$$M(X_\gamma^2) = \frac{3}{2(1-\gamma)^{\frac{3}{2}}} \int_0^{\sqrt{1-\gamma}} x^2(1-\gamma-x^2)dx = \frac{1-\gamma}{5}$$

$$var(X_\gamma) = M(X_\gamma^2) - M(X_\gamma)^2 = \frac{1-\gamma}{5} - \frac{9(1-\gamma)}{64} = \frac{19(1-\gamma)}{320}.$$

$$M(Y_\gamma^2) = \frac{3}{2(1-\gamma)^{\frac{3}{2}}} \int_0^{1-\gamma} y^2\sqrt{1-\gamma-y}\,dy = \frac{8(1-\gamma)^2}{35}.$$

$$var(Y_\gamma) = M(Y_\gamma^2) - M(Y_\gamma)^2 = \frac{8(1-\gamma)^2}{35} - \frac{4(1-\gamma)^2}{25} = \frac{12(1-\gamma)^2}{175}.$$

The covariance of X_γ and Y_γ:

$$\mathrm{cov}(X_\gamma, Y_\gamma) = M(X_\gamma Y_\gamma) - M(X_\gamma)M(Y_\gamma) = -\frac{(1-\gamma)^{\frac{3}{2}}}{40},$$

and we can calculate the probabilistic correlation of the random variables:

$$\rho(X_\gamma, Y_\gamma) = \frac{\mathrm{cov}(X_\gamma, Y_\gamma)}{\sqrt{\mathrm{var}(X_\gamma)}\sqrt{\mathrm{var}(Y_\gamma)}} = -\sqrt{\frac{35}{228}} \approx -0.392.$$

And finally the f-weighted possibilistic correlation of A and B:

$$\rho_f(A, B) = \int_0^1 -\sqrt{\frac{35}{228}} f(\gamma) d\gamma = -\sqrt{\frac{35}{228}}.$$

3   A Transition from Zero to -1/2

Suppose that a family of joint possibility distribution of A and B (where $A(x) = B(x) = (1-x)\cdot\chi_{[0,1]}(x)$, for $x \in \mathbb{R}$) is defined by

$$C_n(x,y) = \begin{cases} 1 - x - \frac{n-1}{n}y, & \text{if } 0 \le x \le 1, x \le y, \frac{n-1}{n}y + x \le 1 \\ 1 - \frac{n-1}{n}x - y, & \text{if } 0 \le y \le 1, y \le x, \frac{n-1}{n}x + y \le 1 \\ 0, & \text{otherwise} \end{cases}$$

In the following, for simplicity, we well write C instead of C_n. A γ-level set of C is computed by

$$[C]^\gamma = \left\{ (x,y) \in \mathbb{R}^2 \mid 0 \le x \le \frac{n}{2n-1}(1-\gamma), 0 \le y \le 1-\gamma-\frac{n-1}{n}x \right\} \cup$$

$$\left\{ (x,y) \in \mathbb{R}^2 \mid \frac{n}{2n-1}(1-\gamma) \le x \le 1-\gamma, 0 \le \frac{n-1}{n}y \le 1-\gamma-x \right\}.$$

The density function of a uniform distribution on $[C]^\gamma$ can be written as

$$f(x,y) = \begin{cases} \frac{1}{\int_{[C]^\gamma} dxdy}, & \text{if } (x,y) \in [C]^\gamma \\ 0 & \text{otherwise} \end{cases} = \begin{cases} \frac{2n-1}{n(1-\gamma)^2}, & \text{if } (x,y) \in [C]^\gamma \\ 0 & \text{otherwise} \end{cases}$$

We can calculate the marginal density functions:

$$f_1(x) = \begin{cases} \dfrac{(2n-1)(1-\gamma-x)}{(n-1)(1-\gamma)^2}, & \text{if } \dfrac{n}{2n-1}(1-\gamma) \leq x \leq 1-\gamma \\[3mm] \dfrac{(2n-1)(1-\gamma-\frac{n-1}{n}x)}{n(1-\gamma)^2}, & \text{if } 0 \leq x \leq \dfrac{n}{2n-1}(1-\gamma) \\[3mm] 0 & \text{otherwise} \end{cases}$$

and

$$f_2(y) = \begin{cases} \dfrac{(2n-1)(1-\gamma-y)}{(n-1)(1-\gamma)^2}, & \text{if } \dfrac{n}{2n-1}(1-\gamma) \leq y \leq 1-\gamma \\[3mm] \dfrac{(2n-1)(1-\gamma-\frac{n-1}{n}y)}{n(1-\gamma)^2}, & \text{if } 0 \leq y \leq \dfrac{n}{2n-1}(1-\gamma) \\[3mm] 0 & \text{otherwise} \end{cases}$$

We can calculate the probabilistic expected values of the random variables X_γ and Y_γ, whose joint distribution is uniform on $[C]^\gamma$ for all $\gamma \in [0, 1]$ as,

$$M(X_\gamma) = \frac{2n-1}{n(1-\gamma)^2} \int_0^{\frac{n(1-\gamma)}{2n-1}} x(1-\gamma-\frac{n-1}{n}x)dx$$
$$+ \frac{2n-1}{(n-1)(1-\gamma)^2} \int_{\frac{n(1-\gamma)}{2n-1}}^{1-\gamma} x(1-\gamma-x)dx = \frac{(1-\gamma)(4n-1)}{6(2n-1)}$$

and

$$M(Y_\gamma) = \frac{(1-\gamma)(4n-1)}{6(2n-1)}.$$

(We can easily see that for $n = 1$ we have $M(X_\gamma) = \dfrac{1-\gamma}{2}$, and for $n \to \infty$ we find $M(X_\gamma) \to \dfrac{1-\gamma}{2}$.)

We calculate the variations of X_γ and Y_γ as,

$$M(X_\gamma^2) = \frac{2n-1}{n(1-\gamma)^2} \int_0^{\frac{n(1-\gamma)}{2n-1}} x^2(1-\gamma-\frac{n-1}{n}x)dx$$
$$+ \frac{2n-1}{(n-1)(1-\gamma)^2} \int_{\frac{n(1-\gamma)}{2n-1}}^{1-\gamma} x^2(1-\gamma-x)dx$$
$$= \frac{(1-\gamma)^2((2n-1)^3+8n^3-6n^2+n)}{12(2n-1)^3}.$$

(We can easily see that for $n = 1$ we have $M(X_\gamma^2) = \dfrac{(1-\gamma)^2}{3}$, and for $n \to \infty$ we find $M(X_\gamma^2) \to \dfrac{(1-\gamma)^2}{6}$.)

Furthermore,

$$\text{var}(X_\gamma) = M(X_\gamma^2) - M(X_\gamma)^2 = \frac{(1-\gamma)^2((2n-1)^3 + 8n^3 - 6n^2 + n)}{12(2n-1)^3}$$

$$-\frac{(1-\gamma)^2(4n-1)^2}{36(2n-1)^2} = \frac{(1-\gamma)^2(2(2n-1)^2 + n)}{36(2n-1)^2}.$$

And similarly we obtain

$$\text{var}(Y_\gamma) = \frac{(1-\gamma)^2(2(2n-1)^2 + n)}{36(2n-1)^2}.$$

(We can easily see that for $n = 1$ we have $\text{var}(X_\gamma) = \frac{(1-\gamma)^2}{12}$, and for n→∞ we

find $\text{var}(X_\gamma^2) \to \frac{(1-\gamma)^2}{18}$.)

And,

$$\text{cov}(X_\gamma, Y_\gamma) = M(X_\gamma Y_\gamma) - M(X_\gamma)M(Y_\gamma)$$

$$= \frac{(1-\gamma)^2 n(4n-1)}{12(2n-1)^2} - \frac{(1-\gamma)^2(1-n)(4n-1)}{36(2n-1)^2}.$$

(We can easily see that for $n = 1$ we have $\text{cov}(X_\gamma, Y_\gamma) = 0$, and for n→∞ we find

$\text{cov}(X_\gamma, Y_\gamma) \to -\frac{(1-\gamma)^2}{36}$.)

We can calculate the probabilistic correlation of the random variables,

$$\rho(X_\gamma, Y_\gamma) = \frac{\text{cov}(X_\gamma, Y_\gamma)}{\sqrt{\text{var}(X_\gamma)}\sqrt{\text{var}(Y_\gamma)}} = \frac{(1-n)(4n-1)}{2(2n-1)^2 + n}.$$

(We can easily see that for $n = 1$ we have $\rho(X_\gamma, Y_\gamma) = 0$, and for $n\to\infty$ we find

$\rho(X_\gamma, Y_\gamma) \to -\frac{1}{2}$.)

And finally the f-weighted possibilistic correlation of A and B is computed as,

$$\rho_f(A,B) = \int_0^1 \rho(X_\gamma, Y_\gamma)f(\gamma)d\gamma = \frac{(1-n)(4n-1)}{2(2n-1)^2 + n}.$$

We obtain, that $\rho_f(A,B) = 0$ for $n = 1$ and if $n\to\infty$ then $\rho_f(A,B) \to -\frac{1}{2}$.

4   A Transition from -1/2 to -3/5

Suppose that a family of joint possibility distribution of A and B (where $A(x) = B(x) = (1-x)\cdot\chi_{[0,1]}(x)$, for $x \in \mathbb{R}$) is defined by

$$C_n(x,y) = (1 - x - y)\cdot\chi_{T_n}(x,y),$$

where

$$T_n = \left\{ (x,y) \in \mathbb{R}^2 \mid x \geq 0, y \geq 0, x+y \leq 1, \frac{1}{n-1}x \geq y \right\} \cup$$

$$\left\{ (x,y) \in \mathbb{R}^2 \mid x \geq 0, y \geq 0, x+y \leq 1, (n-1)x \leq y \right\}.$$

In the following, for simplicity, we well write C instead of C_n. A γ-level set of C is computed by

$$[C]^\gamma = \left\{ (x,y) \in \mathbb{R}^2 \mid 0 \leq x \leq \frac{1}{n}(1-\gamma), (n-1)x \leq y \leq 1 - \gamma - x \right\} \cup$$

$$\left\{ (x,y) \in \mathbb{R}^2 \mid 0 \leq y \leq \frac{1}{n}(1-\gamma), (n-1)y \leq x \leq 1 - \gamma - y \right\}.$$

The density function of a uniform distribution on $[C]^\gamma$ can be written as

$$f(x,y) = \begin{cases} \dfrac{1}{\int_{[C]^\gamma} dxdy}, & \text{if } (x,y) \in [C]^\gamma \\ 0 & \text{otherwise} \end{cases} = \begin{cases} \dfrac{n}{(1-\gamma)^2}, & \text{if } (x,y) \in [C]^\gamma \\ 0 & \text{otherwise} \end{cases}$$

We can calculate the marginal density functions:

$$f_1(x) = \begin{cases} \dfrac{n\left(1 - \gamma - nx + \frac{x}{n-1}\right)}{(1-\gamma)^2}, & \text{if } 0 \leq x \leq \dfrac{1-\gamma}{n} \\ \dfrac{nx}{(1-\gamma)^2(n-1)}, & \text{if } \dfrac{(1-\gamma)}{n} \leq x \leq \dfrac{(n-1)(1-\gamma)}{n} \\ \dfrac{n(1-\gamma-x)}{(1-\gamma)^2}, & \text{if } \dfrac{(n-1)(1-\gamma)}{n} \leq x \leq 1 - \gamma \\ 0 & \text{otherwise} \end{cases}$$

and,

$$
f_2(y) = \begin{cases}
\dfrac{n(1-\gamma-ny+\frac{y}{n-1})}{(1-\gamma)^2}, & \text{if } 0 \le y \le \dfrac{1-\gamma}{n} \\[2mm]
\dfrac{ny}{(1-\gamma)^2(n-1)}, & \text{if } \dfrac{(1-\gamma)}{n} \le y \le \dfrac{(n-1)(1-\gamma)}{n} \\[2mm]
\dfrac{n(1-\gamma-y)}{(1-\gamma)^2}, & \text{if } \dfrac{(n-1)(1-\gamma)}{n} \le y \le 1-\gamma \\[2mm]
0 & \text{otherwise}
\end{cases}
$$

We can calculate the probabilistic expected values of the random variables X_γ and Y_γ, whose joint distribution is uniform on $[C]^\gamma$ for all $\gamma \in [0, 1]$ as,

$$
M(X_\gamma) = \frac{n}{(1-\gamma)^2} \int_0^{\frac{1-\gamma}{n}} x(1-\gamma-nx+\frac{x}{n-1})dx
$$

$$
+ \frac{n}{(1-\gamma)^2} \int_{\frac{1-\gamma}{n}}^{\frac{(n-1)(1-\gamma)}{n}} \frac{x^2}{n-1}dx + \frac{n}{(1-\gamma)^2} \int_{\frac{(n-1)(1-\gamma)}{n}}^{(1-\gamma)} x(1-\gamma-x)dx
$$

$$
= \frac{1-\gamma}{3}.
$$

That is, $M(Y_\gamma) = \dfrac{1-\gamma}{3}$. We calculate the variations of X_γ and Y_γ as,

$$
M(X_\gamma^2) = \frac{n}{(1-\gamma)^2} \int_0^{\frac{1-\gamma}{n}} x^2(1-\gamma-nx+\frac{x}{n-1})dx
$$

$$
+ \frac{n}{(1-\gamma)^2} \int_{\frac{1-\gamma}{n}}^{\frac{(n-1)(1-\gamma)}{n}} \frac{x^3}{n-1}dx
$$

$$
+ \frac{n}{(1-\gamma)^2} \int_{\frac{(n-1)(1-\gamma)}{n}}^{(1-\gamma)} x^2(1-\gamma-x)dx
$$

$$
= \frac{(1-\gamma)^2(3n^2-3n+2)}{12n^2}.
$$

and,

$$
\mathrm{var}(X_\gamma) = M(X_\gamma^2) - M(X_\gamma)^2 = \frac{(1-\gamma)^2(3n^2-3n+2)}{12n^2} - \frac{(1-\gamma)^2}{9}
$$

$$
= \frac{(1-\gamma)^2(5n^2-9n+6)}{36n^2}.
$$

And, similarly, we obtain

$$
\mathrm{var}(Y_\gamma) = \frac{(1-\gamma)^2(5n^2-9n+6)}{36n^2}.
$$

From

$$\text{cov}(X_\gamma, Y_\gamma) = M(X_\gamma Y_\gamma) - M(X_\gamma)M(Y_\gamma) = \frac{(1-\gamma)^2(3n-2)}{12n^2} - \frac{(1-\gamma)^2}{9}.$$

we can calculate the probabilistic correlation of the random variables:

$$\rho(X_\gamma, Y_\gamma) = \frac{\text{cov}(X_\gamma, Y_\gamma)}{\sqrt{\text{var}(X_\gamma)}\sqrt{\text{var}(Y_\gamma)}} = -\frac{-3n^2 + 7n - 6}{5n^2 - 9n + 6}.$$

And finally the f-weighted possibilistic correlation of A and B:

$$\rho_f(A, B) = \int_0^1 \rho(X_\gamma, Y_\gamma)f(\gamma)d\gamma = -\frac{-3n^2 + 7n - 6}{5n^2 - 9n + 6}$$

We obtain, that for n = 2

$$\rho_f(A, B) = -\frac{1}{2}$$

and if if $n \to \infty$ then

$$\rho_f(A, B) \to -\frac{3}{5}$$

We note that in this extreme case the joint possibility distribution is nothing else but the marginal distributions themselves, that is, $C_\infty(x, y) = 0$, for any interior point (x,y) of the unit square.

5   Trapezoidal Marginal Distributions

Consider now the case,

$$A(x) = B(x) = \begin{cases} x, & \text{if } 0 \le x \le 1 \\ 1, & \text{if } 1 \le x \le 2 \\ 3-x, & \text{if } 2 \le x \le 3 \\ 0, & \text{otherwise} \end{cases}$$

for $x \in \mathbb{R}$, that is $[A]^\gamma = [B]^\gamma = [\gamma, 3-\gamma]$, for $\gamma \in [0, 1]$. Suppose that the joint possibility distribution of these two trapezoidal marginal distributions – a considerably truncated pyramid – given by:

$$C(x,y) = \begin{cases} y, & \text{if } 0 \le x \le 3, 0 \le y \le 1, x \le y, x \le 3-y \\ 1, & \text{if } 1 \le x \le 2, 1 \le y \le 2, x \le y \\ x, & \text{if } 0 \le x \le 1, 0 \le y \le 3, y \le x, x \le 3-y \\ 0, & \text{otherwise} \end{cases}$$

Then $[C]^\gamma = \{(x,y) \in \mathbb{R}^2 | \gamma \le x \le 3-\gamma, \gamma \le y \le 3-x \}$. The density function of a uniform distribution on $[C]^\gamma$ can be written as

$$f(x,y) = \begin{cases} \dfrac{1}{\int_{[C]^\gamma} dxdy}, & \text{if } (x,y) \in [C]^\gamma \\ 0 & \text{otherwise} \end{cases} = \begin{cases} \dfrac{2}{(3-2\gamma)^2}, & \text{if } (x,y) \in [C]^\gamma \\ 0 & \text{otherwise} \end{cases}$$

The marginal functions are obtained as

$$f_1(x) = \begin{cases} \dfrac{2(3-\gamma-x)}{(3-2\gamma)^2}, & \text{if } \gamma \le x \le 3-\gamma \\ 0 & \text{otherwise} \end{cases}$$

and

$$f_2(y) = \begin{cases} \dfrac{2(3-\gamma-y)}{(3-2\gamma)^2}, & \text{if } \gamma \le y \le 3-\gamma \\ 0 & \text{otherwise} \end{cases}$$

We can calculate the probabilistic expected values of the random variables X_γ and Y_γ, whose joint distribution is uniform on $[C]^\gamma$ for all $\gamma \in [0, 1]$:

$$M(X_\gamma) = \frac{2}{(3-2\gamma)^2} \int_\gamma^{3-\gamma} x(3-\gamma-x)dx = \frac{\gamma+3}{3}$$

and,

$$M(Y_\gamma) = \frac{2}{(3-2\gamma)^2} \int_\gamma^{3-\gamma} y(3-\gamma-y)dx = \frac{\gamma+3}{3}$$

We calculate the variations of X_γ and Y_γ, with the formula $var(X) = M(X^2) - M(X)^2$:

$$M(X_\gamma^2) = \frac{2}{(3-2\gamma)^2} \int_\gamma^{3-\gamma} x^2(3-\gamma-x)dx = \frac{2\gamma^2+9}{6}$$

and,

$$var(X_\gamma) = M(X_\gamma^2) - M(X_\gamma)^2 = \frac{2\gamma^2+9}{6} - \frac{(\gamma+3)^2}{9} = \frac{(3-2\gamma)^2}{18}.$$

And similarly we obtain

$$var(Y_\gamma) = \frac{(3-2\gamma)^2}{18}.$$

Using the relationship,

$$cov(X_\gamma, Y_\gamma) = M(X_\gamma Y_\gamma) - M(X_\gamma)M(Y_\gamma) = -\frac{(3-2\gamma)^2}{36},$$

we can calculate the probabilistic correlation of the random variables:

$$\rho(X_\gamma, Y_\gamma) = \frac{\text{cov}(X_\gamma, Y_\gamma)}{\sqrt{\text{var}(X_\gamma)}\sqrt{\text{var}(Y_\gamma)}} = -\frac{1}{2}.$$

And finally the f-weighted possibilistic correlation of A and B is,

$$\rho_f(A, B) = -\int_0^1 \frac{1}{2} f(\gamma) d\gamma = -\frac{1}{2}.$$

6   Time Series with Fuzzy Data

A time series with fuzzy data is referred to as fuzzy time series (see [4]). Consider a fuzzy time series indexed by $t \in (0, 1]$:

$$A_t(x) = \begin{cases} 1 - \dfrac{x}{t}, & \text{if } 0 \leq x \leq t \\ 0 & \text{otherwise} \end{cases}$$

and

$$A_0(x) = \begin{cases} 1, & \text{if } x = 0 \\ 0 & \text{otherwise} \end{cases}$$

It is easy to see that in this case, $[A_t]^\gamma = [0, t(1-\gamma)]$, for $\gamma \in [0, 1]$. If we have $t_1, t_2 \in (0, 1]$, then the joint possibility distribution of the corresponding fuzzy numbers is given by:

$$C(x, y) = \left(1 - \frac{x}{t_1} - \frac{y}{t_2}\right) \cdot \chi_T(x, y),$$

where

$$T = \left\{ (x, y) \in \mathbb{R}^2 \mid x \geq 0, y \geq 0, \frac{x}{t_1} + \frac{y}{t_2} \leq 1 \right\}.$$

Then

$$[C]^\gamma = \left\{ (x, y) \in \mathbb{R}^2 \mid x \geq 0, y \geq 0, \frac{x}{t_1} + \frac{y}{t_2} \leq 1 - \gamma \right\}.$$

The density function of a uniform distribution on $[C]^\gamma$ can be written as

$$f(x,y) = \begin{cases} \dfrac{1}{\int_{[C]^\gamma} dx dy}, & \text{if } (x,y) \in [C]^\gamma \\ 0 & \text{otherwise} \end{cases}$$

That is,

$$f(x,y) = \begin{cases} \dfrac{2}{t_1 t_2 (1-\gamma)^2}, & \text{if } (x,y) \in [C]^\gamma \\ 0 & \text{otherwise} \end{cases}$$

The marginal functions are obtained as

$$f_1(x) = \begin{cases} \dfrac{2(1-\gamma-\frac{x}{t_1})}{t_1(1-\gamma)^2}, & \text{if } 0 \le x \le t_1(1-\gamma) \\ 0 & \text{otherwise} \end{cases}$$

and,

$$f_2(y) = \begin{cases} \dfrac{2(1-\gamma-\frac{y}{t_2})}{t_2(1-\gamma)^2}, & \text{if } 0 \le y \le t_2(1-\gamma) \\ 0 & \text{otherwise} \end{cases}$$

We can calculate the probabilistic expected values of the random variables X_γ and Y_γ, whose joint distribution is uniform on $[C]^\gamma$ for all $\gamma \in [0, 1]$:

$$M(X_\gamma) = \frac{2}{t_1(1-\gamma)^2} \int_0^{t_1(1-\gamma)} x(1-\gamma-\frac{x}{t_1})dx = \frac{t_1(1-\gamma)}{3}$$

and

$$M(Y_\gamma) = \frac{2}{t_2(1-\gamma)^2} \int_0^{t_2(1-\gamma)} y(1-\gamma-\frac{y}{t_2})dx = \frac{t_2(1-\gamma)}{3}.$$

We calculate now the variations of X_γ and Y_γ as,

$$M(X_\gamma^2) = \frac{2}{t_1(1-\gamma)^2} \int_0^{t_1(1-\gamma)} x^2(1-\gamma-\frac{x}{t_1})dx = \frac{t_1^2(1-\gamma)^2}{6}$$

and,

$$\text{var}(X_\gamma) = M(X_\gamma^2) - M(X_\gamma)^2 = \frac{t_1^2(1-\gamma)^2}{6} - \frac{t_1^2(1-\gamma)^2}{9} = \frac{t_1^2(1-\gamma)^2}{18}.$$

And, in a similar way, we obtain,

$$\text{var}(Y_\gamma) = \frac{t_2^2(1-\gamma)^2}{18}.$$

From,

$$\text{cov}(X_\gamma, Y_\gamma) = -\frac{t_1 t_2 (1-\gamma)^2}{36}.$$

we can calculate the probabilistic correlation of the random variables,

$$\rho(X_\gamma, Y_\gamma) = \frac{\text{cov}(X_\gamma, Y_\gamma)}{\sqrt{\text{var}(X_\gamma)}\sqrt{\text{var}(Y_\gamma)}} = -\frac{1}{2}.$$

The f-weighted possibilistic correlation of A_{t1} and A_{t2},

$$\rho_f(A_{t_1}, A_{t_2}) = \int_0^1 -\frac{1}{2}f(\gamma)d\gamma = -\frac{1}{2}.$$

So, the autocorrelation function of this fuzzy time series is constant. Namely,

$$R(t_1, t_2) = -\frac{1}{2}$$

for all t_1, $t_2 \in [0, 1]$.

Acknowledgments. We are greatly indebted to Prof. Tamás Móri of Department of Probability Theory and Statistics, Eötvös Loránd University, Budapest, for his long-term help with probability distributions.

References

[1] Carlsson, C., Fullér, R., Majlender, P.: On Possibilistic Correlation. Fuzzy Sets and Systems 155, 425–445 (2005)

[2] Fullér, R., Majlender, P.: On Interactive Fuzzy Numbers. Fuzzy Sets and Systems 143, 355–369 (2004)

[3] Fullér, R., Mezei, J., Várlaki, P.: An Improved Index of Interactivity for Fuzzy Numbers. Fuzzy Sets and Systems (submitted)

[4] Möller, B., Beer, M., Reuter, U.: Theoretical Basics of Fuzzy Randomness - Application to Time Series with Fuzzy Data. In: Augusti, G., Schueller, G.I., Ciampoli, M. (eds.) Safety and Reliability of Engineering Systems and Structures - Proceedings of the 9th Int. Conference on Structural Safety and Reliability, pp. 1701–1707. Millpress, Rotterdam

A Novel Bitmap-Based Algorithm for Frequent Itemsets Mining

János Abonyi

Department of Process Engineering, University of Pannonia
POB. 158, H-8200 Veszprém, Hungary
abonyij@fmt.uni-pannon.hu

Abstract. Mining frequent itemsets in databases is an important and widely studied problem in data mining research. The problem of mining frequent itemsets is usually solved by constructing candidates of itemsets, and identifying those itemsets that meet the requirement of frequent itemsets. This paper proposes a novel algorithm based on BitTable (or bitmap) representation of the data. Data - related to frequent itemsets - are stored in spare matrices. Simple matrix and vector multiplications are used to calculate the support of the potential n+1 itemsets. The main benefit of this approach is that only bitmaps of the frequent itemsets are generated. The concept is simple and easily interpretable and it supports a compact and effective implementation (in MAT-LAB). An application example related to the BMS-WebView-1 benchmark data is presented to illustrate the applicability of the proposed algorithm.

Keywords: frequent itemsets, BitTable.

1 Introduction

Mining frequent itemsets is an important and widely studied problem in the field of data mining research [1]. When the minimum support is small, and hence the number of frequent itemsets is very large, most algorithms either run out of memory or run over of allowed disk space due to the huge number of frequent itemsets. Most of early researches focused on the reduction of the amount of candidate itemsets and the time required to database-scanning. A lot of these algorithms adopt an Apriori-like candidate itemsets generation and support count approach which is a time-demanding process and needs a huge memory capacity.

Among the wide range of the developed algorithms this paper focuses on bitmap based solutions, like MAFIA [3] and BitTableFI [4], where BitTable is used for compressing the database horizontally and vertically for quick candidate itemsets generation and support count. Experiments with both synthetic and real databases show that BitTableFI outperforms Apriori and CBAR which uses ClusterTable for quick support count. Wei Song at al. has developed this concept resulting the Index-BitTableFI algorithm [7].

I.J. Rudas et al. (Eds.): Computational Intelligence in Engineering, SCI 313, pp. 171–180.
springerlink.com

In these algorithms the task of mining frequent itemsets is still solved by construct-ing candidate itemsets, then, identifying the itemsets that meet the frequent itemset requirement. The motivation of this paper is to develop an extremely simple and eas-ily implementable algorithm based on the bitmap-like representation of the frequent itemsets. The key idea is simple: store the data related to a given itemset in a binary vector. Hence, data related to frequent itemsets is stored in spare matrices, simple ma-trix and vector multiplications are used to calculate the support of the potential $k+1$ itemsets. The main benefit of this approach is that only bitmaps of the frequent item-sets are generated based on the elementwise products of the binary vectors corre-sponding the building k-1 *frequent itemsets*. Furthermore, when fuzzy membership values are stored in the bitmap-like matrices, the algorithm can directly be used to generate fuzzy frequent itemsets. The concept is simple and easily interpretable, so it supports the compact and effective implementation of the algorithm (in MATLAB).

The paper is organized as follows. In Section 2 we are going to briefly revisit the problem definition of frequent itemset mining by basic definitions and show the de-tails of the proposed algorithm. In Section 3, an application for web usage mining will be presented. Finally, the results and the advantages of the proposed method will be summarized.

2   The Proposed Algorithm

The problem of finding frequent itemsets can be formally stated by the following way: let $I = \{i_1, i_2, \ldots, i_m\}$ be a set of distinct literals, called items. Let $D = \{T_1, T_2, \ldots, T_N\}$ be a set of transactions, where each transaction T is a set of items such that $T \subseteq I$. A transaction T is said to support an itemset X if it contains all items of X, i.e., $X \subseteq T$. The support of an itemset X is the number (or percent-age) of transactions that support X. An itemset is frequent if its support is greater or equal to a user-specified minimum support threshold, denoted *MinSup*. Frequent item-sets are also known as large itemsets. An itemset X is called k-*itemset* if it contains k items from I.

An illustrative example for D transactional database is shown in Table 1(a). The transactional database can be transformed into a bitmap-like matrix representation, where if an item $i = 1, \ldots, m$ appears in transaction $T_j, j = 1, \ldots, N$, the bit i of the j-th row of the binary incidence matrix will be marked as one (as seen in Table 1).

For mining association rules, non-binary attributes have to be mapped to binary at-tributes. The straightforward mapping method is to transform the metric attributes to k ordinal attributes by building categories (e.g., an attribute income might be trans-formed into a ordinal attribute with the three categories: "low", "medium" and "high"). Then, in a second step, each categorical attribute with categories k is repre-sented by k binary dummy attributes which correspond to the items used for mining. An example application using questionnaire data can be found in [6].

Table 1. Illustrative example for a transactional dataset and its binary incidence matrix representation

Items	
Tid	**Items**
T_1	a, c, d
T_2	b, c, e
T_3	a, b, c, e
T_4	b, e

(a)

	Items				
Tid	**1(a)**	**2(b)**	**3(c)**	**4(d)**	**5(e)**
T_1	1	0	1	1	0
T_2	0	1	1	0	1
T_3	1	1	1	0	1
T_4	0	1	0	0	1
Sum	**2**	**3**	**3**	**1**	**3**

(b)

As the support of an itemset is a percentage of the total number of transactions, the sum of the columns of this $\mathbf{B}^0_{N \times n}$ matrix represent the support of the $j = 1, \ldots, n$ items. (see the bottom of Table 1(b)) Hence, if $\mathbf{b}^0_j$ represents the j-th column of $\mathbf{B}^0_{N \times n}$ which is related to the occurrence of the i_j-th item, then the support of the i_j item can be easily calculated as

$$\sup(X = i_j) = (\mathbf{b}^0_j)^T \mathbf{b}^0_j / N \tag{1}$$

(in this case the result is given in percentage). Similarly, the support of an $X_{i,j} = \{i_i, i_j\}$ itemset can be easily calculated by a simple vector product of the two related bitvectors, since when both i_i and i_j items appear in a given transaction the product of the two related bits can represent the AND connection of the two items:

$$\sup(X_{i,j} = \{i_i, i_j\}) = (\mathbf{b}^0_i)^T \mathbf{b}^0_j / N \tag{2}$$

The matrix representation allows the effective calculation of all of the itemsets:

$$\mathbf{S}^2 = (\mathbf{B}^0)^T \mathbf{B}^0 \tag{3}$$

where the i,j-th element of the $\mathbf{S}^2$ matrix represents the support of the $X_{i,j} = \{i_i, i_j\}$ 2-itemset. Of course, only the upper triangular elements of this, this symmetrical matrix has to be checked, whether the $X_{i,j} = \{i_i, i_j\}$ 2-*itemsets* are frequent or not.

Fuzzy membership values can also be stored in the same matrix structure, where the columns represent the items and the rows the transactions (see Table 2). Hence, beside the analysis of classical transactional datasets, the analysis of fuzzy data is also considered in this paper. In this case let $D = \{t_1, t_2, \ldots, t_N\}$ be a transformed fuzzy dataset of N tuples (data points) with a set of variables $Z = \{z_1, z_2, \ldots, z_n\}$ and let

$c_{i,j}$ be an arbitrary fuzzy interval (fuzzy set) associated with attribute z_i in Z. Use the notation $\langle z_i : c_{i,j} \rangle$ for an *attribute-fuzzy interval pair*, or simply *fuzzy item*, (i.e.: $\langle Age : young \rangle$). For *fuzzy itemsets*, we use expressions like $\langle Z : C \rangle$ to denote an ordered set $Z \subseteq \mathbf{Z}$ of attributes and a corresponding set C of some fuzzy intervals, one per attribute, i.e. $\langle Z : C \rangle$.

Table 2. Example database containing membership values

	Items		
	$\langle Balance : medium \rangle$	$\langle Credit : high \rangle$	$\langle Income : high \rangle$
T_1	0.5	0.6	0.4
T_2	0.8	0.9	0.4
T_3	0.7	0.8	0.7
T_4	0.9	0.8	0.3
T_5	0.9	0.7	0.6

Equation (3) can also be used in case the $\mathbf{B}^0_{N \times n}$ matrix stores fuzzy membership values. A fuzzy support reflects how the record of the dataset supports the itemset. In the literature, the fuzzy support value has been defined in different ways. Some of the researchers suggest the minimum operator as in fuzzy intersection, others prefer the product operator. They can be defined formally as follows: value $T_k(z_i)$ for attribute z_i, then the *fuzzy support* of $\langle Z : C \rangle^2$ with respect to D is defined as

$$FS(Z:C) = \frac{\sum_{k=1}^{N} \prod_{\langle z_i : c_{i,j} \rangle \in \langle Z:C \rangle} T_k(z_i)}{N} \tag{4}$$

The following example illustrates the calculation of the fuzzy support value. Let $\langle X : A \rangle = [\langle Balance : medium \rangle \cup \langle Income : high \rangle]$ be a fuzzy itemset, the dataset shown in Table 2. The fuzzy support of $\langle X : A \rangle$ is given by:

$$FS(X:A) = \\ = \frac{0.5 \cdot 0.4 + 0.8 \cdot 0.4 + 0.7 \cdot 0.7 + 0.9 \cdot 0.3 + 0.9 \cdot 0.6}{5} = 0.364 \tag{5}$$

An itemset $\langle Z : C \rangle$ is called frequent if its fuzzy support value is higher than or equal to a user-defined minimum support threshold σ.

2.2 Mining Frequent Itemsets Based on Bitmap-Like Representation

2.2.1 Apriori Algorithm for Mining Frequent Itemsets

The best-known and most commonly applied frequent pattern mining algorithm *Apriori* was developed by Agrawal et al. [2]. The name is based on the fact that the algorithm uses prior knowledge of the already determined frequent itemsets.

It is an iterative, breadth-first search algorithm, based on generating stepwise longer *candidate* itemsets, and clever pruning of non-frequent itemsets. Pruning possesses the advantage of the so-called *apriori* (or *upward closure*) *property* of frequent itemsets: all subsets of a frequent itemset must also be frequent. Each candidate generation step is followed by a counting step where the supports of candidates are checked and non-frequent ones deleted.

Given a user-specified *MinSup*, Apriori passes multiple times over the database to find all frequent itemsets. In the first pass, Apriori scans the transaction database to count the support of each item and identify the frequent *1-itemsets* marked as L_1. In a subsequent k-th pass, Apriori establishes a candidate set of frequent k-itemsets (which are itemsets of length k) marked as C_k from L_{k-1}. Two arbitrary L_{k-1} join each other, when their first k-1 items are identical. Then, the downward closure property is applied to reduce the number of candidates. This property refers to the fact that any subset of a frequent itemset must be frequent. Therefore, the process deletes all the k-*itemsets* whose subsets with length k - 1 are not frequent. Next, the algorithm scans the entire transaction database to check whether each candidate k-itemset is frequent.

Generation and counting alternate, until at some step all generated candidates turn out to be non-frequent. A high-level pseudocode of the algorithm used for mining fuzzy frequent itemsets based on the *apriori* principle is given in Table 3.

Table 3. Algorithm: Mining Frequent Itemsets (minimum support σ, dataset D)

```
k = 1
(C_k;D_F) = Transform(D)
F_k = Count(C_k, D_F, σ)
while |C_k| ≠ 0 do
inc(k)
C_k = Generate(F_{k-1})
C_k = Prune(C_k)
F_k = Count(C_k, D_F, σ)
F = F ⋃ F_k
end
```

In case of mining fuzzy frequent itemsets the subroutines are outlined as follows:

- *Transform(D):* Generates a fuzzy database D_F from the original dataset D. At the same time the complete set of candidate items C_1 is found.
- *Count(C_k, D_F, σ):* In this subroutine the fuzzy database is scanned and the fuzzy support of candidates in C_k is counted. If this support is not less than minimum support σ for a given itemset, we put it into the set of frequent itemsets F_k.

- *Generate(F_{k-1}):* Generates candidate itemsets C_k from frequent itemsets F_{k-1}, discovered in the previous iteration $k-1$. For example, if
$F_1 = \{\langle Balance: high\rangle, \langle Income: high\rangle\}$, then
$C_2 = \{\langle Balance: high\rangle \cup \langle Income: high\rangle\}$

- *Prune(C_k):* During the prune step, the itemset will be pruned if one of its subsets does not exist in the set of frequent itemsets F.

2.2.2 The Proposed Algorithm for Mining Frequent Itemsets

The proposed algorithm has similar philosophy as the Apriori TID [5], which is does not revisit the original table of data, $\mathbf{B}^0_{N \times n}$, for computing the supports larger itemsets, but transforms the table as it goes along with the generation of the k-itemsets, $\mathbf{B}^1_{N_1 \times n_1} \cdots \mathbf{B}^k_{N_k \times n_k}$ $N_k < N_{k-1} < K < N$.

$\mathbf{B}^1_{N_1 \times n_1}$ represents the data related to the 1-frequent itemsets. This table is generated from $\mathbf{B}^0_{N \times n}$, by erasing the columns related to the non-frequent items, to reduce the size of a Bittable and improve the performance of the generation process, because all non-frequent 1-itemsets are not useful for further analysis. The rows of $\mathbf{B}^k_{N_k \times n_k}$ which do not contain frequent itemsets (the sum of the row is zero) are also deleted from the table. This concept is taken from the Apriori TID algorithm based on using new data structure called *counting_base* to store the transactions which can support the actual list of candidates.

If a column remains, the index of its original position is written into matrix that stores the indexes ("pointers") of the element of the itemsets, $\mathbf{L}^1_{N_1 \times 1}$.

Data related to frequent itemsets are stored in spare matrices. Simple matrix and vector multiplications are used to calculate the support of the potential $k+1$ itemsets:

$$\mathbf{S}^k = \left(\mathbf{B}^{k-1}\right)^T \mathbf{B}^{k-1} \tag{6}$$

where the i,j-th element of the $\mathbf{S}^k$ matrix represent the support of the $X_{i,j} = \{\mathbf{L}^{k-1}_i, \mathbf{L}^{k-1}_j\}$ itemset. Of course, only the upper triangular elements of this symmetrical matrix has to be checked, whether the *itemsets* are frequent or not. The main benefit of this approach is that only the BitTables of the frequent itemsets are generated, by forming the columns of the $\mathbf{B}^k_{N_k \times n_k}$ as the element wise products of the columns of the $\mathbf{B}^{k-1}_{N_{k-1} \times n_{k-1}}$, $\mathbf{b}^k_i$, $\mathbf{b}^k_j$.

The concept is simple and easily interpretable that supports the compact and effective implementation (see the appendix for the MATLAB code).

The above presented approach is related to the lazy version of the algorithm. The performance of the support count can be significantly decreased when only the relevant blocks of the $\mathbf{B}^{k-1}_{N_{k-1}\times n_{k-1}}$ matrix are multiplied by their transpose. When $L^{k-1}_{N_{k-1}\times k-1}$ matrices related to the indexes of the k-1-itemsets are ordered it is easy to follow the heuristics of the apriori algorithm, as only the itemsets L_{k-1} each other, when their first k-1 items are identical (the set of these itemsets form the blocks of the $\mathbf{B}^{k-1}_{N_{k-1}\times n_{k-1}}$ matrix).

3 Application Example

The main benefit of the proposed algorithm is that it can be effectively implemented in tools tailored to perform matrix manipulations. In the appendix of this paper the full implementation of the algorithm is shown as a MATLAB function. This code with 15 lines is optimized to give a reasonable calculation time. Hence, the matrix multiplications are performed in block-wise manner and the unnecessary transactions (rows) are removed from the $\mathbf{B}^{k}_{N_{k}\times n_{k}}$ matrices.

The proposed algorithm and MATLAB code has been applied to the BMS-WebView-1 benchmark problem, [8] where data taken from the www.gazelle.com web portal is analyzed. This database contains 59,602 transactions and 497 items (webpages). The maximal size of a basket is 267, while its average size is 2.5 (length of the average visit of the portal).

The calculation time is shown in Figure 1 is quite reasonable, considering the MATLAB framework is not optimized to calculation speed. The results agree with the results of other applications [8] (see Figure 2 for the number of the mined itemsets). It is interesting to see Figure 3 that nicely illustrates the key element of the proposed approach, the bitmap of the 2th itemset of the studied benchmark data.

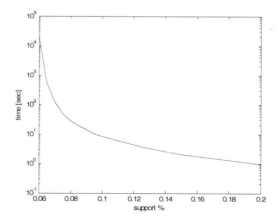

Fig. 1. Time required to mine frequent itemsets with a given support

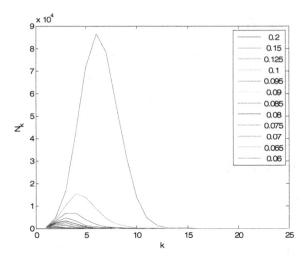

Fig. 2. Number of frequent itemsets related to different support treshold (MinSup). As can be seen, at smaller support values the number of itemsets can be really huge.

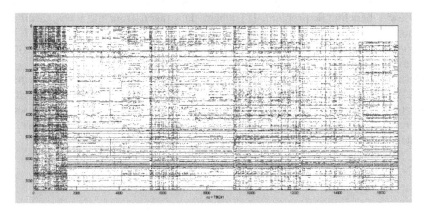

Fig. 3. Data related to frequent itemsets are stored in "Bitmaps" like shown in this figure, where the columns represent the itemsets and the dots in the rows represen the given itemset is in the transaction related to the row of the matrix

4 Conclusions

This paper proposed a novel algorithm for mining frequent itemsets. The key idea is to store the data related to a given itemset in a binary vector. Hence, data related to frequent itemsets are stored in spare matrices and simple matrix and vector multiplications are used to calculate the support of the potential $k+1$ itemsets.

The main benefit of this approach is that only bitmaps of frequent itemsets are generated based on the elementwise products of the binary vectors corresponding the building $k-1$ frequent itemsets, since bitwise AND operation is greatly faster than comparing each item in two frequent itemsets (as at Apriori). Furthermore, when fuzzy membership values are stored in the bitmap-like matrices, the algorithm can directly be used to generate fuzzy frequent itemsets. The concept is simple and easily interpretable, so it supports the compact and effective implementation of the algorithm (in MATLAB). The application example related to the BMS-WebView-1 benchmark problem demonstrated that applicability of the developed compact MATLAB code that can be easily used by medium-sized firms having around 100 000 transactions and several thousand items.

Acknowledgments. The financial support from the TAMOP-4.2.2-08/1/2008-0018 (Livable environment and healthier people – Bioinnovation and Green Technology research at the University of Pannonia, MK/2) project is gratefully acknowledged.

References

[1] Burdick, D., Calimlim, M., Gehrke, J.: MAFIA: A Maximal Frequent Itemset Algorithm for Transactional Databases Department of Computer Science. Cornell University, Ithica
[2] Hastie et al (2001)
[3] Abonyi, J., et al.: Adatbányászat – a hatékonyság eszköze. Computerbooks (2006)
[4] Dong, J., Han, M.: Bit Table FI: An Efficient Mining Frequent Itemsets Algorithm, Science Direct (2006)
[5] Han, J., Kamber, M.: Data Mining Concepts and Techniques. Elsevier, Amsterdam (2001)
[6] Agrawal, R., Srikant, R.: Fast Algorithm for Mining Association Rules in Large Databases. In: Proceedings of 1994 International Conference on VLDB, pp. 487–499 (1994)
[7] Song, W., Yang, B., Xu, Z.: Index-BitTableFI: An Improved Algorithm for Mining Frequent Itemsets. Knowledge-based Systems Journal homepage: elsevier.com/locate/knosys (2008)
[8] Zheng, Z., Kohavi, R., Mason, L.: Real World Performance of Association Rule Algorithms. ACM, New York (2001)

Appendix
The Word's Most Compact Frequent Itemset Miner in MATLAB

```
function [items,Ai]=bittable(A, suppp)
[N,n]=size(A);
items{1}=find(sum(A,1)>=suppp)';
k=1; Ai{1}=A(:,items{1});
while ~isempty(items{k})
  k=k+1; Ai{k}=[]; items{k}=[];
  index=[0;find(sum(abs(diff((items{k-1}(:,1:end-1)))),2)~=0);...
          size(items{k-1},1)];
  for i=1:length(index)-1
    v=[index(i)+1:index(i+1)]; m=Ai{k-1}(:,v)'*Ai{k-1}(:,v);
    m=triu(m,1); [dum1,dum2]=find((m)>suppp);
    for j=1:length(dum1)
      items{k}=[items{k}; [items{k-1}(v(dum1(j)),:)items{k-1}...
                        (v(dum2(j)),end) ] ];
      Ai{k}= [ Ai{k} Ai{k-1}(:,v(dum1(j))).*Ai{k-1}...
                        (:,v(dum2(j))) ];
    end
  end
  [items{k},I]=sortrows(items{k}); Ai{k}=Ai{k}(:,I);
end
```

Neural Networks Adaptation with NEAT-Like Approach

Jaroslav Tuhársky and Peter Sinčák

Center for Intelligent Technologies, Department of Cybernetics and Artificial Intelligence,
FEI TUKE Kosice
jaroslav.tuharsky@gmail.com, peter.sincak@tuke.sk

Abstract. This paper describes experience with NEAT (NeuroEvolution of Augmenting Topologies) method which is based on evolutionary computation and optimization of neural networks structure and synaptic weights. Non-linear function XOR approximation is tested and evaluated with this method with the aim of perspective application in humanoid robot NAO. The experiments show that selected method NEAT is suitable for this type of adaptation of NN, because of its ability to deal with the problems which emerge in TWEAN methods.

Keywords: neural networks, evolutionary algorithms, Neuroevolution, TWEANN, NEAT, genetic algorithms.

1 Introduction

Evolutionary algorithm (EA) is used for solving optimization problems and one of these tasks could be a search for optimal Neural Network (NN) and its topology.

Finding the optimal neural networks by using EA may consist of NN topologies optimization - searching NN topology able to solve the problem and of NN synaptic weights (SW) optimization - search for suitable values of SW. As it is described in [6], neuroevolution (NE), the artificial evolution of neural networks using genetic algorithms (GA), has shown great promise in complex learning tasks.

2 NEAT

The method NeuroEvolution of Augmenting Topologies (NEAT) was created by K. O. Stanley and R. Miikkulainen, from the Texas University in Austin, described in [6]. From the same publication is the following description.

2.1 Genetic Encoding

NEAT's genetic encoding scheme is designed to allow corresponding genes to be easily lined up when two genomes cross over during matting. Genomes are linear representations of network connectivity. Each genome includes a list of connections

I.J. Rudas et al. (Eds.): Computational Intelligence in Engineering, SCI 313, pp. 181–190.
springerlink.com

genes, each of which refers to two node genes being connected. Each connection gene specifies the in-node, the out-node, the weight of the connection, whether or not the connection gene is expressed, and an innovation number, which allows finding corresponding genes, see [6].

2.2 Historical Markings of Genes

Whenever a new gene appears through structural mutation, a global innovation number is incremented and assigned to that gene. The innovation number thus represents a chronology of the appearance of every gene in the system. The historical markings of genes give NEAT new capability. The system now knows exactly which genes match up with which. When crossing over, the matching genes in both genomes with the same innovation numbers are lined up, see [6].

2.3 Protecting Innovations through Speciation

Speciating the population allows organisms to compete primarily within their own niches instead of with the population at large. This way, topological innovations are protected in a new niche where they have time to optimize their structure through competition within the niche. This task appears to be a topology matching problem. In NEAT is the measure of the compatibility distance of a different structures a simple linear combination of the number of excess E and disjoint D genes, as well as the average weight differences of matching genes W, including disabled genes, see (1). [6]

$$\delta = c_1 \cdot \frac{E}{N} + c_2 \cdot \frac{D}{N} + c_3 \cdot \overline{W} \tag{1}$$

$\overline{W}$ – average of SW differences of matching genes
E – number of excess genes
D – number of disjoint genes
N – number of genes in larger genome (for normalization because of its size)
c_1, c_2, c_3 – coefficients
$\tilde{\delta}$ compatibility (gene's) distance

3 Implementation of NEAT Approach

For the implementation of experiments we have proposed and implemented the software in the creation of which we was inspired by the method NEAT, see Chapter 2. The reason why we have decided for NEAT is that it provides the best solutions for the problems associated with TWEANN - see [6, 9] and is using evolutionary calculations, namely the GA to find the simplest topologies with optimized SW for XOR problem.

3.1 Representing Individuals

The population is made up of individuals - NN. Particular individual, which we call the genome, contains a list of "genes of links" [6].

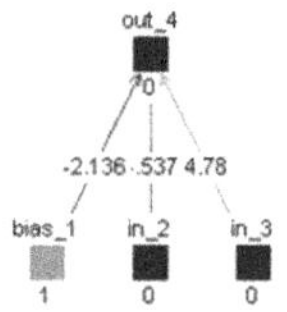

Fig. 1. Initial NN topology – 4-Node NN

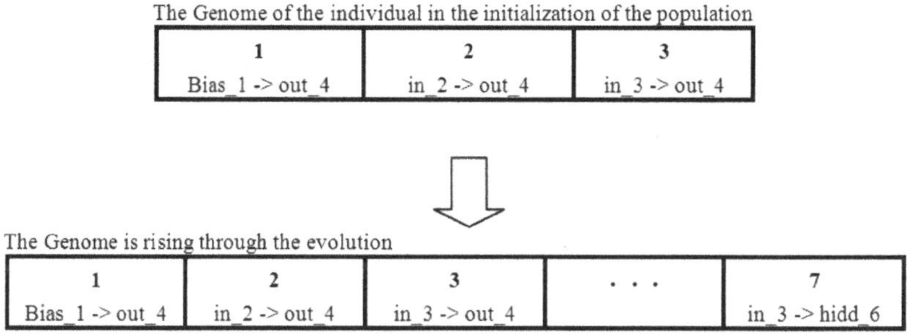

The Genome of the individual in the initialization of the population

1	2	3
Bias_1 -> out_4	in_2 -> out_4	in_3 -> out_4

The Genome is rising through the evolution

1	2	3	. . .	7
Bias_1 -> out_4	in_2 -> out_4	in_3 -> out_4		in_3 -> hidd_6

Fig. 2. The Genome is rising through the evolution

In the initialization of the population are individuals with a minimum NN topology, see Fig. 1, whose structure is made up of only input and output layer, i.e. without hidden layers, and with the philosophy that their structure is rising only when it is appropriate for a given solution, see Fig. 2. Input layer consists of two input nodes, output layer of one output node. Node "bias" was incorporated into the topology so that we can introduce the entry of the external world to all of the neurons.

For each neuron in the NN, we used the same sigmoid activation function.

3.2   Genetic Operators

Using the Genetic Operators (GO) program searches the SW space for an optimal NN, able to solve the required task, which is in our experiment XOR. In the program we have used the following GO: crossover, SW mutation, add node mutation, and add connection mutation. Mutation in NEAT can change both SW and NN topology. SW mutation is performed by generating a random value from the interval of SW's values. Structural mutations occur in two ways as a add node mutation and as a add connection mutation.

3.2.1   Crossover
The probability of crossing is given by parameter CrossProb. Crossover between two compatible genes is indeed in the calculation of SW value, which is inherited by the offspring from the values of his parents SW. Crossover can be done in two ways, either by calculating the average SW value of the parents - AVG parameter or the value of SW is randomly generated from one or the other parent.

Crossover GO is applied only to the individuals of the same species, i.e. can not be a crossing of different species individuals, thus addressing the problem of permutation, see [6, 9]. The number of offsprings for each species is based on a probabilistic relationship, see (2).

$$\lambda_S = \frac{\overline{\Phi}_S}{\sum\limits_{j=1}^{N} \overline{\Phi}_j} \cdot \mu$$

(2)

$\overline{\Phi}_S$ – average fitness of individuals of the species S, see equation (3)
$\tilde{\mu}$ – number of individuals in the population
N – number of species in the population
$\tilde{\lambda_S}$ – number of offsprings of the species S

where

$$\overline{\Phi}_S = \frac{\sum\limits_{i=1}^{\mu_S} \Phi(a_i)}{\mu_S}$$

(3)

$\Phi(a_i)$ – fitness function of individual a_i
μ_S – number of individuals in the species S

3.2.2 Fitness Function

We split the calculation of a Fitness function (FF) into 2 steps. Calculating gross FF (4) from purpose function (5), and calculating adjusted FF (6) based on fitness sharing method. In the experiment, the purpose function is the Sum of Squared Error (SSE) of NNs (5) in solving the XOR task.

$$\Phi(f(a_i)) = (1-\varepsilon) \cdot \frac{f(a_i) - \min\limits_j f(a_j)}{\min\limits_j f(a_j) - \max\limits_j f(a_j)} + 1$$

(4)

Φ – fitness function
$f(a_i)$ – purpose function of individual a_i
ε – the lowest possible value of fitness

where

$$f(a_i) = \sum_{k=1}^{4} e_k^2$$

(5)

e_k^2 – SSE Sum of Squared Error on k-th solution of XOR

$$\Phi'(a_i) = \frac{\Phi(a_i)}{\sum\limits_{j=1}^{\mu} sh(d(a_i, a_j))}$$

(6)

$\Phi(a_i)$ – fitness function of individual a_i
$\Phi'(a_i)$ – adjusted fitness function of individual a_i
$sh(d(a_i,a_j))$ – sharing function

where

$$\sum_{j=1}^{\mu} sh\big(d\big(a_i,a_j\big)\big)= \mu_S \tag{7}$$

μ_S – number of individuals in the species S
S – species in which is the individual who fitness is adjusted

4   Experiments

4.1   Experiment Example No 1

This experiment shows that in 500 generations, the program NEAT created 4-Node, 5-Node, 6-Node and 7-Node NN. Figure 3 shows the number of individuals pertaining to the topology in the certain generation, as well as the emergence and disappearance of species in the population.

In this case, the 7-node NN was created at the end of the experiment, in the 450. gen. there were only 5 such individuals in the whole population, see Fig. 3. Therefore, the program had sufficient time to search SS (State Space) of SW of 6-Node NN and founds its optimization in the 350. gen., i.e. founds the optimal topology and values of NN's SW able to solve XOR task. The 5-Node NN was optimized in the 130.gen. and the search for the 4-Node NN, see Fig. 1 (which we know that is not able to solve XOR task) the program stopped in 260.gen., so that entire 4-Node type was thrown away from the population, see Fig. 3 and charts of SSE (Sum of Squared Error) during the evolution which are shown in Figs. 4-8.

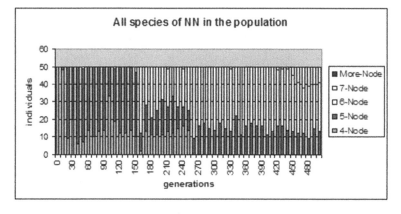

Fig. 3. Number of individuals in the population

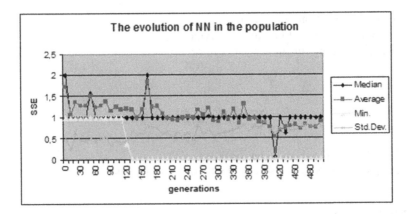

Fig. 4. SSE of NN through evolution in all population

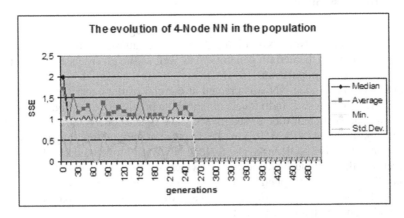

Fig. 5. SSE of the 4-Node NN through the evolution

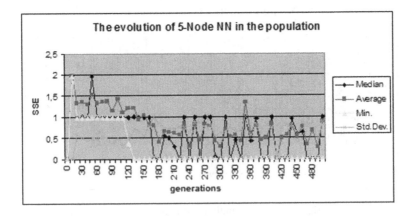

Fig. 6. SSE of the 5-Node NN through the evolution

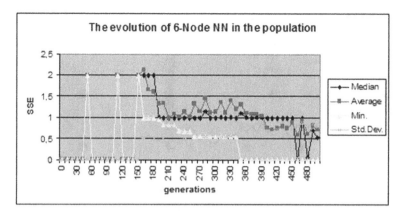

Fig. 7. SSE of the 6-Node NN through the evolution

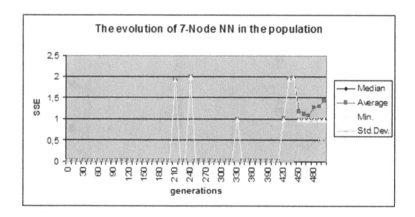

Fig. 8. SSE of the 7-Node NN through the evolution

The reason why the 7-Node NN was not optimized, is due to lack of time (generations) needed for sufficient scan of NN's SW state space.

In most cases to find the simplest (5-Node) topology of NN able to solve XOR task (where SSE = 0) only 100.gen. were needed, but for NN with more complex structure we need more generations for its optimization.

Figures 9-12 shows individuals of particular species which were evolved on the end of the evolution process.

Table 1. SSE of the best individual of each species at the end of evolution

INDIV. ID	SSE	INDIV. ID	SSE
ID_28551	3.09e-05	ID_30748	1.19e-08
ID_28588	0.006266	ID_30647	1

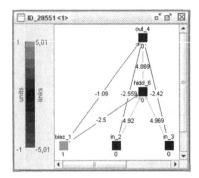

Fig. 9. The individual from species No1

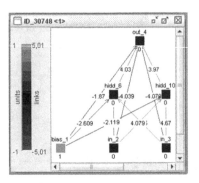

Fig. 10. The individual from species No2

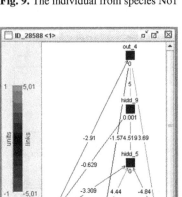

Fig. 11. The individual from species No3

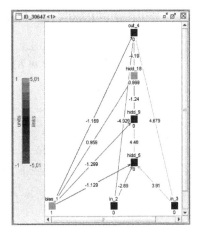

Fig. 12. The individual from species No4

4.2   Experiment Example No 2

This experiment shows that the two phenotypic identical individuals, i.e. NN with the same topology, see Figs. 13 and 14, do have identical genetic codes, i.e. they are not coded in the same genetic sequence, which shows that the program solves the permutations problem, see [6, 9]. That these two individuals are not coded in the same genetic code, proving the node in hidden layer, where one is "hidd_5" and second is "hidd_6". This situation, where two different genes (nodes) coding the same pheno-typic feature, i.e. it is a node on the same position in the topology of NN, occurs in cases where the genes encoding the same phenotypic feature have different historical marking of genes, i.e. this is not the same gene. In this particular case, one node emerged between the nodes "in_2" and "out_4", but the second between "in_3" and "out_4" and finally the gradual addition of more synaptic connections through the evolution was due to the nodes were on the same position in topology of NN.

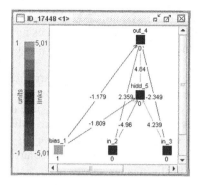
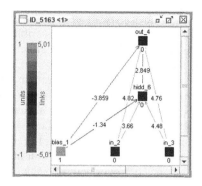

Fig. 13. The individual from species No1   **Fig. 14.** The individual from species No2

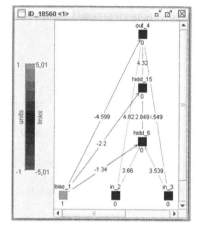
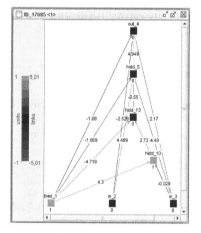

Fig. 15. The individual from species No3   **Fig. 16.** The individual from species No4

Table 2. SSE of the best individual of each species at the end of evolution

INDIV. ID	SSE	INDIV. ID	SSE
ID_17448	2.5348e-005	ID_5163	1
ID_18560	1	ID_17685	4.0929e-005

4.3 Experiments Results

For selected experiments we have demonstrated the functionality of the implemented NEAT method with ability to cope with problems that arise from the TWEANN approaches see [6, 9].

5 Conclusions

The functionality of the NEAT method was tested for its ability to evolve various topologies of NN able to deal with XOR problem.

This work has shown strong and weak points of the system TWEANN, as well as the NEAT system, and outlined possible pitfalls, which can be given when using these systems. Tested approach is very effective for the problems of TWEANN which were easily solved.

Our plan was to create a program that will help us to understand this extremely interesting method in depth which is crucial for its future use or improvement.

This work is seen as an essential step on our way to create a system capable of optimizing the NN topology, together with its SW, so that they were able to solve the challenges associated with control of BOTs in space, which would create a system able to adapt on the situation without human intervention. This intelligent control will be used in the control of real robots in space, or BOTs in video games.

Acknowledgment. This publication is the result of the project implementation Centre of Information and Communication Technologies for Knowledge Systems (project number: 26220120020) supported by the Research & Development Operational Programme funded by the ERDF.

References

[1] Sinčák, P., Andrejková, G.: Neuron Networks (engineering approach) part 1 and part 2, Elfa, Kosice (1996) (in Slovak)

[2] Mach, M.: Evolutionary Algorithms – Elements and Principles. Elfa, Kosice (2009) ISBN 978-80-8086-123-0

[3] Kvasnička, V., Pospíchal, J., Tiňo, P.: Evolutionary Algorithms. STU, Bratislava (2000) (in Slovak) ISBN 80-227-1377-5

[4] Mařík, V., Štěpánková, O., Lažanský, J., et al.: Artificial Intelligence 3. Academia, Praha (2001) (in Czech) ISBN 80-200-0472-6

[5] Mařík, V., Štěpánková, O., Lažanský, J., et al.: Artificial Intelligence 4. Academia, Praha (2003) (in Czech) ISBN 80-200-1044-0

[6] Stanley, K.O., Miikkulainen, R.: Evolving Neural Networks through Augmenting Topologies. The MIT Press Journals (2002)

[7] Stanley, K.O., Miikkulainen, R.: Efficient Evolution of Neural Network Topogies. In: Proceedings of 2002 Congerss on Evolutionary Coputation. IEEE, Piscataway (2002)

[8] Stanley, K.O., Bryant, B.D., Miikkulainen, R.: Real-Time Neuroevolution in the NERO Video Game. IEEE Transactions on Evolutionary Computation (2005)

[9] Tuhársky, J., Reiff, T., Sinčák, P.: Evolutionary Approach for Structural and Parametric Adaptation of BP-like Multilayer Neural Networks. In: Proceedings of the 10th International Symposium of Hungarian Researchers on Computational Intelligence and Informatics, Budapest, Hungary, November 12-14, pp. 41–52 (2009)

Incremental Rule Base Creation with Fuzzy Rule Interpolation-Based Q-Learning

Dávid Vincze and Szilveszter Kovács

Department of Information Technology, University of Miskolc
Miskolc-Egyetemváros, H-3515 Miskolc, Hungary
{david.vincze, szkovacs}@iit.uni-miskolc.hu

Abstract. Reinforcement Learning (RL) is a widely known topic in computational intelligence. In the RL concept the problem needed to be solved is hidden in the feedback of the environment, called rewards. Using these rewards the system can learn which action is considered to be the best choice in a given state. One of the most frequently used RL method is the Q-learning, which was originally introduced for discrete states and actions. Applying fuzzy reasoning, the method can be adapted for continuous environments, called Fuzzy Q-learning. An extension of the Fuzzy Q-learning method with the capability of handling sparse fuzzy rule bases is already introduced by the authors. The latter suggests a Fuzzy Rule Interpolation (FRI) method to be the reasoning method applied with Q-learning, called FRIQ-learning. The main goal of this paper is to introduce a method which can construct the requested FRI fuzzy model from scratch in a reduced size. The reduction is achieved by incremental creation of an intentionally sparse fuzzy rule base. Moreover an application example (cart-pole problem simulation) shows the promising results of the proposed rule base reduction method.

Keywords: reinforcement learning, fuzzy Q-learning, fuzzy rule interpolation, fuzzy rule base reduction.

1 Introduction

Reinforcement learning methods can help in situations, where the task to be solved is hidden in the feedback of the environment, i.e. in the positive or negative rewards (the negative reward is often called a punishment) provided by the environment. The rewards are calculated by an algorithm created especially for expressing the task needed to be solved. Based on the rewards of the environment RL methods are approximating the value of each possible action in all the reachable states. Therefore RL methods can solve problems where a priory knowledge can be expressed in the form what is needed to be achieved, not in how to solve the problem directly. Reinforcement learning methods are a kind of trial-and-error style methods adapting to dynamic environment through incremental iterations. The primary ideas of reinforcement learning techniques (dynamical system state and the idea of 'optimal return', or 'value'

I.J. Rudas et al. (Eds.): Computational Intelligence in Engineering, SCI 313, pp. 191–203.
springerlink.com   © Springer-Verlag Berlin Heidelberg 2010

function) are inherited from optimal control and dynamic programming [3]. A common goal of the reinforcement learning strategies is to gather an optimal policy by constructing the state-value- or action-value-function [20]. The state-value-function $V^{\pi}(s)$, is a function of the expected return (a function of the cumulative reinforcements), related to a given state $s \in S$ as a starting point, following a given policy π. These rewards, or punishments (reinforcements) are the expression of the desired final goal of the learning agent as a kind of evaluation following the previous action (in contrast to the instructive manner of error feedback based approximation techniques, for example the gradient descent optimisation). The optimal policy is basically the description of the agent behaviour, in the form of mapping between the agent states and the corresponding suitable actions. The action-value function $Q^{\pi}(s,a)$ is a function of the expected return, in case of taking action $a \in A_s$ in state s following policy π. In possession of the action-value-function, the optimal (greedy) policy, which always takes the optimal (the greatest estimated value) action in every state, can be constructed as [20]:

$$\pi(s) = \arg \max_{a \in A_s} Q^{\pi}(s,a) \tag{1}$$

For the estimation of the optimal policy, the action-value function $Q^{\pi}(s,a)$ should be approximated. Approximating the latter function is a complex task given that both the number of possible states and the number of the possible actions could be an extremely high value. Evaluating all the possibilities could take a considerable amount of computing resources and computational time, which is a significant drawback of reinforcement learning. However there are some cases where a distributed approach with continuous reward functions can reduce these resource needs [17]. Generally reinforcement learning methods can lead to results in practically acceptable time only in relatively small state and action spaces.

Adapting fuzzy models the discrete Q-learning can be extended to continuous state and action space, which in case of suitably chosen states can lead to the reduction of the size of the state-action space [13].

2   Q-Learning and Fuzzy Q-Learning

Q-learning is a reinforcement learning method which has the purpose of finding the fixed-point solution (Q) of the Bellman Equation [3] via iteration. In the case of discrete *Q-Learning* [25], the action-value-function is approximated by the following iteration:

$$Q_{i,u} \approx \tilde{Q}_{i,u}^{k+1} = \tilde{Q}_{i,u}^k + \Delta \tilde{Q}_{i,u}^{k+1} = \tilde{Q}_{i,u}^k + \alpha_{i,u}^k \cdot \left(g_{i,u,j} + \gamma \cdot \max_{v \in U} \tilde{Q}_{j,v}^{k+1} - \tilde{Q}_{i,u}^k \right) \tag{2}$$

$\forall i \in I, \forall u \in U$, where $\tilde{Q}_{i,u}^{k+1}$ is the $k+1$ iteration of the action-value taking the uth action A_u in the ith state S_i, S_j is the new (j^{th}) observed state, $g_{i,u,j}$ is the observed reward completing the $S_i \rightarrow S_j$ state-transition, γ is the discount factor and

$\alpha_{i,u}^k \in [0,1]$ is the step size parameter (can vary during the iteration steps), I is the set of the discrete possible states and U is the set of the discrete possible actions. There are various existing solutions [1], [4], [5], [6] for applying this iteration to continuous environment by adopting fuzzy inference (called Fuzzy Q-Learning). Most commonly the simplest FQ-Learning method, the 0-order Takagi-Sugeno Fuzzy Inference model is adapted. Hereby in this paper the latter one is studied (a slightly modified, simplified version of the Fuzzy Q-Learning introduced in [1] and [6]). In this case for characterizing the value function $Q(s,a)$ in continuous state-action space, the 0-order Takagi-Sugeno Fuzzy Inference System approximation $\tilde{Q}(s,a)$ is adapted in the following way:

$$\text{If } s \text{ is } S_i \text{ And } a \text{ is } A_u \text{ Then } \tilde{Q}(s,a) = Q_{i,u} , i \in I, u \in U , \tag{3}$$

where S_i is the label of the ith membership function of the n dimensional state space, A_u is the label of the uth membership function of the one dimensional action space, $Q_{i,u}$ is the singleton conclusion and $\tilde{Q}(s,a)$ is the approximated continuous state-action-value function. Having the approximated state-action-value function $\tilde{Q}(s,a)$, the optimal policy can be constructed by function (1). Setting up the antecedent fuzzy partitions to be *Ruspini partitions*, the zero-order Takagi-Sugeno fuzzy inference forms the following approximation function:

$$\tilde{Q}(s,a) = \sum_{i_1,i_2,\cdots,i_N,u}^{I_1,I_2,\ldots,I_N,U} \prod_{n=1}^{N} \mu_{i_n,n}(s_n) \cdot \mu_u(a) \cdot q_{i_1 i_2 \ldots i_N u} \tag{4}$$

where $\tilde{Q}(s,a)$ is the approximated state-action-value function, $\mu_{i_n,n}(s_n)$ is the membership value of the i_n^{th} state antecedent fuzzy set at the nth dimension of the N dimensional state antecedent universe at the state observation s_n, $\mu_u(a)$ is the membership value of the uth action antecedent fuzzy set of the one dimensional action antecedent universe at the action selection a, $q_{i_1 i_2 \ldots i_N u}$ is the value of the singleton conclusion of the $i_1, i_2, \ldots, i_N, u^{-th}$ fuzzy rule. Applying the approximation formula of the Q-learning (2) for adjusting the singleton conclusions in (4), leads to the following function:

$$q_{i_1 i_2 \ldots i_N u}^{k+1} = q_{i_1 i_2 \ldots i_N u}^{k} + \prod_{n=1}^{N} \mu_{i_n,n}(s_n) \cdot \mu_u(a) \cdot \Delta \tilde{Q}_{i,u}^{k+1} =$$

$$= q_{i_1 i_2 \ldots i_N u}^{k} + \prod_{n=1}^{N} \mu_{i_n,n}(s_n) \cdot \mu_u(a) \cdot \alpha_{i,u}^k \cdot \left(g_{i,u,j} + \gamma \cdot \max_{v \in U} \tilde{Q}_{j,v}^{k+1} - \tilde{Q}_{i,u}^k \right) \tag{5}$$

where $q_{i_1 i_2 ... i_N u}^{k+1}$ is the $k+1$ iteration of the singleton conclusion of the $i_1 i_2 ... i_N u^{\text{th}}$ fuzzy rule taking action A_u in state S_i, S_j is the new observed state, $g_{i,u,j}$ is the observed reward completing the $S_i \rightarrow S_j$ state-transition, γ is the discount factor and $\alpha_{i,u}^k \in [0,1]$ is the step size parameter. The $\mu_{i_n}, n(s_n) \cdot \mu_u(a)$ is the partial derivative of the conclusion of the 0-order Takagi-Sugeno fuzzy inference $\tilde{Q}(s,a)$ with respect to the fuzzy rule consequents $q_{u,i}$ according to (4). This partial derivative is required for the applied steepest-descent optimization method. The $\tilde{Q}_{j,v}^{k+1}$ and $\tilde{Q}_{i,u}^k$ action-values can be approximated by equation (4).

3 FRIQ-Learning

The Fuzzy Rule Interpolation based Q-learning (FRIQ-learning) is an extension of the traditional fuzzy Q-learning method with the capability of handling sparse fuzzy rule bases. In the followings the FIVE FRI embedded FRIQ-learning (originally introduced in [23]) will be studied in more details.

2.1 The FRI Method 'FIVE'

Numerous FRI methods can be found in the literature. A comprehensive overview of the recent methods is presented in [2]. FIVE is one of the various existing techniques.

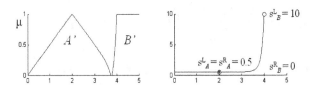

Fig. 1. Approximate scaling function s generated by non-linear interpolation (on the right). On the left hand side the partition is shown as described by the approximate scaling function (A', B').

FIVE is an application oriented FRI method (introduced in [12], [10] and [14]), hence it is fast and serves crisp conclusions directly so there is no need for an additional defuzzification step in the process. Also FIVE has been already proved to be capable of serving the requirements of practical applications [22].

The main idea of the FIVE is based on the fact that most of the control applications serves crisp observations and requires crisp conclusions from the controller. Adopting the idea of the vague environment (VE) [9], FIVE can handle the antecedent and consequent fuzzy partitions of the fuzzy rule base by scaling functions [9] and therefore turn the fuzzy interpolation to crisp interpolation. The idea of a VE is based on the similarity (in other words: indistinguishability) of the considered elements. In VE the fuzzy membership function $\mu_A(x)$ is indicating the level of similarity of x to a specific

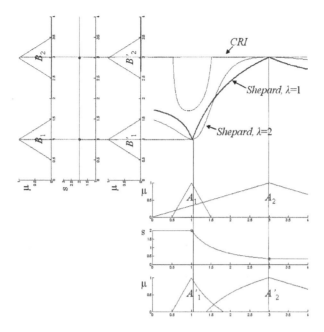

Fig. 2. Interpolation of two fuzzy rules rules (Ri: $A_i{\rightarrow}B_i$), by the Shepard operator based FIVE, and for comparison the min-max CRI with COG defuzzification

element a that is a representative or prototypical element of the fuzzy set $\mu_A(x)$, or, equivalently, as the degree to which x is indistinguishable from a [9]. Therefore the α-cuts of the fuzzy set $\mu_A(x)$ are the sets which contain the elements that are $(1-\alpha)$-indistinguishable from a. Two values in a VE are ε-distinguishable if their distance is greater than ε. The distances in a VE are weighted distances. The weighting factor or function is called scaling function (factor) [9]. If a VE of a fuzzy partition (the scaling function or at least the approximate scaling function [12], [14]) exists, the member sets of the fuzzy partition can be characterized by points in that VE (see e.g. scaling function s in Fig. 1). This way any crisp interpolation, extrapolation, or regression method can be adapted very simply for FRI [12], [14]. FIVE integrates the Shepard operator based interpolation (first introduced in [19]) method (see e.g. Fig. 2) because of its simple multidimensional applicability. Precalculating and caching of the consequent and antecedent sides of the vague environment is straightforward, speeding up the method considerably.

The source code of the FIVE FRI along with other FRI methods is freely available as a MATLAB FRI Toolbox [8]. These can be downloaded from [26] and [27] for free of charge.

3.1 FRIQ-Learning-Based on 'FIVE'

The introduction of FIVE FRI in FQ-learning allows the omission of fuzzy rules (i.e. action-state values in this case) from the rule base and also gains the potentiality of applying the proposed method in higher state dimensions with a reduced rule-base sized describing the action-state space. An example for effective rule base reduction by FRI FIVE is introduced in [16]. Moreover in case the fuzzy model is a function, there are methods which can generate sparse fuzzy rule base for FRI directly from the given input-output data [7].

Substituting the 0-order Takagi-Sugeno fuzzy model of the FQ-learning with the FIVE FRI turns the FQ-learning to FRIQ-learning [23].

The FIVE FRI based fuzzy model in case of singleton rule consequents [11] can be expressed by the following formula:

$$\tilde{Q}(s,a) = \begin{cases} q_{i_1 i_2 \ldots i_N u} & \text{if } \mathbf{x} = \mathbf{a}_k \text{ for some } k, \\ \left(\sum_{k=1}^{r} q_{i_1 i_2 \ldots i_N u} / \delta_{s,k}^{\lambda} \right) \Big/ \left(\sum_{k=1}^{r} 1/\delta_{s,k}^{\lambda} \right) & \text{otherwise.} \end{cases} \tag{6}$$

where the fuzzy rules R_k have the form:

If $x_1 = A_{k,1}$ **And** $x_2 = A_{k,2}$ **And** ... **And** $x_m = A_{k,m}$ **Then** $y = c_k$, (7)

$\delta_{s,k}$ is the scaled distance:

$$\delta_{s,k} = \delta_s(\mathbf{a}_k, \mathbf{x}) = \left[\sum_{i=1}^{m} \left(\int_{a_{k,i}}^{x_i} s_{X_i}(x_i) dx_i \right)^2 \right]^{1/2}, \tag{8}$$

and s_{X_i} is the ith scaling function of the m dimensional antecedent universe, $\mathbf{x}$ is the m dimensional crisp observation and $\mathbf{a}_k$ are the cores of the m dimensional fuzzy rule antecedents A_k.

The application of the FIVE FRI method with singleton rule consequents (6) to be the model of the state-action-value function results in the following:

$$\tilde{Q}(s,a) = \begin{cases} q_{i_1 i_2 \ldots i_N u} & \text{if } \mathbf{x} = \mathbf{a}_k \\ & \text{for some } k, \\ \sum_{i_1, i_2, \ldots, i_N, U}^{I_1, I_2, \ldots, I_N, U} \prod_{n=1}^{N} (1/\delta_{s,k}^{\lambda}) \Big/ \left(\sum_{k=1}^{r} 1/\delta_{s,k}^{\lambda} \right) \cdot q_{i_1 i_2 \ldots i_N u} & \text{otherwise} \end{cases} \tag{9}$$

where $\tilde{Q}(s,a)$ is the approximated state-action-value function.

The partial derivative of the model consequent $\tilde{Q}(s,a)$ with respect to the fuzzy rule consequents $q_{u,i}$, required for the applied fuzzy Q-learning method (5) in case of the FIVE FRI model from (9) can be expressed by the formula above (according to [15]):

$$\frac{\partial \tilde{Q}(s,a)}{\partial q_{i_1 i_2 \ldots i_N u}} = \begin{cases} 1 & \text{if } x = a_k \text{ for some } k, \\ (1/\delta_{s,k}^{\lambda}) \Big/ \left(\sum_{k=1}^{r} 1/\delta_{s,k}^{\lambda} \right) & \text{otherwise.} \end{cases} \tag{10}$$

where $q_{u,i}$ is the constant rule consequent of the kth fuzzy rule, $\delta_{s,k}$ is the scaled distance in the vague environment of the observation, and the kth fuzzy rule antecedent, λ is a parameter of Shepard interpolation (in case of the stable multidimensional extension of the Shepard interpolation it equals to the number of antecedents according to [21]), x is the actual observation, r is the number of the rules.

Replacing the partial derivative of the conclusion of the 0-order Takagi-Sugeno fuzzy inference (5) with the partial derivative of the conclusion of FIVE (10) with respect to the fuzzy rule consequents $q_{u,i}$ leads to the following equation for the Q-Learning action-value-function iteration:

if $\mathbf{x} = \mathbf{a}_k$ for some k:

$$q_{i_1 i_2 \ldots i_N u}^{k+1} = q_{i_1 i_2 \ldots i_N u}^{k} + \Delta \tilde{Q}_{i,u}^{k+1} =$$

$$= q_{i_1 i_2 \ldots i_N u}^{k} + \alpha_{i,u}^{k} \cdot \left(g_{i,u,j} + \gamma \cdot \max_{v \in U} \tilde{Q}_{j,v}^{k+1} - \tilde{Q}_{i,u}^{k} \right)$$

$$\text{(11)}$$

otherwise

$$q_{i_1 i_2 \ldots i_N u}^{k+1} = q_{i_1 i_2 \ldots i_N u}^{k} + \prod_{n=1}^{N} \left(1/\delta_{s,k}^{\lambda}\right) / \left(\sum_{k=1}^{r} 1/\delta_{s,k}^{\lambda} \right) \cdot \Delta \tilde{Q}_{i,u}^{k+1} =$$

$$= q_{i_1 i_2 \ldots i_N u}^{k} + \prod_{n=1}^{N} \left(1/\delta_{s,k}^{\lambda}\right) / \left(\sum_{k=1}^{r} 1/\delta_{s,k}^{\lambda} \right) \cdot \alpha_{i,u}^{k} \cdot \left(g_{i,u,j} + \gamma \cdot \max_{v \in U} \tilde{Q}_{j,v}^{k+1} - \tilde{Q}_{i,u}^{k} \right)$$

where $q_{i_1 i_2 \ldots i_N u}^{k+1}$ is the $k+1$ iteration of the singleton conclusion of the $i_1 i_2 \ldots i_N u^{th}$ fuzzy rule taking action A_u in state S_i, S_j is the new observed state, $g_{i,u,j}$ is the observed reward completing the $S_i \to S_j$ state-transition, γ is the discount factor and $\alpha_{i,u}^{k} \in [0,1]$ is the step size parameter.

As in the previous chapter the $\tilde{Q}_{j,v}^{k+1}$ and $\tilde{Q}_{i,u}^{k}$ action-values can be approximated by equation (11). This way the FIVE FRI model is used for the approximation of the mentioned action-value function.

In multidimensional cases to slightly reduce the computational needs it is a good practice to omit updates on rules which have a distance (d_r) considered far away from the actual observation (for example a predefined limit $\varepsilon_d : d_r > \varepsilon_d$) in the state-action space.

4 Incremental Fuzzy Rule-Base Construction

Towards achieving the fuzzy rule base size-reduction, an incremental rule base creation is suggested. This method (also introduced in [24]) simply increases the number of the fuzzy rules by inserting new rules in the required positions (for an example see Fig. 3). Instead of initially building up a full rule base with the conclusions of the rules (q values) set to a default value, only a minimal sized rule base is created with 2^{N+1} fuzzy rules at the corners of the $N+1$ dimensional antecedent (state-action space) hypercube. Similarly like creating Ruspini partitions with two triangular shaped fuzzy sets in all the antecedent universes (see Fig. 3/1). In cases when the action-value function update (11) is high (e.g. greater than a preset limit ε_Q: $\Delta\tilde{Q} > \varepsilon_Q$), and even the closest existing rule to the actual state is farther than a preset limit ε_s, then a new rule is inserted to the closest possible rule position (see Fig. 3/1). The possible rule positions are gained by inserting a new state among the existing ones ($s_{k+1} = s_k$, $\forall k > i$,

$s_{i+1} = \dfrac{s_i + s_{i+2}}{2}$, see e.g. in Fig. 3/2). In case if the update value is relatively low

($\Delta\tilde{Q} \le \varepsilon_Q$), or the actual state-action point is in the vicinity of an already existing fuzzy rule, then the rule base remains unchanged (only the conclusions of the corresponding rules will be updated). The next step is to update the q value, done regarding to the FRIQ-Learning method according to the equation (11) as it was discussed earlier.

Selecting a new rule position could be difficult in case of having many dimensions. To simplify the new rule position calculation it is practical to handle the dimensions separately and choose a position not in the n-dimension space, but in simple vectors for each of the n dimensions. Then use the selected values in the vectors for the position in the n-dimensional space.

This way the resulting action-value function will be modeled by a sparse rule base which contains only the fuzzy rules which seem to be most relevant in the model. Applying the FIVE FRI method, as stated earlier, allows the usage of sparse rule bases which could result in saving a considerable amount of computational resources and reduced state space.

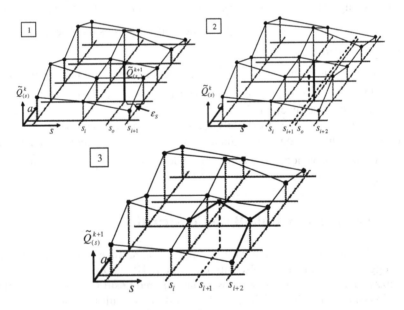

Fig. 3. 1) The next approximation of Q at s_o: $\tilde{Q}_{(s_o)}^{k+1}$.

2) A new fuzzy rule should be inserted at s_{i+1}.

3) Next approximation, with a new rule inserted, and value updated according to (11)

5 Application Example

The notorious Q-learning demonstration application, the cart-pole (reversed pendulum) problem is used for demonstration purposes in this paper as a benchmark for the proposed incremental rule-base building FRIQ-learning method. A cart pole simulation implementation by José Antonio Martin H. which uses SARSA [18] (a Q-learning method) in discrete space is available at [28]. In order to make the benchmarking simple, this latter implementation was modified to make use of the introduced FIVE FRI based rule-base size reduction method. For the easier comparability purposes, in the application example the discrete Q-learning and the proposed reduced rule-base FRIQ-learning had the same resolution of state-action space. Resolution in case of a discrete model means the number of the discrete cases, in the fuzzy model case these are the number of the cores of the fuzzy sets in the antecedent fuzzy partitions.

Table 1. Sample rules from the initial Q value calculation rule base

R#	s_1	s_2	s_3	s_4	a	q
01	N	N	N12	N	AN10	0
16	N	P	P12	P	AP10	0
30	P	P	P12	N	AP10	0
31	P	P	P12	P	AN10	0
32	P	P	P12	P	AP10	0

The example program runs through episodes, where an episode means a cart pole simulation run. The goal of the application is to move the cart to the center position while balancing the pole. Maximum reward is gained when the pole is in vertical position and the cart is on the center position mark. An episode is considered to be successfully finished (gains positive reinforcement in total) if the number of iterations (steps) reaches 1000 while the pole stays up without the cart crashing into the walls. Otherwise the episode is considered to be failed (gains negative reinforcement in total).

Sample rules from the initial rule base used for calculating the q values (set to zero initially) can be seen in Table 1. The rule antecedent variables are the following: s_1 – shift of the pendulum, s_2 – velocity of the pendulum, s_3 – angular offset of the pole, s_4 – angular velocity of the pole, a – compensation action of the cart. Five antecedent variables in total, hence the 32 rules (corners of a 5-dimensional hypercube). The linguistic terms used in the antecedent parts of the rules are: Negative (N), Zero (Z), Positive (P), the multiples of three degrees in [-12,12] degree interval (N12, N9, N6, N3, Z, P3, P6, P9, P12) and for the actions: from negative to positive in one tenth steps (AN10-AP10, Z), 21 in total.

For the sake of simpler comparison purposes the above resolution of the state-action spaces are the same as the resolution of the discrete Q-learning implementation selected as the reference example (see [28]). The consequents (q) are initialized with zero values. These values will be then updated while learning according to (11), when the difference of the current and previous q value is considered small. In the opposite case, when the difference is considered large, then the program calculates a new rule position where the new q value should be stored. If the new rule position does not already exits, then it will be inserted into the rule base, otherwise the consequent of the rule and its surrounding rules will be updated like in the case when the q difference is considered small.

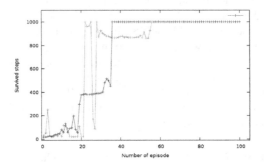

Fig. 5. Survived steps per episode: original discrete method – red line, FRIQ-learning method – green line

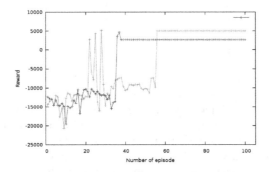

Fig. 6. The cumulative rewards per episode: original discrete method – red line, FRIQ-learning method – green line

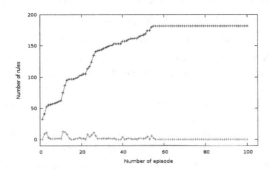

Fig. 7. The number of rules per episode – blue line, the difference in rule numbers per episode – magenta line

Calculation of the new position of a rule is done in the following manner: possible positions are stored separately for each state space and the action state. These vectors are initialized with the possible state/action values. Minimally the start and the end positions of the interval are required, but additional hints can be given to get faster results. When an observation falls close to an existing rule then that existing rule will be updated. Else when the observation falls into the vicinity of the midst of the segment connecting two neighbouring values, a new possible state is registered in the midst of the segment, and that will be the place of the new rule in the corresponding dimension.

The fuzzy rules are defined in the following form:

$$R_i: \tag{12}$$

If $s_1 = A_{1,i}$ **and** $s_2 = A_{2,i}$ **and** $s_3 = A_{3,i}$ **and** $s_4 = A_{4,i}$
and $a = A_{5,i}$ **Then** $q = B_i$

The built-in randomized action selection (see [28]) (state-action space exploration) was disabled temporarily for testing and validation reasons.

The comparison of the results of the original discrete method simulation and the FRIQ-learning method using incremental rule-base creation is shown in Figs. 5 and 6. Fig. 5 shows the number of iteration steps the cart could survive both with the original discrete method (red) and the FRIQ-learning method (green) while Fig. 6 shows the cumulated rewards in each iteration. These figures show that both versions learn a suitable state-action-value function, and the FRIQ-learning version gives better results at first, but converges slower. The blue curve in Fig. 7 shows the number of fuzzy rules in the rule base. Starting with the initial 32 rules, the size of the rule base quickly rises through the episodes, stabilizing at 182 rules. A higher rate of rule insertion can be observed (magenta coloured curve in Fig. 7) till the first successful episode, then the insertion rate drops approximately to the third of the previous rate, then stays constant while all the episodes will be successful. For describing the original discrete states 2268 rules would be needed, but the same result can be achieved with only 182 rules.

The MATLAB source code of the extended cart pole simulation program can be accessed at [27] free of charge.

5 Conclusions

With the introduction of the FRIQ-learning, continuous spaces can be applied instead of the originally discrete state-action spaces. FRIQ-learning also allows the application of sparse fuzzy rule bases, which means that rules which are considered unimportant can be left out from the rule base. A possible method is introduced for automatically creating a reduced size rule base. The targeted reduced rule base size is achieved by incremental creation of an intentionally sparse fuzzy rule base. The fuzzy rule base is incrementally built up from scratch and will contain only the rules which seem to be most relevant in the model, meanwhile the existing rules are also updated when required. This way the real advantages of the FIVE based FRIQ-learning method could be exploited: reducing the size of the fuzzy rule base has the benefits not only in decreasing the computing resource requirements, but having less rules (optimizable parameters), it also speeds up the convergence of the FRIQ-learning. The application example of the paper shows the results of the method, instead of 2268 rules, 182 is sufficient for achieving the same task, which is a significant difference. A smaller rule base considerably speeds up the FRIQ-learning process itself and also the whole application.

Acknowledgments. This research was partly supported by the Hungarian National Scientific Research Fund grant no: OTKA K77809 and by the ETOCOM project (TÁMOP-4.2.2-08/1/KMR-2008-0007) through the Hungarian National Development Agency in the framework of Social Renewal Operative Programme supported by EU and co-financed by the European Social Fund.

References

[1] Appl, M.: Model-based Reinforcement Learning in Continuous Environments. Ph.D. thesis, Technical University of München, München, Germany, dissertation.de, Verlag im Internet (2000)

[2] Baranyi, P., Kóczy, L.T., Gedeon, T.D.: A Generalized Concept for Fuzzy Rule Interpolation. IEEE Trans. on Fuzzy Systems 12(6), 820–837 (2004)

[3] Bellman, R.E.: Dynamic Programming. Princeton University Press, Princeton (1957)

[4] Berenji, H.R.: Fuzzy Q-Learning for Generalization of Reinforcement Learning. In: Proc. of the 5th IEEE International Conference on Fuzzy Systems, pp. 2208–2214 (1996)

[5] Bonarini, A.: Delayed Reinforcement, Fuzzy Q-Learning and Fuzzy Logic Controllers. In: Herrera, F., Verdegay, J.L. (eds.) Genetic Algorithms and Soft Computing, Studies in Fuzziness, 8, pp. 447–466. Physica-Verlag, Berlin (1996)

[6] Horiuchi, T., Fujino, A., Katai, O., Sawaragi, T.: Fuzzy Interpolation-based Q-learning with Continuous States and Actions. In: Proc. of the 5th IEEE International Conference on Fuzzy Systems, vol. 1, pp. 594–600 (1996)

[7] Johanyák, Z.C.: Sparse Fuzzy Model Identification Matlab Toolbox - RuleMaker Toolbox. In: IEEE 6th International Conference on Computational Cybernetics, November 27-29, pp. 69–74. Stara Lesná, Slovakia (2008)

[8] Johanyák, Z.C., Tikk, D., Kovács, S., Wong, K.W.: Fuzzy Rule Interpolation Matlab Toolbox – FRI Toolbox. In: Proc. of the IEEE World Congress on Computational Intelligence (WCCI 2006), 15th Int. Conf. on Fuzzy Systems (FUZZ-IEEE 2006), Vancouver, BC, Canada, July 16-21, pp. 1427–1433. Omnipress (2006)

[9] Klawonn, F.: Fuzzy Sets and Vague Environments. Fuzzy Sets and Systems 66, 207–221 (1994)

[10] Kovács, S., Kóczy, L.T.: Approximate Fuzzy Reasoning Based on Interpolation in the Vague Environment of the Fuzzy Rule base as a Practical Alternative of the Classical CRI. In: Proceedings of the 7th International Fuzzy Systems Association World Congress, Prague, Czech Republic, pp. 144–149 (1997)

[11] Kovács, S.: Extending the Fuzzy Rule Interpolation FIVE by Fuzzy Observation. In: Reusch, B. (ed.) Advances in Soft Computing, Computational Intelligence, Theory and Applications, pp. 485–497. Springer, Germany (2006)

[12] Kovács, S.: New Aspects of Interpolative Reasoning. In: Proceedings of the 6th International Conference on Information Processing and Management of Uncertainty in Knowledge-based Systems, Granada, Spain, pp. 477–482 (1996)

[13] Kovács, S.: SVD Reduction in Continuos Environment Reinforcement Learning. In: Reusch, B. (ed.) Fuzzy Days 2001. LNCS, vol. 2206, pp. 719–738. Springer, Heidelberg (2001)

[14] Kovács, S., Kóczy, L.T.: The Use of the Concept of Vague Environment in Approximate Fuzzy Reasoning. In: Fuzzy Set Theory and Applications, Tatra Mountains Mathematical Publications, Mathematical Institute Slovak Academy of Sciences, vol. 12, pp. 169–181. Slovak Republic, ratislava (1997)

[15] Krizsán, Z., Kovács, S.: Gradient-based Parameter Optimisation of FRI FIVE. In: Proceedings of the 9th International Symposium of Hungarian Researchers on Computational Intelligence and Informatics, Budapest, Hungary, November 6-8, pp. 531–538 (2008) ISBN 978-963-7154-82-9

[16] Kovács, S.: Interpolative Fuzzy Reasoning in Behaviour-based Control, Advances in Soft Computing. In: Reusch, B. (ed.) Computational Intelligence, Theory and Applications, vol. 2, pp. 159–170. Springer, Germany (2005)

[17] José Antonio Martin, H., De Lope, J.: A Distributed Reinforcement Learning Architecture for Multi-Link Robots. In: 4th International Conference on Informatics in Control, Automation and Robotics, ICINCO 2007 (2007)

[18] Rummery, G.A., Niranjan, M.: On-line Q-learning Using Connectionist Systems. In: CUED/F-INFENG/TR, vol. 166, Cambridge University, UK (1994)

[19] Shepard, D.: A Two Dimensional Interpolation Function for Irregularly Spaced Data. In: Proc. 23rd ACM Internat. Conf., pp. 517–524 (1968)

[20] Sutton, R.S., Barto, A.G.: Reinforcement Learning: An Introduction. MIT Press, Cambridge (1998)

[21] Tikk, D., Joó, I., Kóczy, L.T., Várlaki, P., Moser, B., Gedeon, T.D.: Stability of Interpolative Fuzzy KH-Controllers. Fuzzy Sets and Systems 125(1), 105–119 (2002)

[22] Vincze, D., Kovács, S.: Using Fuzzy Rule Interpolation-based Automata for Controlling Navigation and Collision Avoidance Behaviour of a Robot. In: IEEE 6th International Conference on Computational Cybernetics, Stara Lesná, Slovakia, November 27-29, pp. 79–84 (2008) ISBN: 978-1-4244-2875-5

[23] Vincze, D., Kovács, S.: Fuzzy Rule Interpolation-based Q-learning. In: SACI 2009, 5th International Symposium on Applied Computational Intelligence and Informatics, Timisoara, Romania, May 28-29, pp. 55–59 (2009) ISBN: 978-1-4244-4478-6

[24] Vincze, D., Kovács, S.: Reduced Rule Base in Fuzzy Rule Interpolation-based Q-learning. In: Proceedings of 10th International Symposium of Hungarian Researchers on Computational Intelligence and Informatics, November 12-14, pp. 533–544. Budapest Tech, Hungary (2009)

[25] Watkins, C.J.C.H.: Learning from Delayed Rewards. Ph.D. thesis, Cambridge University, Cambridge, England (1989)

[26] The FRI Toolbox is available at, http://fri.gamf.hu/

[27] Some FRI applications are available at,
     http://www.iit.uni-miskolc.hu/~szkovacs/,
     http://www.iit.uni-miskolc.hu/~szkovacs/

[28] The cart-pole example for discrete space can be found at:
     http://www.dia.fi.upm.es/~jamartin/download.htm

Protective Fuzzy Control of Hexapod Walking Robot Driver in Case of Walking and Dropping

István Kecskés[1] and Péter Odry[2]

[1] Singidunum, Beograd, Serbia
   kecskes.istvan@gmail.com
[2] Subotica Tech, Subotica, Serbia
   odry@vts.su.ac.rs

Abstract. The new hexapod walking robot is assembly phase. It was design to overcome rough terrain using the latest numerical tests on the model. One of the likelihoods of walking on rough terrain is falling over. This posed the requirement that the robot had to be able to continue walking even after multiple falls. One of the goals was to create a control mechanism in the engine layer that will ensure optimal walk as well as protection from breaking down. In order to achieve the best results, it would be necessary to ensure the feedback of the torque, but there was no possibility for that. Thus the only solution was to feedback the engine power; however, such feedback will input a delay into the feedback branch anyway. The suitable Fuzzy rules are being sought for, which will help minimize the effects of the delay in the control branch. The fitness function is defined for determining the optimal hexapod walk algorithm. During the test the worst cases were tested with the lower arm of the robot closing in a 90 degree angle with the upper arm due to the fall, this is when the robot structure is endures maximum load.

Keywords: hexapod walking robot modeling, fuzzy control, drop test, fitness function for determination of optimal walking function.

1 Introduction

The detailed analysis of the previously created robot and the robot to be built, focusing on the optimal walk algorithm is given in [1, 2]. In [1] the reference lists all publications related to the "Szabadka-robot". [3] presents the previously built hexapod walking robots and gives a detailed description of the walking robot called "Hamlet". The control in the engine layer has been the subject of a whole range of publications. The use of PID control circuit algorithm for hexapod walking devices has been detailed in the publication [4, 5]. Moreover, [5] presents some solutions in connection with determining the optimal parameters for the stable operation of the manipulator. A wide range of literature deals with the issue of Fuzzy controls, this works specially mentions some of them: [6] in the field of servo-engine drive, [7, 8] focusing on fuzzy optimization issues in detail. Several papers have been written in the field of

I.J. Rudas et al. (Eds.): Computational Intelligence in Engineering, SCI 313, pp. 205–217.
springerlink.com   © Springer-Verlag Berlin Heidelberg 2010

determination of the walking robot's structural elements, including the sizes of arms, most represented in [9]. This publication came to the following conclusions: distribution of legs around the body to decrease maximum foot forces against the ground, which greatly affects the determination of the robot shape and actuator size. A detailed analysis of the construction of the "Szabad(ka)" robot can be found in [10]. Measuring the robot drive quality has resulted in several functions, published in [12].

The built robot was created for the purpose of developing the navigational algorithm, for accessing labyrinths. The labyrinths were constructed without an even ground but with obstacles to make progress in the labyrinth more difficult. The plan is to equip the robot with a reliable walking algorithm which will 'know' what to do on uneven ground and around obstacles. In order for the robot to be able to solve the given tasks 'on its own', the hexapod manipulator has to feed back the body's acceleration from a 3D accelerometer, as well as the force values from the end of the robot's arms. Fig. 1 shows the entire model structure of the walking robot, starting from the entire structure of the driver to the completion of the given course. The relation of the 2nd trajectory and 3rd walking layer presents the feedback mechanism using data from the previously mentioned 3D accelerometer and the force values at the end of the arms.

In realization the new robot is important problem is building and testing the full kinematic and dynamic model. The first step when in constructing a real robot is the insight from the simulation results. Therefore we should build a perfect model. Our research is a part of the process of building the new walking hexapod robot. Our robot architecture has 18 degree of freedom, and about 5 kg weight. The issue of the robot's drive soft-computing using small power intake microcontrollers is discussed in [6].

The low-power DC micro motors are widely used, especially in the field of robotics and mechatronics. We choose this micro motor from FaulHaber [11].

The marginal parameters affecting the quality of robot drive is discussed in detail in this article.

In order for this problem to be solved we first must make sure that the damaged caused by the fall will be corrected. Thus on the lowest level, on the engine control layer such an engine control must be found that will ensure maximum protection for the system compared to the well-chosen PID controller, meaning that it will yield protection against sudden torque and power surges. A sudden great change in torque will lead to gear damage, while the great value of current will lead to damage in the engine. In order to avoid such damages fuzzy control must be found that will function reliably within the operating domain, just as a PID optimally chosen for the given structure, but in extreme cases it is much softer, the system controls in a finer way. Several drop tests have been performed on the model to determine the effect on the robot structure and the driver components.

In order to be able to model the above-described, first an optimal PID controller has to be chosen. This near-perfect PID controller is used as a reference system when creating the fuzzy structure. In the course of problem solution another aspect has to be taken into consideration, namely that the lower three layers of the robot are to be solved using low-power intake embedded tools. This requirement will limit the choice of model.

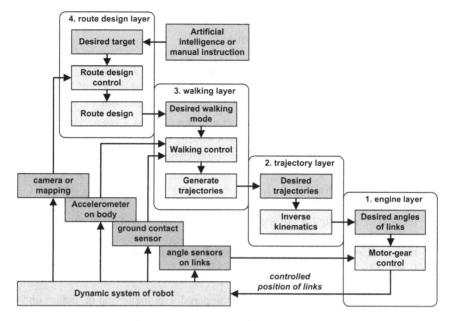

Fig. 1. The hexapod walking robot controlling hierarchy

2 Simulation Model of Hexapod Walker

The current simulation model is acceptable if the results carry information even given a certain error. Often it is enough if the range of signals or the spatial and temporal range can be seen, then it is possible to draw conclusions, which direction to move towards.

This system has many parameters, but basically these can be divided into hardware and software parameters.

The HW parameters for a hexapod robot can further be classified into:

- structural parameters
  - the shape and size of the robot's body and leg joints
  - the suspension points of the legs on the body
  - rotation end values of the legs
  - choice of spring and damper part

- engine parameters
  - choice of DC engines
  - choice of gears
  - choice of encoders

- electronics parameters
  - choice of microcontrollers
  - capacity of the batteries

Parameters of the control electronic software

- 1st engine layer – engine control parameters
  - sampling control
  - types of engine control (Fuzzy, Fuzzy-Neural, PID, PI, hybrid)
  - control parameters: in case of PID (P, I, D values) and in case of Fuzzy (limits and shapes of membership functions, rules, calculation methods, etc...)

Fig. 2 illustrates the engine layer of the robot in the form of a block diagram. Fig. 2 shows the mechanical structure of the robot and its lowest level of control divided into the robot's legs and body. Dependencies are illustrated globally and broken down in detail for one leg.

- 2nd trajectory layer - walking
  - speed of walking (in step/second)
  - method of walking (out of the 6 legs how many legs and lifted simultaneously, and in what order)
  - the arc shape of the lifted part (spline, ellipse, polyline)
  - length of a step
  - height of a step
  - temporal ration of the lifted period compared with the period on the ground
- 3rd walking layer – control of walking. The parameters of the system can only be chosen correctly, in fact optimized with the help of simulation, if one or more goals are set. Generally these goals are:
  - achieving maximum speed of walking taking up as minimal electric energy as possible
  - make the torques on the joints and gears, and the currents of the engines discursive and spiky as little as possible, and let their variation be as small as possible
  - while walking, let the robot's body acceleration be minimal in all 3 dimension directions
  - make the robot function the same even with a greater load (a load of at least its own body weight)

Since the change of these parameters will influence the optimal values of the other parameters – that is, these are not independent parameters – that's why the optimal parameter set is to be sought in the multi-parameter space. Naturally, these can be grouped, meaning that not all parameters have to be monitored at the same time, only those that have a significant influence on each other.

Trying all the combination would take a very long time, and since our current Simulink model is about 1500 times slower than running it on an ordinary computer, this means that simulating 1 second of robot walk takes about 25 minutes. If, thus, in one group there are 10 independent parameters, and for each we take 5 values, then it would take 12000 years to run all the combinations. Later on in the research the aim is to find suitable starting points and destinations, in order to run the optimizing algorithm we need greater performance equipment than we now have available.

Currently we are forced to test each parameter based on intelligent thinking. For example, it is not difficult to figure out that by rounding the leg-end walking path the sudden jolts in the joint can be decreased. Or a further example, if the fuzzy controller

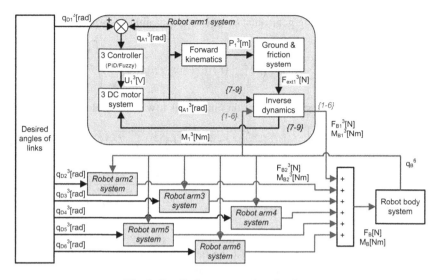

Fig. 2. Detailed structure of engine layer

tries to suppress the change in current, the result will probably be an even softer controller, thus the change in current will also be smaller. Naturally, we must pay attention that this does not hurt our other goals, such as that the speed of the robot will decrease significantly, or it can't bear the weight, etc.

The walking algorithms determine the robot's movement, the load of the robot engine, the dynamics of power consumption, torque of the separate reductors, etc. In the case of hexapod items there are several basic types of walking. One is when during the walk 3 legs are always on the ground, and 3 are in the air, another type is when 4 legs are constantly on the ground and 2 legs are in the air, and the third type is when 5 legs are continuously on the ground, and only 1 leg is in the air. The following section will focus on three-legged walk.

2.1 The Aims of Choosing a Controller Algorithm

In the process of tuning fuzzy it must be clear what kind of controller we aim for, i.e. what kind of walk we aim for. The following issues have to be considered when determining the quality of walk:

- the walk algorithm is to ensure the longest possible distance for the robot using the least possible power

$$E_{walk}\left[\frac{J}{m}\right] = \frac{\int \sum_{i=1}^{18} P_{motor}^i(t)dt}{D_X(t_{end})} \tag{1}$$

- when walking, its average speed must be the highest possible

$$\overline{V}_X\left[\frac{m}{s}\right] = \frac{D_X(t_{end})}{t_{end}} \tag{2}$$

- torque on the load bearing gear must be as low as possible (its squared average)

$$E_{Mgear}[Nm] = rms\left(\sum_{i=1}^{18} abs\left(M_{gear}^{i}(t)\right)\right), \text{ when } rms(x) = \sqrt{\frac{\sum_{i=1}^{n} x_i^2}{n}} \qquad (3)$$

- the robot's body and angle acceleration must be as low as possible

$$E_{abody}\left[\frac{m}{s^2}\right] = rms\left(\sqrt{a_X(2)^2 + a_Y(2)^2 + a_Z(2)^2}\right) \qquad (4)$$

$$E_{\alpha body}\left[\frac{rad}{s^2}\right] = rms\left(\sqrt{\alpha_X(2)^2 + \alpha_Y(2)^2 + \alpha_Z(2)^2}\right) \qquad (5)$$

- the robot's body must maintain its initial height (firstly, it must not collapse or lower, but it must not rise, either). At the beginning an offset of 1cm is chosen to increase the value so as to make the fitness function less sensitive

$$D_{bodyZ}[m] = abs(D_Z) + 0.01 \qquad (6)$$

- As a result, the following fitness function must be created:

$$Fitness = \frac{E_{walk} \cdot E_{Mgear} \cdot E_{abody} \cdot E_{\alpha body} \cdot D_{bodyZ}}{\overline{V}_X} \qquad (7)$$

The minimum of the *fitness function* provides the parameters of the optimal walk. Fig. 3 shows the results of the calculation of the fitness function and auxiliary function used for calculating the fitness function for various values of the Fuzzy parameters. On the fitness function's result curve the minimum of the fitness function can be clearly seen which provides optimal value of the Fuzzy parameters for robot construction and assumptions regarding the robot's forward movement.

2.2 Choosing the Fuzzy Rules

Three rules groups were built into the Fuzzy model structure. This structure contains the following three inputs:

- **Angle error input:** negative feedback has to ensure and change the output voltage into the direction so that the link's angle error will decrease.
- **Driver current input:** the rule group which takes the current value into consideration. The task of this rule group is to decrease the fuzzy in case of great power, move the values of the first group into the reverse direction. This rule makes sure there is no excessive current value.
- **Driver current change input:** this group protects against sudden torque and current surges, namely for a great change in current it gives a reverse direction voltage so that the great torque caused by outside effects can be decreased immediately.

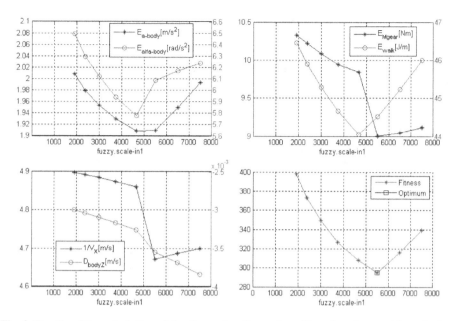

Fig. 3. Results of the calculation of the fitness function and auxiliary function used for calculating the fitness function for various values of the Fuzzy parameters

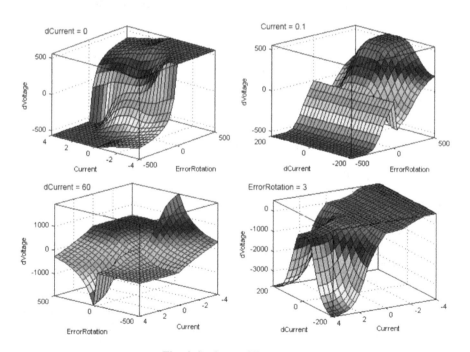

Fig. 4. Surfaces of Fuzzy rules

The third rule group is the most critical, the least linear, thus its integration into the system is the most critical. Sudden great torque value will cause the change in current to be very positive, and if the system reacted to this with a strong negative voltage, then current would change direction and decreased significantly. This will turn off the reaction, and again lead to torque increase. Sudden reactions in the controller system lead to oscillations, resulting in strong current and torque oscillation. This oscillation is critical because the frequency is very high, close to the Fuzzy controller's frequency.

In the test case the robot was dropped from a height of 6 cm's. Due to the fall a 10 ms-long strong torque appears on the gears. Compared with the torque, the current change lags behind by 1ms. This time equals the drivers' time constant. If the direct torque were fed back, it would be easier to find a controller solution, but this is not possible in the current robot building set up. With the current created robot only the current values can be fed back.

The speed of the fuzzy cycle is important when choosing the microprocessor controlling the driver. This system uses the MSP430F2xxx processor functioning with a 16 MHz clock signal. The clock signal of this processor family is 16 MHz; this speed equals the speed of command. The arms of the future robot will be driven by six such processors. Based on the above described aspects a 1600 per second instruction cycle will be needed for the calculation of a fuzzy drive.

3 Presentation of Tuning Process Results of General Robot Parameters

The principles of choosing parameters for a robot model are described in detail in [1]. The robot motion and all dynamic results are observed within a given value of the robot's body mass. As it has been stated earlier in this work, the robot has to be able to carry twice its own body weight. The weight of the robot is 1.5 kg; the 6 arms weigh all together 3 kg, so the robot weighs a total of 4.5 kg at least. If this weight is loaded on the body, the maximum body weight to be calculated with is 1,5+4,5=6 kg. With maximum load the robot's entire weight is 6+3=9 kg.

Following the optimization of the previous parameter using the best values, and maximum weight the following results were reached:

- Robot moving in direction X along with start under 1 second: X = 0.214 [m];
- The taken total current $I_{average}$ = 3.11 [A];
- When the robot is walking horizontally, it takes P_e = 9 [W] average performance, and moves forward 1 meter using energy: E_{walk} = 42 [J/m]. So the 24 V, 3300 [mAh] battery with 80% utilization can sustain about 7 hour of motion.
  These results indicate the following things:
- if the taken electrical energy is transformed into equivalent potential energy, then it would reach a height of 0,114 [m] - Equivalent Z rise ($E_{pot} = m \cdot g \cdot h$):   Z = 0.147 [m]
- horizontally it would reach 88% farther than if the equivalent electrical energy had been transformed into height - Rise-moving ratio: Z/X = 188%. This value is also significant because this is how we express numerically the initial goal, namely to achieve the fastest possible walk using the least possible energy.

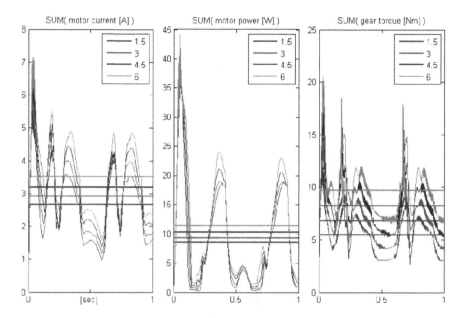

Fig. 5. The effect of body mass on the total current and total power of robot

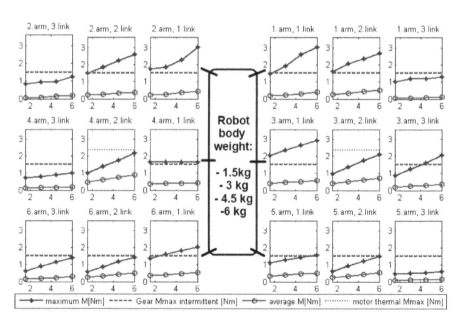

Fig. 6. The maximum and average value for the torque in connection with the mass of robot's body

Simulations have been performed for four types of robot body weight: 1.5, 3, 4.5, 6 [kg]. The robot's forward motion a given direction and its speed are shown in the Figs. 6 and 7.

In Fig. 6 we can see the maximum and average value for the torque in connection with the mass of robot's body. The critical load of gears is drawn using horizontal lines for lifting, and the broken lines show the engine load in torque.

It is seen how the load of the second link is almost linear, but mostly on the two middle arms (3rd and 4th arms). The torque's temporal form shows that when walking on 3 arms, the two mentioned links are constantly under a 1.5 [Nm]. Here the robot's body weight is 3 kg, which is the value for no extra load Fig. 7.

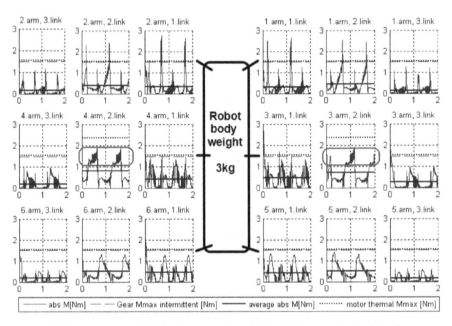

Fig. 7. The torque of links in connection with the 3 kg mass of robot's body

As it can be seen from the previous sections, if there is a working model, then by running the simulation with different parameters, both the hardware as well as the software parameters can be optimized.

The model was mainly helpful in building the robot itself, as the choice of engine parts (motor, gear, encoder, and battery) was vital. They had to endure the load, but it ought to be built using the smallest and lightest parts. The model showed that the carrier links of the central 2 arms have to have the stronger motor-gear pair, as they have to be able to lift the weight of half the robot when walking on three arms. Further, the simulation results were compared for calculating the walk's pathway, and so the most suitable parameters were chosen.

Naturally, not nearly all possibilities have been testes, e.g. if the robot does not use 3 but only 2 arms for walking. Numerous parameter combinations can be analyzed, many new ideas tested. These options are all open to us; only a small segment was realized and described in this article.

4 Tuning the Fuzzy Parameters for Drop Test

The chosen PID controller turned out to be linear; its three parameters have an optimal domain for which there is no need for further tuning. If it is increased, then the advantages will equal the disadvantages, different changes in walk and load will always call for different values depending on the situation.

The fuzzy controller is created using whichever membership functions and controllers are deemed fit. However, this type of controller then has to be tuned, as it is impossible to determine in advance which inputs and outputs will have which optimal domain, when shape the optimal membership functions will be, and what effect the controls and their weights will have on the model.

Unlike with the PID controller, using the fuzzy controller it is possible to define such a rule which will cause a sudden great torque to give a sudden inverse output, which will move the link into the direction of the outside effect despite retaining the required angle. On the other hand, it is also possible to realize a softer system in the higher power domain, and a stronger system in the lower power domain. The effect is

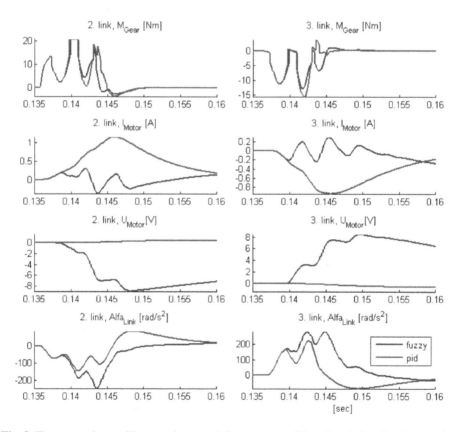

Fig. 8. The upper pictures illustrate the essential parameters of the robot during the drop test for the case of PID and fuzzy control

as if the scale of the PID was tuned depending on the current. When building the model we opted for a walk with having three arms on the ground at all times. In this case the middle load bearing links of the mid-section arms (arms 3 and 4) endure more load than the rest of the links. After tuning the fuzzy controller the current on these links was lower compared with the PID controller, and in all other cases the current was higher, as seen in Fig. 8.

5 Conclusions

In this material we have been looking for a solution in the engine layer in case of the fall of the robot simultaneously enabling quality control without control signals (acceleration or gyro signals from the robot's body) from the upper layers (from the trajectory or walking layers 1; Figure 1).

In the third chapter some simulation results of the robot's quality motion are given which are discussed in more detail in [1]. Chapter 4 contains the simulation results of a theory related to the quality estimation of protection from damage in case of the robot's fall (drop test).

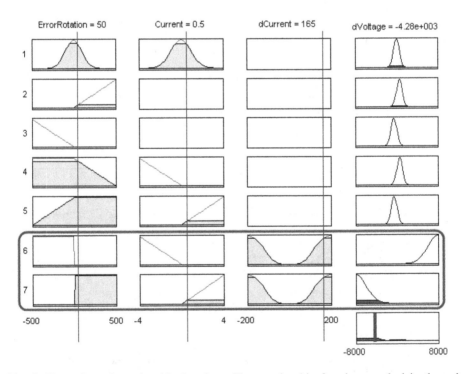

Fig. 9. Illustration of membership functions; The membership functions marked in the red frame are those responsible for the drop test

The suggested solution of mechanism control (in one moment the quality movement, in the other the anti-drop protection of the robot or its careful movement over an unknown terrain) lies in the turning on or turning off of some membership functions in the fuzzy control.

The membership functions responsible for action during the drop test are marked with red color. Changing the weight of the rules in the control algorithm marked with red we can modify the characteristics of the controller so as to be optimal in the case of drop test and walking as well.

Acknowledgments. The realization of the project, which includes building the new Hexapod robot, is aided by the *FAULHABER Motors Hungaria Ltd.* for which we are very grateful. Besides this company we are thankful to the *College of Dunaújváros*, as well as the *Appl DSP Ltd.* for their kind support.

References

[1] Kecskés, I., Odry, P.: Walk Optimization for Hexapod Walking Robot. In: Proceedings of 10th International Symposium of Hungarian Researchers on Computational Intelligence and Informatics, Budapest, Hungary, November 12-14, pp. 265–277 (2009) ISBN: 978-963-7154-96-6

[2] http://www.szabadka-robot.com

[3] Fielding, M.R., Dunlop, G.R.: Omnidirectional Hexapod Walking and Efficient Gaits Using Restrictedness. The International Journal of Robotics Research 23, 1105 (2004), The online version of this article can be found at:
http://ijr.sagepub.com/cgi/content/abstract/23/10-11/1105

[4] Meza, J.L., Santibanez, V., Campa, R.: An Estimate of the Domain of Attraction for the PID Regulator of Manipulators. International Journal of Robotics and Automation 22(3) (2007)

[5] Alvarez, J., Cervantes, I., Kelly, R.: PID Regulation of Robot Manipulators: Stability and Performance. Systems and Control Letters 41, 73–83 (2000)

[6] Odry, P.: Sz. Diveki, A. Csasznyi, N. Burany: Fuzzy Logic Motor Control with MSP430x14x, Texas Instruments, slaa 235 (2005),
http://focus.ti.com/lit/an/slaa235/slaa235.pdf (included program package)

[7] Odry, P., et al.: Fuzzy Control of Brush Motor - Problems of Computing. In: Proceedings of 2nd Serbian-Hungarian Joint Symposium on Intelligent Systems (SISY 2004), Subotica, Serbia and Montenegro, October 1-2, pp. 37–46 (2004) (invited paper)

[8] Victor, P., Vroman, A., Deschrijver, G., Kerre, E.E.: Real-Time Constrained Fuzzy Arithmetic. IEEE Transactions on Fuzzy Systems 17, 630–640 (2009)

[9] Gonzalez de Santos, P., Garcia, E., Estremera, J.: Improving Walking-Robot Performances by Optimizing Leg Distribution. Auton Robot 23, 247–258 (2007)

[10] Burkus, E., Odry, P.: Autonomous Hexapod Walker Robot Szabad(ka). Acta Polytechnica Hungarica 5, 69–85 (2008)

[11] http://www.faulhaber-group.com

[12] Xu, C., Ming, A., Mak, K., Shimojo, M.: Design for High Dynamic Performance Robot based on Dynamically Coupled Driving and Joint Stops. International Journal of Robotics and Automation 22, 281–293 (2007)

Survey on Five Fuzzy Inference-Based Student Evaluation Methods

Zsolt Csaba Johanyák

Institute of Information Technologies, Kecskemét College
Izsáki út 10, H-6000 Kecskemét, Hungary
johanyak.csaba@gamf.kefo.hu

Abstract. In case of non-automated examinations the evaluation of students' academic achievements involves in several cases the consideration of impressions and other subjective elements that can lead to differences between the scores given by different evaluators. The inherent vagueness makes this area a natural application field for fuzzy set theory-based methods aiming the reduction of the mentioned differences. After introducing a criterion set for the comparison the paper surveys five relevant fuzzy student evaluation methods that apply fuzzy inference for the determination of the students' final score.

Keywords: fuzzy student evaluation, rules based system, fuzzy inference, fuzzy rule interpolation.

1 Introduction

The evaluation of students' answerscripts containing narrative responses or assignments that cannot be rated fully automatically is from nature vague, which can lead to quite different scores given by different evaluators. This problem usually is solved by defining scoring guides that become more and more complex after the developers face new and new cases that seemed to be unimaginable previously. The more specific the guides are the more tedious they become, which leads to inconsistency in their application and increases the time need of the scoring. Owing to the increased complexity and hard-to-learn character of the comprehensive scoring guides evaluators often use ad hoc inference methods that lack a formal mechanism. Beside the demand on consistency of the evaluation the easy-to-explain/confirm character is also important not only for the teachers but also for other interested parties, like students, parents, etc.

A completely new approach appeared in the late 90s in field of evaluation methods by emerging the fuzzy set theory based evaluation techniques, which make possible a good trade-off between the demand on quick evaluation and high consistence of the results. Biswas [2] proposed a particular (FEM) and a generalized (GFEM) method that were based on the vector representation of fuzzy membership functions and a special aggregation of the grades assigned to each question of the student's answerscripts. Chen and Lee [4] suggested a simple (CL) and a generalized (CLG) method

I.J. Rudas et al. (Eds.): Computational Intelligence in Engineering, SCI 313, pp. 219–228.
springerlink.com

that produced improvements by applying a finer resolution of the scoring interval and by including the possibility of weighting the four evaluation criteria. Wang and Chen [18] extended the CL/CLG method pair by introducing the evaluator's optimism as a new aspect, and by using type-2 fuzzy numbers for the definition of the satisfaction. Johanyák suggested a fuzzy arithmetic based simple solution (FUSBE) in [7] for the aggregation of the fuzzy scores. Nolan [13] introduced a fuzzy classification model for supporting the grading of student writing samples in order to speed up and made more consistent the evaluation. Bai and Chen [1] developed a method for the ranking of students that obtained the same total score during the traditional evaluation. They used a three-level fuzzy reasoning process. Saleh and Kim [16] enhanced the BC method by excluding some subjective elements and applying Mamdani [12] type inference. Rasmani and Shen [15] introduced a data driven fuzzy rule identification method. Johanyák suggested a low complexity fuzzy rule interpolation based method (SEFRI) in [9].

The fuzzy student evaluation techniques can be classified in two main groups depending on their algorithm: (1) methods applying fuzzy inference (e.g. [1] [9] [13] [15] [16]), and (2) methods applying "only" fuzzy arithmetic (e.g. [2] [4] [7] [18]). The advantage of the first approach is that the rules are close to the traditional human thinking, they are easily readable and understandable. Their drawback is however that they usually require a tedious preparation work done by human expert graders. Besides, such a system is usually task/subject specific, i.e. minor modifications in the aspects can lead to a demand on a completely redefinition of the rule base. This feature makes the system rigid. Another problem arises from the fact that in general the rule based systems can only operate with a low number of fuzzy sets owing to the exponentially growing number of necessary rules in multidimensional cases if a full coverage of the input space should be ensured.

The advantage of the second approach is its simplicity and easy adaptability. Furthermore the methods based on it can operate with a higher resolution of the input space. However, as its disadvantage one should mention the lack of the humanly easy-to-interpret rules. The rest of this paper is organized as follows. Section 2 introduces the criterion set considered as relevant for fuzzy student evaluation methods. Section 3 gives a short survey on fuzzy inference based evaluation methods. The conclusions are drawn in Section 4.

2   Criteria for Comparison of Fuzzy Evaluation Methods

In this section, we introduce a criterion set [10] for fuzzy methods aiming the evaluation of the students' academic performance. We consider these requirements as properties that help the reader to compare the overviewed methods. The criteria are the followings.

1   The method should not increase the time needed for the assessment compared to the traditional evaluation techniques.
2   The method should help the graders to express the vagueness in their opinion.
3   The method should be transparent and easy to understand for both parties involved in the assessment process, i.e. the students and the graders.

4 The method should ensure a fair grading, i.e. it should be beneficial for all students.
5 The method should allow the teacher to express the final result in form of a total score or percentage as well as in form of grades using a mapping between them.
6 The method should be easy implementable in software development terms.
7 The method should be compatible with the traditional scoring system, i.e. when the grader provides crisp scores for each response the total score and the final grade should be identical with the one calculated by the traditional way.

3 Fuzzy Inference-Based Student Evaluation Methods

3.1 Evaluation Based on Fuzzy Classification

Nolan published in [13] the development and successful application of a fuzzy rule based evaluation method aiming the rating of writing samples of fourth grade students. Previously in course of the evaluation the teachers used a comprehensive scoring guide that defined which skills have to be measured by the evaluator and which ones have to be determined from them.

The rule base was created from this scoring guide involving the participation of a group of expert evaluators. In order to reduce the complexity of the rule base they defined input partitions with a quite low resolution. In course of the evaluation the rater measures skills like character recognition, text understanding, understanding elements of the plots, and understanding ideas. The system infers the evaluation of skills like reading comprehension. For example a rule of the system is

IF understanding=*high* AND character-recognition=*strong* AND elements-of-plot=*all* AND generates-ideas=*expand* THEN reading-comprehension=*high*.

The main advantage of the method compared to the traditional evaluation form was that it reduced the time necessary for the learning of the scoring technique and the difference between the scores given by different evaluators decreased significantly. The drawback of the method is that it does not support the fuzzy input; the evaluators can express their opinion only in form of crisp values, which will be fuzzyfied later by the method. Based on the description given in the literature we can summarize that the method fulfils the criteria 1, 3, 4, and 6. Furthermore, it surely does not fulfill criteria 2 and 5.

3.2 Bai-and-Chen's Method

In order to reduce the subjectivism in student evaluation Bai and Chen (further on we will refer to it as BC method) suggested a quite complex solution in [1]. However, their method addresses only a part-task of the evaluation, namely the ranking of the students that obtained the same total score.

The BC method is applied as a follow-up of a conventional scoring technique. First, in case of each student (S_j, $1 \leq j \leq n$) each question (Q_i, $1 \leq j \leq m$) is evaluated independently by an accuracy rate a_{ij}, where $a_{ij} \in [0,1]$. Then, the evaluator calculates a total score for the student by

$$TS_j = \sum_{i=1}^{m} a_{ij} \cdot g_i \,, \qquad (3.1)$$

where g_i is the maximum achievable score assigned to the question Q_i ($\sum_{i=1}^{m} g_i = 100$).

In order to rank the students having the same total score Bai and Chen propose an adjustment of their scores. The adjustment is based on introduction of new aspects in the evaluation, i.e. the importance and the complexity of the questions, which are based on fuzzy sets determined by the evaluator or by domain experts. The measurement part of the evaluation is also extended by including the time necessary for answering the individual questions divided by the maximum time allowed to solve the question (answer-time rate, $t_{ij} \in [0,1]$).

Although it is used only in cases when two or more students achieve the same total score, the answer-time rate has to be measured for each student during the exam because it cannot be obtained posterior.

The modified scores are determined in six steps applying a three-level fuzzy reasoning process whose block diagram is presented in Figure 3.1. After calculating the average of the accuracy rates ($\overline{a_i}$) and the average of the answer-time rates ($\overline{t_i}$) for each question these are fuzzyfied by calculating their membership values in the corresponding predefined partitions resulting in the fuzzy grade matrices $[fa_{ik}]$ and $[ft_{ik}]$.

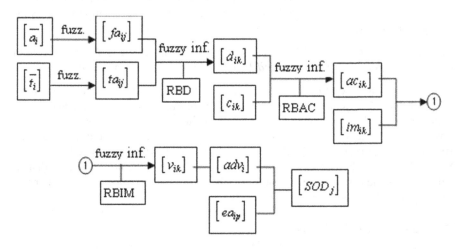

Fig. 3.1. Block diagram of the BC method

In the second step of the method one determines the fuzzy difficulty ($[d_{ik}]$) of each question using a special kind of fuzzy reasoning applying a predefined rule base (*RBD*) and a weighted average of the previously calculated membership values. The third step of the method concentrates on the calculation of the answer-cost of each question (a_{ik}) from the difficulty and the complexity values. The complexity of each question (c_{ik}) is expressed as membership values in the five sets of the predefined complexity partition. The $[c_{ik}]$ matrix is defined by domain experts. This step uses the same fuzzy inference model as the previous one applying a predefined rule base (*RBAC*).

The fourth step of the method calculates the adjustment values (v_{ik}) of each question from the answer-cost and the importance values. The importance of each question (im_{ik}) is expressed as five membership values in the five sets of the predefined importance partition. The $[im_{ik}]$ matrix is defined by domain experts. This step uses the same fuzzy inference model as the previous one applying a predefined rule base (*RBIM*). Next, one calculates the final adjustment value (adv_i) for each question as a weighted average of the individual adjustment values (v_{ik}) corresponding to the question.

In step 5 a new grade matrix ($[ea_{ip}]$) is constructed that contains only those k columns of the original accuracy rate matrix, which correspond to the students having the same total score.

The modified score values of each student (SOD_j, $1 \leq j \leq n$) are calculated in the last step by

$$ SOD_j = \sum_{p=1}^{k} \left[\left[\sum_{\substack{i=1 \\ i \neq j}}^{m} \left(ea_{pj} - ea_{pi} \right) \right] \cdot g_p \cdot \left(0.5 + adv_p \right) \right]. \qquad (3.2) $$

The main advantages of the method are that it does not increase the time needed for the evaluation and it allows the evaluators to make a ranking among students achieving the same score in the traditional scoring system. However, one has to pay a too high price for this result. In course of the exam preparation two matrices have to be defined by domain experts, one describing the complexity $[c_{ik}]$ and one describing the importance $[im_{ik}]$ of each question. It introduces redundancy in the evaluation process because these aspects presumably already have been taken into consideration in course of the definition of the vector $[g_i]$.

Thus it is hardly avoidable the occurrence of cases when the achievable score of a question is not in accordance with its complexity and importance evaluation. Besides, the level of subjectivity is also increased by the fact that the seven weights have to be determined by domain experts and there is no formalized way to determine their optimal values. Another drawback of the method is that it does not allow the evaluator to express the evaluation using fuzzy sets.

The real novel aspect of the evaluation is the answer-time rate. However, it is not clear how the base time for each question is defined. Besides, it seems not too efficient to measure the answer time for each student for each question and then to use it in case of students having the same total score unless it can be done by software automatically. Thus the BC method is not applicable in case of non computer-based exams. We can summarize that it fulfils criteria 1, 4, 5, and 6.

3.3 Saleh-and-Kim's Method

In order to alleviate some shortcoming of the BC method Saleh and Kim [16] suggested the so called *Three node fuzzy evaluation system* (TNFES) that applies Mamdani type fuzzy inference and COG defuzzyfication. Similar to the BC method

TNFES works with five inputs, namely the original grade vector ($[g_i]$), the accuracy grade matrix ($[a_i]$), the time rate matrix ($[t_i]$), the complexity matrix ($[c_{ik}]$), the importance matrix ($[im_{ik}]$), as well as with three rule bases, one for the difficulty (*RBD*), one for the effort (*RBE*), and one for the adjustment (*RBA*). The accuracy rate and answer time rate matrices are results of the examination. The complexity and importance matrices as well as the rule bases are defined by domain experts. The output of the system is a new grade vector, which contains the adjusted score values.

TNFES defines three fuzzy nodes (difficulty, effort and adjustment) that attain a three level fuzzy inference schema as follows:

- Difficulty node: $D = I([a_i], [t_i], RBD)$,
- Effort node: $E = I(D, [c_{ik}], RBE)$,
- Adjusment node: $W = I(E, [im_{ik}], RBA)$,

where *I* represents the Mamdani type fuzzy inference.

Each of the nodes contains a fuzzy logic controller with two scalable inputs and one output. The scalable inputs make possible the weighting of the different aspects, however, the authors do not use this possibility, they consider each input of equal influence. Each node consists of three steps (fuzzyfication, inference, defuzzyfication), which modularity can be also considered as a drawback owing to the redundancy introduced by the consecutive defuzzyfications and fuzzyfications. The result of the third node ($W=[w_i]$) is used for the calculation of the adjusted grade vector $[ga_i]$ by

$$ga_i = g_i \cdot (1 + w_i),\qquad(3.3)$$

followed by a scaling operation

$$ga_i = g_i \cdot \frac{ga_i \cdot \sum_{j=1}^{m} g_j}{\sum_{j=1}^{m} ga_j},\qquad(3.4)$$

where *m* is the number of the questions. The final total score is determined by

$$TS = [a_i]^T \cdot [ga_i].\qquad(3.5)$$

Owing to the similarity between TNFES and the BC approaches, the advantages and the drawbacks of the method are also similar to the features of BC. We can summarize that it fulfils criteria 1, 4, 5, and 6.

3.4 Student Evaluation Based on Fuzzy Rule Interpolation

The method *Student evaluation based on fuzzy rule interpolation* (SEFRI) [9] offers a solution using a rule base containing only the most relevant rules. The method takes into consideration three aspects, namely the accuracy of the response, the time necessary for answering the questions, and the correct use of the technical terms. In course of the preparation the 100 achievable marks are divided between the questions. They are the weights associated to the questions.

In case of the second aspect one works with the total time necessary for answering all of the questions, which is determined automatically and reported to the allowed total response time. The resulting relative time is fuzzyfied (*TR*) using singleton type fuzzyfication.

The characteristics "the accuracy of the response" (*AC*), and "the correct use of the technical terms" (*CU*) are measured by the evaluator with separate fuzzy marks (fuzzy numbers) for each question. The scoring scale is in both cases the unit interval. After assigning the two fuzzy marks for each question one calculates an average *AC* and *CU* value ($\overline{AC}$ and $\overline{CU}$) for the student as a weighted average of the individual values.

Next one determines from the three fuzzy values ($\overline{AC}$, *TR*, and $\overline{CU}$) the general evaluation of the student using fuzzy inference. In order to reduce the complexity of the rule base a fuzzy rule interpolation-based reasoning method called LESFRI [4] is used. Thus the underlying rule base requires only 64 rules in contrast with the 125 rules of the dense rule base owing to the fact that each input dimension contains five fuzzy sets.

The fuzzy inference results the general fuzzy evaluation of the student (*GFE*) that is defuzzyfied using Center Of Area method in order to get the total score (*TS*). Finally the grade of the student is determined using the standardized mapping of the university. For example a possible mapping is presented in Table 3.1.

Similar to the previous techniques this method can only applied in practice when a software support is present. Its advantage is that it contains only one-level inference with a relatively transparent rule base. The drawback of the method is that owing to the sparse character of the rule base it applies a bit complex inference technique that could require more software development work. We can summarize that the method satisfies 1, 2, 3, 4, 5, and 6.

Table 3.1. Mapping between scores and grades [9]

Score intervals	Grades
0 - 50	Unsatisfactory
51 - 60	Satisfactory
61 - 75	Average
76 - 85	Good
86 - 100	Excellent

3.5 Rasmani-and-Shen's Method

Rasmani and Shen proposed in [15] a special fuzzy inference technique and the use of a data driven fuzzy rule identification method that also allowed the addition of expert knowledge. Their main aim was to obtain user comprehensible knowledge from historical data making also possible the justification of any evaluation. The suggested inference technique is the so called weighted fuzzy subsethood based reasoning, which was developed for multiple input single output (MISO) fuzzy systems that apply rules of form

$$IF \ A_1 \ is \ \left[w(E_i, A_{11}) \cdot A_{11} \ OR \ w(E_i, A_{12}) \cdot A_{12} \ OR...OR \ w(E_i, A_{1j}) \cdot A_{1j} \ OR \ ... \right.$$
$$\left. OR \ w(E_i, A_{1n_1}) \cdot A_{1n_1} \right] AND$$
$$A_2 \ is \ \left[w(E_i, A_{21}) \cdot A_{21} \ OR \ w(E_i, A_{22}) \cdot A_{22} \ OR...OR \ w(E_i, A_{2j}) \cdot A_{2j} \ OR \ ... \right.$$
$$\left. OR \ w(E_i, A_{2n_2}) \cdot A_{2n_2} \right] AND \ ... \ AND$$
$$A_k \ is \ \left[w(E_i, A_{k1}) \cdot A_{k1} \ OR \ w(E_i, A_{k2}) \cdot A_{k2} \ OR...OR \ w(E_i, A_{kj}) \cdot A_{kj} \ OR \ ... \right.$$
$$\left. OR \ w(E_i, A_{kn_k}) \cdot A_{kn_k} \right] AND \ ... \ AND \tag{3.6}$$
$$A_m \ is \ \left[w(E_i, A_{m1}) \cdot A_{m1} \ OR \ w(E_i, A_{m2}) \cdot A_{m2} \ OR...OR \ w(E_i, A_{mj}) \cdot A_{mj} \ OR \ ... \right.$$
$$\left. OR \ w(E_i, A_{mn_m}) \cdot A_{mn_m} \right] THEN \ B \ is \ E_i$$

where m is the number of antecedent dimensions, A_k $k \in [1,m]$ are the antecedent linguistic variables, n_k is the number of linguistic terms in the kth antecedent dimension, B is the consequent linguistic variable, E_i $i \in [1,N]$ is the ith consequent linguistic term, N is the number of consequent linguistic terms, and $w(E_i, A_{kj})$ is the relative weight of the antecedent linguistic term A_{kj}. The weight expresses the influence of the set A_{kj} towards the conclusion drawn. One determines the weight as a result of the normalization of the fuzzy subsethood value of the set

$$w(E_i, A_{kj}) = \frac{S(E_i, A_{kj})}{\max\limits_{l=1..n_k} S(E_i, A_{kl})} \ . \tag{3.7}$$

The fuzzy subsethood value S represents in this case the degree to which the fuzzy set A_{kj} is the subset of a the fuzzy set E_i. It is calculated as

$$S(E_i, A_{kj}) = \frac{\sum_{x \in U} \nabla(\mu_{E_i}(x), \mu_{A_{kj}}(x))}{\sum_{x \in U} \mu_{E_i}(x)} \ , \tag{3.8}$$

where U is the universe of discourse, μ is the membership function, and ∇ is an arbitrary t-norm.

The rule base contains only one rule for each consequent linguistic term. The first step of the fuzzy inference is the calculation of the overall weight of each rule by applying the arbitrary disjunction and conjunction operators [5] to the antecedent side. Next, one selects the rule having the highest weight, whose consequent will represent the final score of the student.

One identifies the rule base in the following steps:

1 Create the input and output partitions.
2 Divide the training dataset into subgroups depending on the output linguistic terms.
3 Calculate fuzzy subsethood values for each subgroup.
4 Calculate weights for each linguistic term.
5 Create rules of form (3.6).
6 Test the rule base using a test dataset.

The main advantage of the method proposed by Rasmani and Shen is that it requires a rule base with a low number of rules, which number is equal with the number of output linguistic terms. Besides, it allows the evaluation of a question/test to be made by fuzzy numbers. However, it is not clear how the antecedent and consequent are determined and what is the meaning of the fuzzy subsethood values in case of the evaluation of the students' academic performance. We can summarize that the method satisfies 1, 2, 4, 5, and 6.

4 Conclusions

Fuzzy student evaluation methods can be a very useful tool supporting the evaluator in handling the uncertainty that is often present in the opinion of the rater in cases when the evaluation process is not fully defined, i.e. when it cannot be fully automated. Fuzzy inference based solutions offer a transparency owing to the humanly interpretable character of the rule base.

However, their disadvantage is their rigidity and the implicit weighting. A small change in the aspects or in the weighting could require a completely redefinition of the underlying rule base. Besides, owing to the implicit weighting the importance of the different aspects is not clear visible.

We can summarize that none of the overviewed methods fulfils all the previously defined criteria. The lack of the compatibility with the traditional methods proved to be a common drawback of them, which probably could be solved using automatic fuzzy rule base identification methods [3] [14] [17]. The application of other fuzzy inference techniques like the methods presented in [6] and [11] could also contribute to the development of evaluation techniques that better fit the applied criteria. Despite of the fuzzy character of the methods only the last two methods (SEFRI and the method proposed by Rasmani and Shen) allow the fuzzy expression of the evaluator's opinion. As a positive evaluation one can state that all the methods satisfy criteria 1, 4, and 6.

Acknowledgments. This research was supported by Kecskemét College GAMF Faculty grant no: 1KU31, and the National Scientific Research Fund Grant OTKA K77809.

References

[1] Bai, S.M., Chen, S.M.: Evaluating Students' Learning Achievement Using Fuzzy Membership Functions and Fuzzy Rules. Expert Systems with Applications 34, 399–410 (2008)

[2] Biswas, R.: An Application of Fuzzy Sets in Students Evaluation. Fuzzy Sets and System 74(2), 187–194 (1995)

[3] Botzheim, J., Hámori, B., Kóczy, L.T.: Extracting Trapezoidal Membership Functions of a Fuzzy Rule System by Bacterial Algorithm. In: Proceedings of the 7th Fuzzy Days, Dortmund 2001, pp. 218–227. Springer, Heidelberg (2001)

[4] Chen, S.M., Lee, C.H.: New Methods for Students Evaluating Using Fuzzy Sets. Fuzzy Sets and Systems 104(2), 209–218 (1999)

[5] Fodor, J.: Aggregation Functions in Fuzzy Systems, Aspects of Soft Computing. In: Fodor, J., Kacprzyk, J. (eds.) Intelligent Robotics and Control, Studies in Computational Intelligence, pp. 25–50. Springer, Heidelberg (2009)

[6] Hládek, D., Vaščák, J., Sinčák, P.: Hierarchical Fuzzy Inference System for Robotic Pursuit Evasion Task. In: Proceedings of the 6th International Symposium on Applied Machine Intelligence and Informatics (SAMI 2008), Herľany, Slovakia, January 21-22, pp. 273–277 (2008)

[7] Johanyák, Z.C.: Fuzzy Set Theory-based Student Evaluation. In: Proceedings of the IJCCI 2009 - International Joint Conference on Computational Intelligence, ICFC 2009 - International Conference on Fuzzy Computation, Funchal-Madeira, Portugal, October 5-7, pp. 53–58 (2008)

[8] Johanyák, Z.C., Kovács, S.: Fuzzy Rule Interpolation by the Least Squares Method. In: Proceedings of the 7th International Symposium of Hungarian Researchers on Computational Intelligence (HUCI 2006), Budapest, November 24-25, pp. 495–506 (2006)

[9] Johanyák, Z.C.: Student Evaluation Based on Fuzzy Rule Interpolation. International Journal of Artificial Intelligence (to be published, 2009)

[10] Johanyák, Z.C.: Survey on Three Fuzzy Inference-based Student Evaluation Methods. In: Proceedings of the 10th International Symposium of Hungarian Researchers – CINTI 2009, Budapest, November 12-14, pp. 185–192 (2009)

[11] Kovács, S.: Extending the Fuzzy Rule Interpolation FIVE by Fuzzy Observation. In: Reusch, B. (ed.) Advances in Soft Computing, Computational Intelligence, Theory and Applications, pp. 485–497. Springer, Germany (2006)

[12] Mamdani, E.H., Assilian, S.: An Experiment in Linguistic Synthesis with a Fuzzy Logic Controller. International Journal of Man Machine Studies 7, 1–13 (1975)

[13] Nolan, J.R.: An Expert Fuzzy Classification System for Supporting the Grading of Student Writing Samples. Expert Systems With Applications 15, 59–68 (1998)

[14] Precup, R.-E., Preitl, S., Tar, J.K., Tomescu, M.L., Takács, M., Korondi, P., Baranyi, P.: Fuzzy Control System Performance Enhancement by Iterative Learning Control. IEEE Transactions on Industrial Electronics 55(9), 3461–3475 (2008)

[15] Rasmani, K.A., Shen, Q.: Data-driven Fuzzy Rule Generation and its Application for Student Academic Performance Evaluation. Appl. Intell. 25, 305–319 (2006)

[16] Saleh, I., Kim, S.: A Fuzzy System for Evaluating Students' Learning Achievement. Expert Systems with Applications 36, 6236–6243 (2009)

[17] Škrjanc, I., Blažič, S., Agamennoni, O.E.: Interval Fuzzy Model Identification Using l-norm. IEEE Transactions on Fuzzy Systems 13(5), 561–568 (2005)

[18] Wang, H.Y., Chen, S.M.: New Methods for Evaluating the Answerscripts of Students Using Fuzzy Sets. In: Ali, M., Dapoigny, R. (eds.) IEA/AIE 2006. LNCS (LNAI), vol. 4031, pp. 442–451. Springer, Heidelberg (2006)

Fuzzy Hand Posture Models in Man-Machine Communication

András A. Tóth[2,3], Balázs Tusor[2,3], and Annamária R. Várkonyi-Kóczy[1,3]

[1] Institute of Mechatronics and Vehicle Engineering, Óbuda University, Budapest, Hungary
[2] Budapest University of Technology and Economics, Hungary
andras.attilio.toth@gmail.com
[3] Integrated Intelligent Systems Japanese-Hungarian Laboratory
koczy.annamaria@bgk.uni-obuda.hu, balazs.tusor@gmail.com

Abstract. Ever since the assemblage of the first computer, efforts have been made to improve the way people could use machines. This ambition is still present nowadays: indeed, intuitively operated systems are currently under intensive research. Intelligent Space (or iSpace) based systems are good examples: they strive to be comfortable and easy to use, even without demanding technical knowledge from their users. However, their aim is not limited to this: in fact, their ultimate goal is to achieve an intelligent environment for higher quality, natural, and easy to follow lifestyle. The system described in this chapter can be used to create a new, intuitive man-machine interface for iSpace applications. The solution exploits one of the basic human skills, namely the ability to assume various hand postures. The proposed system first processes the frames of a stereo camera pair and builds a model of the hand posture visible on the images and then classifies this model into one of the previously stored hand posture models, by using neural networks and fuzzy reasoning.

Keywords: Ubiquitous Computing, Intelligent Space, Smart Environments, Image Processing, Intuitive User Interface, Soft Computing.

1 Introduction

Today, with the spreading of machine intelligence, "smart environments" are becoming a popular tool for humans to collect information, form the environment, get assistance etc. *Intelligent Space* (*iSpace*) [1], which can be any smart area, such as a room, a railway station, an underpass, road crossing, or even a town, etc. equipped with intelligent sensors and agents, is a special intelligent implementation of the so called *Ubiquitous Computing* paradigm [2]. The main feature of the iSpace is that the intelligence itself is not present in the agents but it is distributed in the whole space. Thus, the architecture of the artificial agents, like robots, is quite simple as they are coordinated by the intelligent sensors. The goal of iSpace is to improve the comfort and safety of everyday life as well as to achieve personalised healthcare and independent living for disabled persons.

I.J. Rudas et al. (Eds.): Computational Intelligence in Engineering, SCI 313, pp. 229–245.
springerlink.com   © Springer-Verlag Berlin Heidelberg 2010

In this chapter, based on the ideas detailed [3], authors introduce a new hand posture and – gesture modelling and identification system, which can be used to develop an intuitive, hand sign driven human-machine interface for iSpace applications. The system described here consists of an image processing and an intelligent modelling component. The former can build a three dimensional model of a hand in various postures using a pre-calibrated stereo camera pair, while the latter learns and classifies these hand postures. The rest of this chapter is organized as follows: in Section 2, an overview of the features and the architecture of the iSpace is given. Section 3 describes the preparation and the classification of the hand postures. In Section 4 experimental results are reported. Finally, Section 5 is devoted to the conclusions and possible improvements are mentioned.

2   The Intelligent Space

Intelligent Space (iSpace) is an intelligent environmental system originally developed at Hashimoto Lab, University of Tokyo, Japan. The main and ultimate goal of the Intelligent Space is to build an environment that comprehends human intentions and satisfies them [4]. This means that the system should be easy to use for the people in it: they should be able to express their will through intuitive actions and there should be no need for them to learn how the system is to be used. Beyond supporting humans, the iSpace should provide an interface also to its artificial agents, e.g. to the robots offering physical services to the humans using the iSpace.

The most characteristic feature of the iSpace is that the intelligence is distributed in the whole space, not in the individual agents around. The other fundamental property of the iSpace is that it is able to monitor what is going on in the space and to build models based on this knowledge. The system can also react with its environment and provide information or physical services to its users.

The iSpace is an active information space because information can be requested by the clients and provided by the system through active devices. The iSpace is thus a so called soft environment: it has the capability to adapt itself to its clients [1].

ISpace applications, existing and under development, (see e.g [4] and [5]) aim at such tasks as the monitoring of physiological functions of humans, the positioning and tracking of humans, the localization and control of mobile robots, finding paths for them by using itineraries taken by people, etc.

There are several requirements that the hardware and software architecture of the iSpace has to satisfy. As stated in [5] these are *modularity, scalability, ease of integration, low cost,* and *ease of configuration and maintenance.*

Software tasks are implemented as distributed individual processes and can be categorized into three types: sensor and actuator servers, intermediate processing, and application processes.

The task of the sensor and actuator servers is to pre-process information and to deliver it to the network. Sensor fusion, temporal integration, and model building occurs at intermediate processing level. These tasks might require real-time capabilities, thus components performing them should be located near the sensor. Finally, the application processes are those that provide the actual functionality of the iSpace. At this level, only low amount of data is processed and a slower reaction time is sufficient. Functions at this level are required to be easily portable and maintainable by the user.

The architecture satisfying all these requirements is described in [4]. The basic elements of the iSpace are the artificial clients and the distributed intelligent sensors. The formers include robots, which provide the physical services to the human users of the iSpace, and the monitors, which furnish them with information.

DIND (Distributed Intelligent Networked Device) is the term for the intelligent sensor, which is the basic building element of the iSpace. This device consists of three main components: a sensor which monitors the dynamic environment, a processor, whose task is to deal with sensed data and to make decisions, and a communication part through which the device is able to communicate with other DINDs or with the artificial agents of the iSpace.

The advantage of the architecture based on the DINDs is that the iSpace can easily be constructed and modified by using them. Indeed, an existing space can be turned into an Intelligent Space just by installing DINDs. The modularity of the architecture makes the renewal of functions easy, and thanks to the network an update can be effectively applied to all DINDs. The other advantage of the network is that it facilitates resource sharing.

3 Fuzzy Hand Posture Classification

The comfortability of the way of communication between humans and the iSpace is of vital importance from the point of view of usability of the system. The more natural the interaction is, the wider the applicability can be. Because of this, we decided to use one of the basic human talents, coordinated complex moving of the hands, for communication. The idea is that after that the sensors of the iSpace detect and separate the hands of the humans, the intelligence of the system determines the postures and movements and translates them to desired actions.

This section describes the subsystem extracting and classifying various hand postures from the images of a stereo camera pair, which is achieved in two steps: first, the frames of the cameras are processed, yielding a series of three dimensional point locations. Then, these spatial points are processed by a soft computing based intelligent modelling component (Fig. 1).

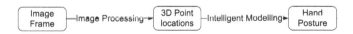

Fig. 1. Overview of the proposed system

At current stage of the development, some simplifications were applied: it is assumed that the whole field of view of the sensors is filled with a homogeneous background and the hand of the user is held before this background. No other parts of the body or other objects are visible on the image delivered by the sensor; however the absence of the hand is permitted. A further assumption is that stretched fingers are well separated and that the hand stays parallel to the background.

In this work, we only focus on hand posture and gesture recognition, however the tracking of slowly moving hands and recognition of predefined movements can also be solved by using a single camera. For details, see [3].

3.1 The 3D Hand Model

Processing the frames of the stereo rig should yield a constant number of spatial points, irrespective of the actual hand posture. These points can then be processed by the intelligent modelling component. In order to achieve this, a solution based on the previous work of the authors (see [6]) has been adopted. The novelty of the solution presented here, namely the subdivision of the whole contour, is motivated by the requirement of a constant count of spatial points.

The image processing component requires two offline preparation steps. On one hand, the hue-saturation histogram of the skin of the hand must be computed. On the other hand, the calibration of the stereo camera pair is needed, which means determining the relative position and orientation of the cameras and other, so called intrinsic parameters. The most important notions are briefly summarized here; for theoretical background please refer to [7].

The aforementioned parameters are represented by the so called *camera matrix*, which, assuming the pin-hole camera model, relates spatial points $\mathbf{X}$ to image points $\mathbf{x}$ in the following way:

$$\mathbf{x} = \mathbf{P} \cdot \mathbf{X} = \mathbf{K} \cdot \mathbf{R}[\mathbf{I}|-\mathbf{C}], \tag{1}$$

where $\mathbf{x}$ and $\mathbf{X}$ are 3- and 4-element vectors, $\mathbf{P}$ is the 3x4 camera matrix, $\mathbf{R}$ is the 3x3 *rotation matrix* between the camera coordinate frame and the world coordinate frame, $\mathbf{I}$ is a 3x3 identity matrix, $\mathbf{C}$ is the 4-element vector representing the *centre of the camera coordinate frame* and $\mathbf{K}$ is the *camera calibration matrix*:

$$\mathbf{K} = \begin{bmatrix} \alpha_x & s & p_x \\ 0 & \alpha_y & p_y \\ 0 & 0 & 1 \end{bmatrix}, \tag{2}$$

where (p_x, p_y) stand for the image center coordinates, α_x and α_y represent the horizontal and vertical size of pixels, and s denotes the skew. If the camera matrices $\mathbf{P}$ and $\mathbf{P}'$ of the left and right cameras are known, the 3x3 *fundamental matrix* $\mathbf{F}$, relating the two images, can be computed:

$$\mathbf{F} = (\mathbf{P}'\cdot\mathbf{C}) \times \mathbf{P}'\cdot\mathbf{P}^+. \tag{3}$$

After the off-line preparation, the image processing steps are executed (Fig. 2). Contour extraction involves more sub-steps: first the hue-saturation histogram H is back-projected on the HSV representation of the input frame (see also [8]):

$$I_{bp}(x, y) = H[I_h(x, y), I_s(x, y)], \tag{4}$$

where I_{bp}, I_h, and I_s denote the single-channel back-projection image, hue plane and saturation plane, respectively. Pixels, having too high or low brightness, according to [8], are discarded. Next, the back-projection is thresholded, assigning the maximal intensity to pixels above a given threshold, and 0 to others. Connected components are then located, on the thresholded back-projection image and those having an area below a predefined value, are discarded, thus a single connected component, representing the hand should remain. The contour of this component is further processed, as described in [9] and [10].

First, the so-called *k-curvature*, θ_i is computed for each contour point C_i:

$$\Theta_i = \arccos\left[\frac{\mathbf{C}_{i-k} - \mathbf{C}_i}{\|\mathbf{C}_{i-k} - \mathbf{C}_i\|} \cdot \frac{\mathbf{C}_{i+k} - \mathbf{C}_i}{\|\mathbf{C}_{i+k} - \mathbf{C}_i\|}\right], \tag{5}$$

where k is a constant number, and the contour point indices are to be computed modulo the count of contour points. After this, contour points whose k-curvature exceeds a given threshold are retained, and non-maximal suppression is performed, locating thus feature points at the curvature extrema of the contour (see Fig. 3). These feature points are separated into peaks and valleys, depending on the convexity c_i of the contour:

$$c_i = \operatorname{sgn}\left[(\mathbf{C}_{i-k} - \mathbf{C}_i) \times (\mathbf{C}_{i+k} - \mathbf{C}_i)\right] \tag{6}$$

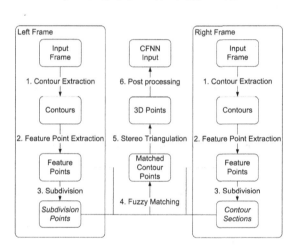

Fig. 2. Image processing steps

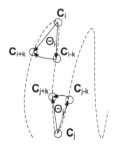

Fig. 3. Computation of k-curvature and convexity

The following procedure is performed only for the left frame: Each section, defined by two consecutive feature points, is subdivided by one or more subdivision points into subsections. The subsections have the same length within one section, but not necessarily all of them are the same length (here, the length of a section means the number of contour points it includes). The count of subdivision points, for each section, is proportional to the length of that section. The total number of subdivision points depends on the number of feature points: these two values, together, should yield a constant number, which equals the number of inputs of the Circular Fuzzy Neural Networks (CFNNs), described in the next subsection. In case of complete lack of feature points, the whole contour is subdivided equidistantly.

After this, feature points (i.e. peaks and valleys) on both frames are matched, using the *Fuzzy Matching Algorithm*, described in [11]. The algorithm computes the epipolar line l' of each feature point **x** on the left frame, using the fundamental matrix **F**, defined in Eq. (3):

$$\mathbf{l'} = \mathbf{F} \cdot \mathbf{x} .\tag{7}$$

For each right feature point **x'**, that lies (in a fuzzy sense) on **l'**, it computes the weighted difference of neighborhood:

$$\sum_{\substack{(x,y)\in w \\ (x',y')\in w'}} \left| I(x, y) - I'(x', y') \right| \cdot \mu_A(x - x_0) \cdot \mu_B(y - y_0) .\tag{8}$$

Here $I(x, y)$ and $I'(x', y')$ stand for the intensities of the gray-scale representation of the left and righ-frame, respectively, μ_A and μ_B are fuzzy membership functions of the environment, w and w' are the rectangular neighborhoods of x and x', and (x_0, y_0) represents the upper left corner of w. The best match is considered to be the pair for which this value is minimal. The whole process is illustrated in Fig. 4.

Subdivision points of the left frame are then matched against contour sections on the right frame, again, using the Fuzzy Matching Algorithm (Fig. 5). In case of some hand postures, there are less than two feature points: when this happens, first some selected subdivision points on the left frame are matched against the whole contour on the right frame, and these points are then treated in the same way, as feature points obtained considering Eq. (5).

When subdivision- and feature points have been matched, the computation of three dimensional coordinates is achieved by the Direct Linear Transformation (DLT), described in [7]. Finally, in order to be used as an input for the CFNN, the set of points is translated, so that its centroid lies at the origin of the world coordinate frame, and the coordinates are scaled to the interval [-1; 1].

3.2 Fuzzy Hand Posture Recognition

In order to efficiently distinguish different hand postures we have developed the *Fuzzy Hand Posture Model* (FHPM). It describes the human hand by fuzzy hand feature sets.

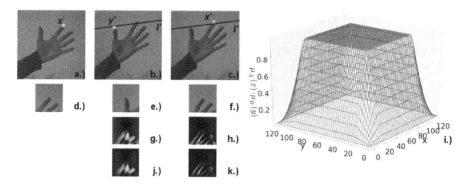

Fig. 4. The fuzzy matching algorithm
    a) Left frame with feature point x
    b) Right frame, epipolar line l' of x near a non-matching feature point, y'
    c) Right frame, epipolar line l' near the matching feature point x'
    d) e) f) Neighborhoods of $x.$, y'. and x', respectively
    g) Absolute difference of neighborhoods of x and y'
    h) Absolute·difference of neighborhoods of x and x'
    i) Environment membership function
    j) Weighted absolute difference of neighborhoods of x and y'
    k) Weighted absolute difference of neighborhoods of x and x'

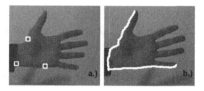

Fig. 5. Subdivision points to contour section matching. a.) Left frame, subdivision points. b.) Right frame, contour section.

3.2.1 Fuzzy Hand Posture Model

Three different types of fuzzy feature sets had been appointed, each one describing a certain type of features of the given hand posture. The first set consists of four fuzzy features; they describe the distance between the fingertips of each adjacent finger. Both the second and the third sets consist of five five fuzzy features: former describes the bentness of each finger, while the latter describes the relative angle between the bottom finger joint and the plane of the palm of the given hand.

Each feature is marked by a linguistic variable that can only have one of the three following values: small, medium, or large. Thus, every feature set type is an answer to a certain question:

- How far are the finger X and the finger Y from each other?
- How big is the angle between the lowest joint of finger W and the plane of the palm?
- How bent is the finger Z?

For each FHPM 14 linguistic variables (which are used to describe the 14 features introduced above) are stored in the *ModelBase*. With this model we can theoretically distinguish among 3^{14} different hand postures, which is more than enough even for the most complex sign languages.

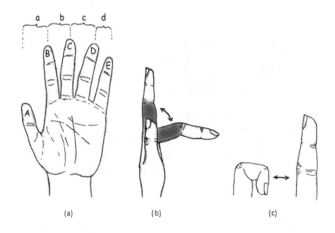

Fig. 6. The three type of fuzzy feature sets

In the following, an example is presented to demonstrate how a hand posture can be described by the FHPM features: Fig. 7 shows the well-known victory sign. The relative position of the fingers from each other and their bentness is clearly visible. From that, we can determine the value of all the 14 linguistic variables. Table 1 shows the values of the variables, divided into the three different feature groups.

Table 1. The features for hand posture "Victory"

Feature group	Feature	Value
	a	Large
Relative distance between adjacent fingers	b	Medium
	c	Small
	d	Small
	A	Medium
	B	Small
Relative angle between the lowest joint of each finger and the plane of the palm	C	Small
	D	Large
	E	Large
	A	Medium
	B	Large
Relative bentness of each finger	C	Large
	D	Small
	E	Small

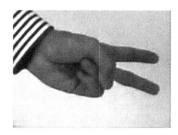

Fig. 7. The classic hand posture for "victory"

3.2.2   The Hand Posture and Gesture Recognition System

For hand posture recognition, the proposed system uses fuzzy neural networks and fuzzy inference. The idea is based on transforming the coordinate model of the detected hand transformed into a *Fuzzy Hand Posture Model* by *Circular Fuzzy Neural Networks* and then to identify it by applying fuzzy inference.

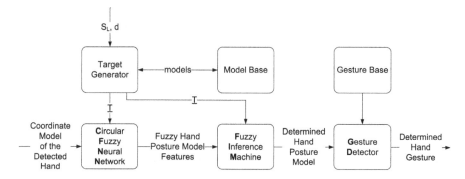

Fig. 8. The block diagram of the Hand Posture and Gesture Recognition system

Fig. 8 shows the architecture of the system. The system receives the coordinate model of the detected hand as input, transforms it into a *Fuzzy Hand Posture Model* using CFNNs, then determinates the hand's shape from the FHPM with the usage of the Fuzzy Inference Machine (FIM) module. The *Gesture Detector* module observes the output of the FIM, searches for matches with pre-defined patterns in the Gesture-Base. The hand gesture is identified in case of fuzzy matching with any stored hand gesture or is refused if the hand gesture is unknown for the GestureBase.

The features for each model are stored in the *ModelBase*, as linguistic variables (with possible values: *small, medium, large*). One of the easiest realizations of the *ModelBase* can be e.g. using XML files. Fig. 9 shows an example for two pre-defined models.

The *GestureBase* contains the pre-defined hand gestures. Each hand gesture is identified by its name and is described by up to 5 string values, which are existing *Fuzzy Hand Posture Model* names. The string values are listed in the *sequence* parameter. The parameter *quantity* denotes how many *FHPMs* the given hand gesture model consists of. Fig. 10 shows an example for the realization of the *GesturBase*.

```
<base>
    <model>
        <name> Open_hand </name>
        <value_a> large large large large </value_a>
        <value_b> small small small small small </value_b>
        <value_c> large large large large large </value_c>
    </model>
    <model>
        <name> Fist </name>
        <value_a> small small small small </value_a>
        <value_b> medium large large large large </value_b>
        <value_c> medium small small small small </value_c>
    </model>
    <model>
        <name> Three </name>
        <value_a> medium medium medium small </value_a>
        <value_b> small small small large large </value_b>
        <value_c> large large large small small </value_c>
    </model>
</base>
```

Fig. 9. An example for the realization of the ModelBase with three defined models

```
<base>
    <gesture>
        <name> Classic </name>
        <quantity> 3 </quantity>
        <sequence> Open_hand Fist Three </sequence>
    </gesture>
    <gesture>
        <name> Neo </name>
        <quantity> 3 </quantity>
        <sequence> Point Thumb-up Victory </sequence>
    </gesture>
</base>
```

Fig. 10. An example for the realization of the GestureBase with two defined gestures

The *Target Generator* is used to calculate the desired target parameters for the *Circular Fuzzy Neural Networks* and the *Fuzzy Inference Machine* using the *ModelBase*. The input parameters of the *Target Generator* module are:

- d - identification value (ID) of the model in the ModelBase. It can simply be the name of the given model or an identification number.
- S_L - a linguistic variable for setting the width the fuzzy triangular sets, thus tuning the accuracy of the system.

In out experiments, we used triangular shaped fuzzy sets with centers set to the following numerical values: *small = 0.3*; *medium = 0.5*; *large = 0.7*.

The other parameter of the fuzzy sets is their width. This parameter is explicitly settable with a linguistic variable S_L. In case of S_L chosen to be small, medium or large, the widths of the fuzzy feature sets equal to 0.33, 0.66, and 1, respectively.

Equation (9) is used to compute the ends of the interval that represents the given fuzzy feature set.

$$(a, b) = (center - 0.2 \cdot width, \; center + 0.2 \cdot width) \tag{9}$$

where a, b denotes the lower and upper ends of the interval, *center* corresponds to the numerical value of the given feature, and *width* equals to the numerical value of linguistic variable S_L.

For the task of converting the coordinate model of the given hand to a Fuzzy Hand Posture Model, we have developed *Circular Fuzzy Neural Networks* based on fuzzy neural networks with interval arithmetics proposed by [15]. The novelty of our network lies in its modified topology.

The networks have fuzzy numbers in their weights, biases, and in the output of each neuron. Each fuzzy number is represented by the interval of its alpha cut at value 0, i.e. by the base of the triangular fuzzy set. Thus, each fuzzy number is determined by two real numbers: the lower and the upper ends of the given interval.

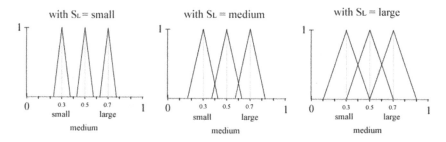

Fig. 11. The possible values of the triangular fuzzy feature sets

Since the coordinate model consists of 15 three dimensional coordinates, the neural networks have 45 inputs. In order to increase the robustness, instead of using one network with 14 output layer neurons, three different networks are used with the only difference in the number of the output layer neurons. The first network computes the first group of fuzzy features, having 4 neurons in the output layer. The second and third networks have 5-5 neurons in the output layer, similarly. Furthermore, all the networks consist of one hidden layer with 15-15 neurons.

In order to enhance the speed of the training session and to take advantage of the fact that the input data consists of coordinate triplets, the topology of the network has been modified. Originally, each input was connected to all hidden layer neurons. In the modified network topology only 9 inputs (corresponding to 3 adjacent coordinate triplets) are connected to each. This way every hidden layer neuron processes the data of 3 adjacent coordinates. The topology between the hidden layer and the output layer neurons has not been changed. This way, the network has been realigned into a *circular* network, see Fig. 12. For the sake of better visibility not every connection is shown. In the outer circle layer the input coordinates can be found, in the middle circle the hidden layer neurons are placed, and in the inner circle the output layer neurons of the network are located.

These reductions cause a dramatic decrease in the required training time, while the precision and accuracy of the networks are not affected.

The last step of the hand posture identification procedure is the fuzzy reasoning based classification. This part compares the output of the CFNNs to all fuzzy hand posture models stored in the *ModelBase* and chooses the model that corresponds the most to the model presented by the CFNNs.

The algorithm works as follows: For each model in the *ModelBase*, we calculate the intersection value between their fuzzy feature sets and the fuzzy feature sets of the

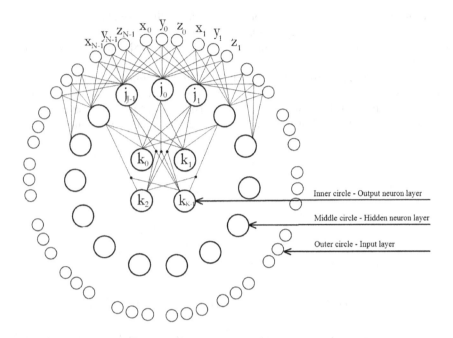

Fig. 12. The topology of Circular Fuzzy Neural Networks

given FHPM, thus gaining β_i for each of its features. The minimum of β_i shows the correspondence ratio between each model and the given FHPM. The maximum value of the correspondence ratios indicates which model is the most corresponding to the detected hand posture.

The ID of each recognized hand posture model is put in a queue, which is monitored by the *Gesture Detector* module. It searches for hand gesture patterns predefined in the *GestureBase*, and in case of matching it gives the ID (in our realization: the name) of the detected hand gesture as the output of the system, i.e. it identifies the gesture.

The reliability of the identification system can be improved by applying a "fault tolerant" sign language, i.e. when there is a $d_H > 2$ Hamming distance among the meaningful hand posture series. In this case, a detected unknown hand posture sequence caused by false detection(s) can be corrected to the nearest known hand gesture.

3.2.3 The Training Session

For increasing the speed of the training of the CFNNs, we have developed a new training procedure. In order to decrease the quantity of the training samples, instead of directly using the training data in the training phase, they are first clustered and the CFNNs are trained by the centers of the obtained clusters (that are still 45 real numbers or 15 spatial coordinate points).

The clustering procedure is based on the k-means method with a simple modification: comparing a given sample to the clusters one by one, the sample gets assigned to the first cluster with the distance of its center to the given sample less than an arbitrary value. If there is no such cluster then a new cluster is appointed with the given sample assigned to it. The clustering is used on only one type of hand model at one time, so the incidental similarity between different type of hand models will not cause problems in the clustering phase.

4 The Experimental System

4.1 Image Processing Component

We have built an experimental set up for testing and analyzing the performance of the image processing component. The hardware part of the testing system consists of two web cameras (Genius Look 316) connected to an average PC (CPU: AMD Athlon 64 4000+, 1 GB DDR RAM, Windows XP+SP3 operating system). The cameras are located at one end of a table. At the other end a homogeneous background is mounted (Fig. 13). The image processing algorithms running on the PC and are implemented in C++ using *OpenCV* [12]. For camera calibration the *Camera Calibration Toolbox for Matlab* has been adopted [14].

Throughout the tests, the following parameters were used: the input frames were 8 bits/channel static RGB images, at a resolution of 640x480 pixels. The working images (i.e. the back-projection and the gray scale representation for fuzzy matching), were single-channel images, with the same parameters. After back-projection, only pixels having brightness (i.e. "value", in the HSV color space) between 10 and 200 were retained. The thresholding parameter, before contour component extraction, was 60. In equation (5), k was equal to 40. In the Fuzzy Matching Algorithm, the weighted difference was calculated over an area of 121x121 pixels, having the maximal weight at the central area of 80x80 pixels. Points lying within 20 pixels of the epipolar lines were considered to be matching candidates, others were discarded. The experimental setup and the processed hand postures are shown in Fig. 13.

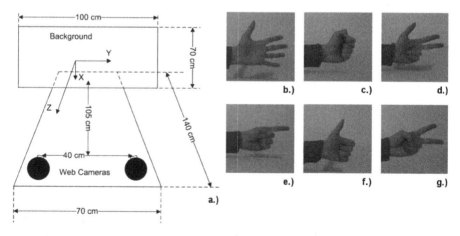

Fig. 13. The experimental setup and hand postures: a) The experimental setup. b) Open hand. c) Fist. d) Three. e) Point. f) Thumb up. g) Victory.

In order to assess the quality of the spatial model, 100 uniform samples of each hand posture were captured at various distances from the homogeneous background (starting at $d=0$ cm, i.e. hand adjacent to the background, until $d=40$ cm, stepping by 5 cm), which is a total of 5400 samples. The yielded spatial models were evaluated in a formal and an informal way. The *formal approach* consisted of fitting a plane to the spatial model (which, in this case was neither translated nor scaled), and calculating the average distance of spatial points from this plane (here, the fitting plane is defined as the plane for which the squared sum of distances to the spatial points is minimal).

The *informal solution* consisted of manually examining the spatial model, by converting it to the format of a third-party program, *Anim8or* [15]. Every 20th sample was examined in this way, and an integer score between 1 (worst) and 5 (best) was assigned to them. The results of both methods are shown in Fig. 14. In general, it is true, that sets having a lower average distance were found to be of "better" quality by the informal examination as well. For instance, models for posture "*Three*" at $d=5$ cm contained more inaccuracies, than those at $d=10$ cm. In the case of posture "*Three*", at $d=35$ cm and $d=40$ cm one point was in a wrong position in the right frame, yielding a considerable displacement also in the spatial model. In another case, however, distance from the fitting plane was not useful for detecting inaccuracy. All the examined samples of posture "*Open Hand*" at $d=5$ cm, and 3 of them at $d=25$ cm, exhibited a

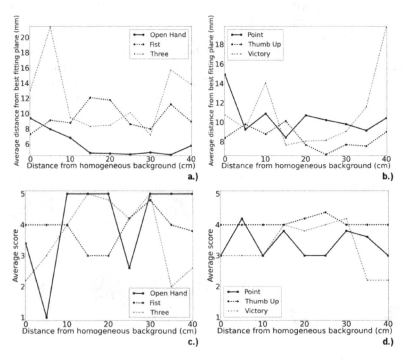

Fig. 14. Average distances and scoring of each measurement set: a) Distance for postures "Open Hand", "Fist" and "Three". b) Distance for postures for postures "Point", "Thumb up" and "Victory". c) Scoring of postures "Open Hand", "Fist" and "Three". d) Scoring of postures "Point", "Thumb up" and "Victory".

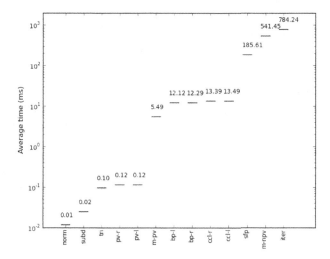

Fig. 15. Average times for various steps. norm – normalization of spatial points, subd – contour subdivision, tri – stereo triangulation, pv-{l, r} – peak and valley location on the left and right frame, m-pv – peak and valley matching, bp-{l, r} – back projection and discarding pixels by brightness (on the left and right frame respectively), ccl-{l, r} – connected component localization after thresholding (on the left, and the right frame respectively), sfp – selecting subdivision points as feature points, m-npv – matching subdivision points to contour sections (i.e., non-peak and valley matching), iter – time for a complete iteration.

particular error: there were no spatial points representing the contour between ring and little finger, causing the two fingers to "blend together". However, since all of them were lying near the fitting plane, a low average distance was reported.

The time needed for each step has been measured by instrumenting the source code before and after the main steps and also at the beginning and at the end of the whole iteration (i.e., all the steps shown in Fig. 2). Then, this iteration was run 200 times on 10 different samples of each posture, and the total time spent during each type of step was averaged. The results are shown in Fig. 15. As expected, it has been found that the bottleneck of the procedure is constituted by steps, which need to match distinct points against whole sections. Due to these operations, the average time needed by an iteration is about 0,78 s, and the maximal value was as high as 1,23 s. In order to use this system in real time applications, ways to optimize the performance must be investigated.

4.2 FHPM-Based Hand Recognition

The experimental options for the training session of the CFNNs (and clustering) were the following:

- Learning rate: 0.8
- The coefficient of the momentum method: 0.5
- The error threshold: 0.1
- S_L: small
- Clustering distance: 0.5

Table 2 summarizes the identification results of a simple initial experimental system which is able to differentiate among six hand models (Open hand, Fist, Three, Thumb-up, Point, and Victory). For better visualization, Fig. 10 shows these hand postures. The set of the generated coordinate model samples has been split into two separate, training and testing, sets. The training set consists of 20 samples from each of the three FHPMs. The clustering procedure reduced the quantity of the training sets, thus instead of 120 only 42 coordinate models were trained to the CFNNs. The networks are trained with choosing S_L=small. The FIM could identify the models in the testing set (with 80 samples of each models) with an average accuracy of 94.8%. (The testing set does not contain the training set.).

While the CFNNs were trained with S_L = small value, in the inference phase S_L = large has been used, which increased the accuracy of the *Fuzzy Inference Machine*.

Table 2. The results of training 6 different type of hands

Name of the model	Quantity of chosen samples	Quantity of clustered training set	Quantity of testing set	Correctly identified	%
Open hand	20	10	80	75	93.75
Fist	20	12	80	78	97.5
Three	20	4	80	78	97.5
Thumb-up	20	7	80	78	97.5
Point	20	4	80	78	97.5
Victory	20	5	80	68	85

5 Conclusions

In this chapter, the concept and implementation of a hand posture and gesture modeling and recognition system are introduced. The system consists of a stereo camera set up and an intelligent modeling part and it is able to classify different simple hand postures. It can be used as a novel interface that makes human users able to control Intelligent Space by hand postures.

In our future work we plan to improve the presented system. E.g. the most limiting factor of the image processing part, the need of the homogeneous background could be solved. A more effective implementation of point matching against a whole section is also required.

The FHPM-based hand gesture recognition method also could be improved, since the actual implementation of the Gesture Detector does not take into account the position of the hand, only its shape.

Acknowledgments. This work was sponsored by the Hungarian National Scientific Fund (OTKA 78576) and the Hungarian-Portuguese Intergovernmental S&T Cooperation Program.

References

[1] Lee, J.-H., Hashimoto, H.: Intelligent Space. In: Intell Robot and Syst. Proc. (2000), doi:10.1109/IROS.2000.893210
[2] Weiser, M.: The Computer for the Twenty-First Century. Sci. Am (1991)
[3] Tóth, A.A., Tusor, B., Várkonyi-Kóczy, A.R.: New Possibilities in Human-Machine Interaction. In: Proc. of the 10th International Symposium of Hungarian Researchers on Computational Intelligence and Informatics 2009, Budapest, Hungary, November 12-14, pp. 327–338 (2009)
[4] Lee, J.-H., Morioka, K., Ando, N., Hashimoto, H.: Cooperation of Distributed Intelligent Sensors in Intelligent Environment. IEEE/ASME Trans on Mechatron (2004)
[5] Appenzeller, G., Lee, J.-H., Hashimoto, H.: Building Topological Maps by Looking at People: An Example of Cooperation Between Intelligent Spaces and Robots. In: Proc. of the Int. Conf. on Intell Robot and Syst. (1997)
[6] Várkonyi-Kóczy, A.R., Tóth, A.A.: ISpace – a Tool for Improving the Quality of Life. J of Autom. Mob. Robot & Intell Syst (2009)
[7] Hartley, R., Zisserman, A.: Multiple View Geometry in Computer Vision, 2nd edn. Cambridge University Press, Cambridge (2003)
[8] Bradski, G.R.: Computer Vision Face Tracking For Use in a Perceptual User Interface. Intel. Techn. J (1998)
[9] Segen, J., Kumar, S.: Shadow Gestures: 3D Hand Pose Estimation Using a Single Camera. In: Proc. of the IEEE Comput Soc. Conf. on Comput. Vis. and Pattern Recognit. (1999)
[10] Malik, S.: Real-time Hand Tracking and Finger Tracking for Interaction. CSC 2503F. Project Report (2003)
[11] Várkonyi-Kóczy, A.R.: Autonomous 3D Model Reconstruction and Its Intelligent Application in Vehicle System Dynamics. In: Proc. of the 5th Int Symp. on Intell. Syst. and Inform. (2007)
[12] Bradski, G., et al.: Intel Open CV Library (2009),
     http://opencv.willowgarage.com/wiki
[13] Ishibushi, H., Tanaka, H.: Fuzzy Neural Networks with Fuzzy Weights and Fuzzy Biases. In: Proc. IEEE Neural Network Conf., San Francisco, vol. 3, pp. 1650–1655 (1993)
[14] Bouguet, J.Y. (2008), Camera Calibration Toolbox for Matlab,
     http://www.vision.caltech.edu/bouguetj/calib_doc
[15] Glanville, R.S. (2008), Anim8or, http://www.anim8or.com

Computational Qualitative Economics
Computational Intelligence-Assisted Building, Writing and Running Virtual Economic Theories

Ladislav Andrášik

Institute of Management, Slovak University of Technology
Vazovova 5, 512 43 Bratislava I, Slovakia
ladislav.andrasik@stuba.sk

Abstract. In steeply evolving global knowledge based society and in authentic economies too are up-and-coming the wide collection of complex problems and tasks that can't be understand and solved only by conventional approaches, methods and tools. The deep understanding such new complex phenomena are asking for better fitting methods. Fortunately the progress in ICT, Internet and first of all in Applied Informatics (AI) and Computational Intelligence (CI) can help to overrule several of such problems. Actually the possibility for using advanced ICT, new generations of web products and services but mainly for utilization of CI are giant jump ahead in deeper understanding complex economic phenomena without historical precedence. In this chapter the author is focusing his attention to new evolving Scientific Program (SP) in General Economics (GE) area called Computational Qualitative Economics (CQE). He shows that CI effectively helps for deeper understanding also conventional complex economic problems even with rather simple CI devices. The great advantage of CI products and services laying in option for assistance not only in knowledge acquisition and in theory building process but can assist in writing of codified theories and in performance of different digital happenings and in compositions and programming of digital scene for authentic performance of evolutionary stories. Among others this chapter will explore the potential advantages and disadvantages of Computational Qualitative Economics (CQE) so for the analysis of real economies, economic systems, models and artificial entities as for the study of codified conventional economic theories and/or economic theories of main stream. The author is also hints at recent debates about the potential methodological costs and benefits of ICT, Internet and CI use in qualitative research, modelling, simulation, and about the relationship between methodological approaches on the one hand and CI assistance methods of qualitative research on the other. It is not consent to argument that the connection between certain ICT, Internet and CI aided strategies and methodological approaches are far looser than is often assumed. He not means that they can panacea potentials but the gains from using them are significant in wide area of socio-economic research. The Qualitative theory building, writing, endosomatic rebuilding and using by authentic subjects assisted by ICT, Internet and CI contribute significant progress in cogitation and so in deeper understanding of economies living in global knowledge based

I.J. Rudas et al. (Eds.): Computational Intelligence in Engineering, SCI 313, pp. 247–261.
springerlink.com   © Springer-Verlag Berlin Heidelberg 2010

society. This chapter is not invented for Computational Quantitative Economics, and/or Econometric or Prognostic Simulations and was limited only to the intentional sphere of positive economics. Similarly it has to be say – from the point of existence of double-fold possibility to methodical approaches the subject that is from up to down (phenomenological approach) and from bottom to up – that this chapter is rest only on using the first one. Primary Classification: J. Computer Applications; → Social and Behavioural sciences; → Subjects: Economics. Additional Classification: I. Computing Methodologies; → Computational (Artificial) Intelligence; → Subjects: Applied products and services of CI utilizable in CQE.

Keywords: Built-in Software: Theaters, Stages and Scenes of CQE; Causal Reasoning; Computational Intelligence; Digital Fables; Digital Storytelling; Economic theory building Engine; Experimentation in Virtual Laboratories; Happenings of Economics Episodes; Knowledge Discovery in running Virtual Experiments; Qualitative: Reasoning, Modeling, Simulation and Theorizing.

1 Introduction

The contemporary economics is very dense branched collection of different economic theories, branches and schools. Among them the major role are played by so-called Main Stream Economics (MSE). The second main school in general economics is so-called Evolutionary Economics (EE). The first is evolved from Classical Political Economy, transformed later to Neoclassical Economy associated by Austrian School and after II. World War created together with Keynesian Economics the so-called Neoclassical Synthesis. The second main stream of economic thinking prefers evolutionary approaches but similarly as first one in contemporary form is methodically based on intensive using of analytical mathematical methods. From this point of view both pointed out schools are primordial methodical base for CQE because the buildings of different computational economic entities, systems and/or models are constructively created on the pedestal of mathematical formalism. But the distinguishing feature of CQE resides in non quantitative identification of result of appropriate computational procedures with those constructs. On the contrary the identification of those results is qualitative in the sense of original H. Poincaré's qualitative theory of complex motions. This is a little different notion of adjective "Qualitative" staying in front of subject "Economics" as in conventional economics. The significant group of traditional economist community believed that a considerable body of sensible economic phenomena can be proposing simply by expression "Qualitative" when this is used as name for the change of sign of sufficient parameter of economic system alone. In other word their simple believe is residing in the statement about form in which the algebraic sign of some economic act and/or effect is predicted from knowledge of the signs only of the relevant structural parameters of the system. But in more realistic view on the nature of Qualitative Economic System (QES) the content of that adjective is fairly wealthier. The difference can be characterizing on the one hand by topological notion of lost structural stability of the system and/or lost their topological equivalence against former state which is caused by exceeding the critical value of parameter (parameters). On the other hand that difference can be clarified from the point of view of variables that is changed loci of starting point of evolution in the area of existing. Both reasons are lead to the notion of perturbation and/or bifurcation in the system caused by them. Decisive role of CI products and services is resting in the potential for creation artificial perturbation on the need of researcher or student. At that base they can find the answer of the question: "What could be happen when…?" So coming from the outside and/or inner self-created perturbation of parameter values or values of starting point coordinates can decides on qualitative nature of coming in situation even of whole future regime of evolution. From other point of view the participant of CI assisted fable can ask the question: "If I take this action now, what might happen in the future?" This can be identified as important step to anticipatory learning by experimentation in virtual laboratory (virtual laboratory in this case is understand as cognitive computational subject, agent and/or engine). The result of those approaches is resides in such outcome that the activities of learning subjects in CQE environments can be better intrinsically motivated than learning activities in conventional economics one. That is such learning subject can above of receiving part of codified knowledge in passive form come to active understanding of considering matter being directly draws to living dialog with process running in virtual laboratory or deal with vital digital scene and via Internet to the vital dialog with other participants of parallel learning process[1]. Finally it has to be settled that the results of conventional research and learning are considerably more perishable than result achieved by CQE.

2 The Object, Subject, Scope and Method of CQE

It is obviously to understand economics and to them belonging partial economic theories as a complex of views on real economy written in economic essays, monographs and textbooks. But on the other hand we can to imagine on economics as a whole and on partial economic theories too as of virtual economy (economies) living in the consciousness of an individual and or in the social consciousness yet. In the last meaning economics and partial economic theories are some eternally evolving complex of knowledge living as special part of world views in different social, cultural and/or religious environment. In the era of ICT, Internet and quickly advancing CI such and similar complex of knowledge can living in another than biological consciousness that is in not biological noosphere as Internet today really is.

The CQE is one of the legitimate parts of Computational Economics (CE) as a whole, see Chart 1. The second such part is Computational Quantitative Economics (CQnE). The disciplinary cluster of corresponding and cooperating scientific branches and hardware-software devices and tool are very complex, Chart2. The object of CQE is double-folded. The principal object of CQE is vital objective economic reality in the age of global knowledge-based society. The secondary object of CQE is the complex of entire system of former (conventional, traditional) economics and partial economic theories. On these two objects the CQE identify the subjects of their scientific interest. The principal interest is focused to causalities and qualitative shapes of economic behaviour of vital object. The secondary interest is touching the question of former economics and

---

[1] By intrinsically motivated activities we understand such behaviour of learning subject when he/her does it for its own sake rather than to receive specific rewards and/or avoid specific penalties. For example student in common education process for credits, grades, counting, etc.

partial economic theories validities by reconstruction them by CI assistance for the purpose of building for them virtual laboratories and in that base that is in virtual reality experimentally confirm or expel validity of former on the real dates based and speculative ends and statements of theorizing. So the scope of CQE is making possible to interesting subjects gaining necessary information, imaginations and qualitative knowledge on complex economic behaviours founded on building, rebuilding, communicating with other human and virtual subject former mental models by assistance of ICT, Applied Informatics (ApI), CI and Cognitive Sciences (CI). That process of those gaining is dominantly realised in the layer of intrinsically motivated human mind. Such intrinsically motivated behaviour is vitally important because the various actions of participants of markets are about their interacting in everyday of life, trying to understand the past, construct the future and make sense of the present. CQE improving very effectively of intersubjectiveness of understanding in a vide community of economists. On the other hand thanks to better prepared single subjects via CQE consequently profiting from these all society again on the base of ICT, ApI and CI assistance because in such way is rising of Knowledge Capital (KC) of all society[2].

The CQE as expected is using the whole collection of reliable conventional methods and toolboxes used in economics but above them are more intensive is using advance methods of applied mathematics than before. The dominant methodical approaches are however products and services delivering and proposing by ICT, ApI and CI. Among others in the area of phenomenological methodical approach is broadly based opinion in using virtual experimentation in economic models built is such software as STELLA, VENSIM, SWARM, SIMULINK, iDmc and without surprise in Excel too. Those approaches is useful in all three level of building and using CQE: in building theory, in writing codified one, in projecting architecture of theatres, scenes and their realisation, in programming digital storytelling's, in communication with other human and artificial subject in social networks[3] of knowledge and so on.

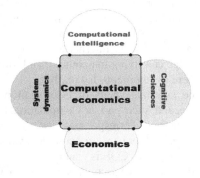

Chart 1. The emerging of Computational economics as device helping to deeper understand new phenomena in global knowledge base economies

---

[2] The KC is not the synonym of the Knowledge Stock of society. The KC is intangible entity living in the network of entire population mind.
[3] Every participant of the network is as node endowed with several sheaves of links coming in from- and going out to- other nodes like a neuron endowed with several synapses and axons.

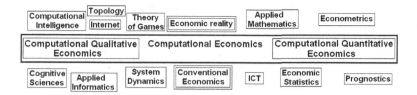

Source: Author's creation

Chart 2. The cluster of correspondent scientific disciplines, cooperating devices and tools to Computational (Qualitative & Quantitative) Economics

On the contrary the surprise may be in the factual experience that relatively simple software enables to build virtual laboratories that help in very deep understanding via dialogue with running experiment. But on the higher level of CQE methods is direct using of CI for building authentic complete theatre of matter at hand economic theory[4]. What is regarding the second methodical approach of investigation and building theories from bottom to up there are a broad collection of approaches, devices and tools in our time. Among such there are well-known devices as are cellular automata, percolation theory, Petri nets, Markov fields, artificial neural networks, genetic and evolutionary algorithms and programs, Bayesian method, theory of random graphs, classification theory, and multi-agent systems, etc.

3   CI-Assisted Theory Building in Economics

Theory building (TB) in economics so as in other social sciences is first of all a process of creativity and imagination realised by authentic scholar. The result of such activities of human mind is becomes of some germs from one side for perturbation impacts to former theory if such exists, and from other side for creation of very new theory if such one nil. So the theory building is actually the composed activity of human mind united destruction and creation. This integration of two polar opposite activities we are using for founding our conclusion that theory building has a nature of creative perturbation arising in former structural stability of economic theory, that is bit by bit vanishing the former structural stability and equivalently, that is gradual creating of, and/or step by step moving to, new structural stability. Using topological approach the former theory after upper descript activities is forced to such change that the successive form of theory looks as if former one lost his topological equivalence. On the other sides of view, because the theory building consist several further activities the job is more complex than it looks in first site. Actually, among other distinctive features, the theory is some kind of story. So the theory building is also special story telling which is parallel with other activities in this process. This complex activity is also on learning and understanding by scrawling and/or featuring and design models and algorithms assisted by CI. The novelties, in decrypted complex activities as theory building is, connected with computational intelligence (CI) assistance are lying in its potential character that may be very intelligent, ready and efficient

---

[4] In this chapter these approaches aren't in the main focus of explanation.

collaborator in qualitative economic analysis and in theory building. The theory builder may acting in perpetual very vital dialog with virtual co-worker and interposes it as interface into gap between own mind and the objective economic reality. We are put emphasis on our belief that this connection between authentic human intelligence (which is some function upon brain) and authentic computational intelligence (which is some function upon computer and used software) isn't the same as connection brain-computer. Apart from that we have to say that so human intelligence as computational intelligence are disposing with imputed intelligence of third subjects products and using them not only in own but naturally also in mutual cooperative cognition process. On the side of virtual partner of theory builder such third that is human persons are CI professionals.

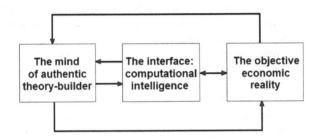

Chart 3. Elementary interactions among the modules of triadic complex

Actually, the vital interaction processes among three entities displayed in Chart 3 is much more complex, but that is another story than one we interesting in this paper. For better understanding our own problem it is suitable to bring some familiar economic examples, because our focus is on economic theory building. Before it is fitting to note that for scholar in meant process there is much to be gained from questioning the assumptions of a former well established theory. Those gains may become the starting point for preliminary destruction of former theory. For example we use known fact that traditional economic theory (in majority named as neoclassical economy) is based on few key assumptions such as perfect information of agents, stable equilibrium (and/or proclivity to homeostasis) of the firms and of whole economy, perfect competitions, etc. So the theory builder can successfully benefit greatly from relaxing these assumptions. The strong destructing perturbation in former theory, for example is releasing the out of this world assumption on stable equilibrium not only in economic systems (that is in constructions of mind), but first of all in objective economic reality. Because the reverse is right: the real economy being in motions far from equilibrium and only time to time is approaching the near neighbor loci, or area close to equilibrium: real economy demonstrates some pattern of discrete form of cyclical behaviour. Not at all, some of cyclical behaviour shows complex crowd modes of behaviour. But this change from equilibrium to non-equilibrium assumptions isn't fallen from heaven. It is result of former theoretical investigation of that matter.

It must be draw attention to the fact that theory building in economics also requires careful reflection on the importance and uniqueness of the phenomenon at hand, the questions explored, and the context of the research. Because theories is ought to serve as signposts that tell us what is important, why it is important, what determines this importance, and what outcomes should be expected. We are familiar also with assertion that theories in several cases is guide the subjects looking for information and knowledge through what was found and why it enriches or even challenges his/her understanding. Theoretically grounded studies pay particular attention to the context of their research and account for its complexity, uniqueness and richness. Such requirements are also very natural screened-off area of theory building. All of said before is several times more confused in the condition of evolving global knowledge society. On the other hand novelties arriving in the same content of that one not surprisingly becoming strong tools helping scholars in solving emerging problems connected with theory building upon new evolving phenomena. Namely, it is to be said that in economics, i. e. in early qualitative analysis and quick to fasten on their result theory building, the major role may play, and/or is really played by computational intelligence at present. Naturally it is clear today, that after first mass produced IBM PC entered to the life of scientist and/or researchers the ICT devices begin to play ceaseless increasing role in scientific activities and in theory building too.

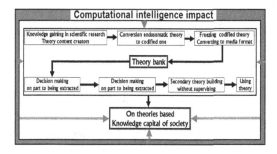

Chart 4. The role of CI in creation of economic knowledge in codified and intangible forms

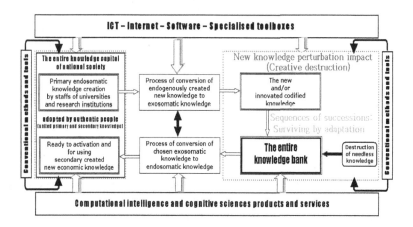

Chart 5. The economic knowledge creation by ICT-Internet-CI-Software-Toolboxes assistance

It must be draw attention to the fact that theory building in economics also requires careful reflection on the importance and uniqueness of the phenomenon at hand, the questions explored, and the context of the research. Because theories is ought to serve as signposts that tell us what is important, why it is important, what determines this importance, and what outcomes should be expected. We are familiar also with assertion that theories in several cases is guide the subjects looking for information and knowledge through what was found and why it enriches or even challenges his/her understanding.

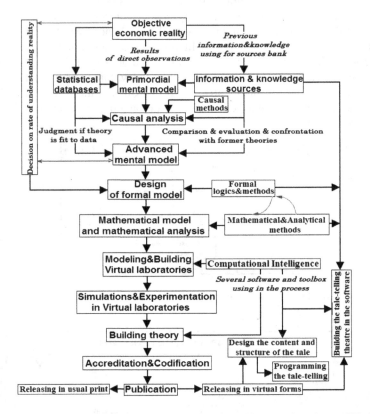

Chart 6. The creative process from objective economic reality to tacit and to codified theory

Theoretically grounded studies pay particular attention to the context of their research and account for its complexity, uniqueness and richness. Such requirements are also very natural screened-off area of theory building. All of said before is several times more confused in the condition of evolving global knowledge society. On the other hand novelties arriving in the same content of that one not surprisingly becoming strong tools helping scholars in solving emerging problems connected with theory building upon new evolving phenomena. Namely, it is to be said that in economics, i.e. in early qualitative analysis and quick to fasten on their result theory building, the major role may play, and/or is really played by computational intelligence at present.

Naturally it is clear today, that after first mass produced IBM PC entered to the life of scientist and/or researchers the ICT devices begin to play ceaseless increasing role in scientific activities and in theory building too. These are the reasons that in narrower community such processes get the name computer-assisted theory building. But in this name is some way and in contents too misunderstanding. From the one hand it is true, that for researcher the PC and/or notebook are their every day assistant at present. But on the other hand greater truth is that real assistant is born on the base created by computational intelligence. This is the reason why we preferred the term computational intelligence assisted theory building. By the way, it is clear that the difference for example between every day using of PC and in opposite using PC and/or in excellence cases also multi-parallel platforms for building virtual partners for dialog like complex virtual laboratories and/or theatres, scenes, digital symposiums are colossal.

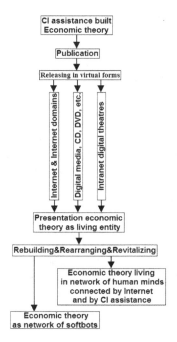

Chart 7. The process of delivering CQE

However in the area of contemporary complex economic systems there arises several difficulties in theory building because of high level requirements for mathematical knowledge's and skill for using mathematical tools. In this situation the scholars can addressee great expectation to product of computational intelligence. Fortunately the devices and tools usable for this purpose isn't put very high requirements to potential computational intelligence product propose in preliminary early steps in the process of theory building. It is interesting and surprising yet, that also relatively simple devices and tools can effectively serve for problem solving in meant area of science. On the other hand the realisation of more advanced theory building is putting extraordinary requirements on tools prepared by computational intelligence. Our purpose in this paper is not so ambitious. We are focusing our attention to more simple tasks, and so we will show only a few examples demonstrating this conclusion for early stages of theory building. Among others there are such subjects those in common economic textbook are prearranged as easy for understanding, but in reality the opposite is true. One from the set of such subjects is broadly familiar problem of monopoly. Intimately spoken new ICT products and services coupling with them entering to market perturbed the former structural stability of that one. So in one hand the manufacturer is immediately becoming real monopolist searching for price and product amount on the market, i.e., he must challenge against uncertainty of demand curve (he must looking for shape and location of that one). On the other hand, the demand side, that is, consumers must react to new situation not only in the market but they are forced them to step by step changing their former economic, social, cultural etc. behaviour, shortly, he must change his former life style. So the new product entering on market is at least two valued entity perturbed the structural stability of market from both sides. But the principal impact is on whole global knowledge society progression. The complex behaviour of that one masterly helps to solve overtly progression in computational intelligence. Similar economic market issues are the problems of duopoly and oligopoly, and on the opposite side of market there is the entity of market type: perfect competition.

4 Rebuilding Common Economic Theories by CI Assistance

For easy ingoing to the area under discussion of this paper we begin with rearranging well-known, but a little more complex economic theories as are frequent with assistance of simple software available for wide community of economist's. Among other the offer in this field is wide enough, for example products of distributed CI, mentioned above as iDmc, STELLA and partially by Excel. In this way we wont to avoid great palette of problems attached with modelling and simulation of social system, because with the alive (perpetually living) economy we may to act and work as with complex social system. In short, to act with complex social system, modelling it, and simulating with them is difficult. As interested scholars known, it is difficult, basically, for

a) Two creator mental reasons: syntactic and semantic complexity, and

b) Two pragmatic reasons: the need of managing clusters of interconnected models and necessitates constraining model possibilities. So we are purposefully resigned to this job in this paper.

For, it is not need to emphasize that theory building is a process of deep creativity and imagination. It demands careful reflection on the importance and uniqueness of the phenomenon at hand, the questions explored, and the context of the research. Theories serve as signposts that tell us what is important, why it is important, what determines this importance, and what outcomes should be expected. Theories also guide the reader through what was found and why it enriches or even challenges our understanding. Theoretically grounded studies pay particular attention to the context of their research and account for its complexity, uniqueness and richness. These studies also offer compelling arguments, provide a fair test of these arguments, and use findings to refine and enrich the theory they have invoked.

4.1 Economic Theories: iDmc and Excel-Assisted Building the Cases of Modified Monopoly Model

Conventional economic theory is arranged so, that deals only with two basic market form: perfect competitive market and market of unitary product, in other word monopoly. From ordinary textbooks of economics occasionally come up deception that monopolist is in advantageous position because he can deliberately choose the level of price and amount of his product for yielding uppermost profit. For realisation of this pleasant goal he must know at least the entirely demand curve established in the market. But this is very unrealistic assumption indeed in very general monopolistic situation. Much harder is that problem with new product monopoly because the demand curve is beginning to emerge as far as new product entering to market. But the monopolist decision making must be done before this, with fairly long advance.

4.1.1 Model of Monopolist's Searching with Linear Demand Function

For demonstration we are using subsequent simple task appropriate for description monopolist possible behaviour by entering future market with new ICT product. Supposing that monopolist just knows a few point on the linear demand function, recently visited in its more or less erratic search of maximum profit. It is however unrealistic assumption but it is convenient for demonstration the problems. The model is based on subsequent formal statements:

Demand curve function is: $P_D = 5 - 2Q$. Revenue curve function is: $R = 5Q - 2Q^2$. Cost curve function is: $C = 2Q - Q^2 + 0.7Q^3$. Profit function is: $\Pi(Q) = 3Q - Q^2 - 0.7Q^3$, because in economics is: $\Pi(Q) = R - C = 5Q - 2Q^2 - 2Q + Q^2 - 0.7Q^3 = 3Q - Q^2 - 0.7Q^3$. And after derivation we are: $\Pi'(Q) = 3 - 2Q - 2.1Q^2$, and for simulation we shall dealing with quadratic equation: $3 - 2x - 2.1x2 = 0$, which after solution given plot of quadratic parabola, and their two roots are: $x_1 = -1.7627862939148853, x_2 = 0.8104053415339328$. Marginal revenue function is: $MR = 5 - 4Q$. Marginal cost function is: $MC = 2 - 2Q + 2.1Q^2$, ($x = y = 0.7378364083397065$).

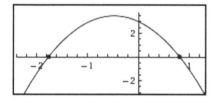

That model we are realised as virtual laboratory in iDmc and subsequent snapshots show some important feature of the monopolist behaviour in uncertain market.

Fig. 1. The simple profit graph

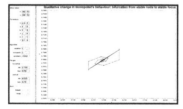 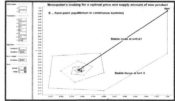

Fig. 2a. Bifurcation of the unstable fixed point from reaching stable node, to reaching stable focus because the longitude of searching step was raised; **b.** The trajectories tracking to stable equilibrium (to fixed point)

4.1.2 Model with Cubic Demand Function Built on Ideas Proposed by T. Puu

On the ideas of T. PUU and on his formal relations [15, p.115-119] we are built virtual laboratory for proposed model with cubic demand function of monopolist's searching in STELLA and in iDmc too.

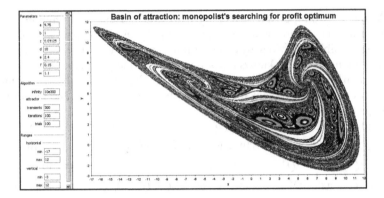

Fig. 3. The sensitivity of the model to starting point values

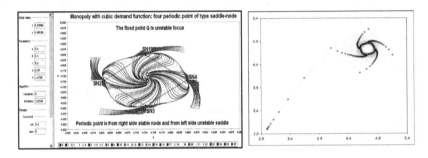

Fig. 4a. By trajectories visualised qualitative nature of four periodic points; **b.** Evolution to saddle-node points via focus trajectory with four branches

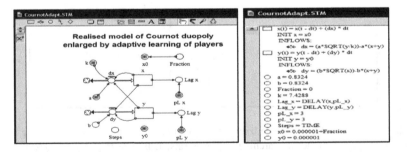

Fig. 5a. Basic scheme of Cournot duopoly enlarged model realised in map/model level of STELLA 8.1; **b.** Principles of basic scheme (the fillings of building boxes)

4.2 STELLA-Assisted Building and Telling Story of Cournot Duopoly

The duopoly model of A. A. Cournot is other very suitable case for demonstration CI assisted economic theory building. More we can to say: it is perfectly fitting for opening phase of CI assisted building economic theory. From authors opinion A. Cournot was the first mathematician and economist who applied mathematics to the field of economics, not necessarily to produce numerical precision in a predictive fashion, but rather to provide clearer formulation of economic relationships. That is the approach I called qualitative computational economics, because parallel with mathematics the computational intelligence have to dominate in economic theory nowadays.

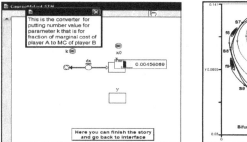 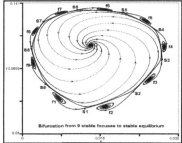

Fig. 6a. Several first steps of story; **b.** Nine attractive focuses, repellent invariant curve and O

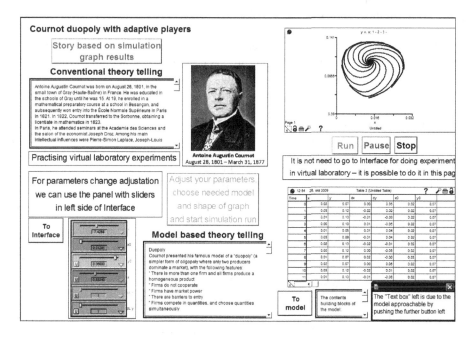

Fig. 7. Reading story telling on duopoly theory (the progress of story watching is about at half)

4.2.1 Building Virtual Laboratory for Duopoly in STELLA

4.2.2 Building Stories on Duopoly Theories

For building entire theory for communication in wider community of scholars it is important to convert tentative theory into for readable for others. In such activities CI can efficiently helps. One from simple approach to this task is to build story telling in programmed form by, for example STELLA software. Only for visual imprint we show two snapshots made in this devices, when the story run by hand pushing mode via spacebar.

5 Concluding Remarks

An economic theory in general is neither simply reflection nor sheer description of chosen by researcher concrete part of real economy. It is more complex, endogenously structured entirety of statements that answers not only to the questions on the superficial character of phenomena but shows deep reason why that phenomena is such, what motives drifts that partial economy to observed behaviours. Shortly, the economic theory generally and CQE principally isn't merely perception and knowing but understanding. However the CQE understanding is higher lever of quality and durability than any other forerunner economic theories.

Methods and tools founded on ICT, Internet and CI are playing important roles in economic research, in building codified theories and telling scientific stories, in education and in ordinary learning too. But it is henceforward valid that the human mind is the Lord and the complex of the ICT, Internet and CI is their Servant.

The CQE is never ended intangible entity parallel living in the authentic subjects consciences creating multilayer net united also by nets in Internet and/or other devices and methods like videoconferences etc., that is community which are furnished not only with vital human subjects but with living in CI environment population of homunculus's like softbot's etc.

Acknowledgments. This chapter was written on the base of paper prepared and presented in the 10th International Symposium (CINTI), in Budapest, November 2009, [3]. Both activities and author's fundamental research are granted by VEGA, No. 1/0877/08 of Ministry of Education of the Slovak Republic. His research is also sponsored by BOAT a. s. (Joint Stock Company) Bratislava.

References

[1] Agliari, A., Bischi, G.I., Dieci, R., Gardini, L.: Global Bifurcations of Closed Invariant Curves in Two-Dimensional Maps: a Computer Assisted Study. International Journal of Bifurcation and Chaos 15, 1285–1328 (2005)
[2] Agliari, A., Dieci, R., Gardini, L.: Homoclinic Tangles in a Kaldor-like Business Cycle Model. Journal of Economic Behaviour & Organization 62, 324–347 (2007), http://www.elsevier.com/locate/econbase
[3] Andrasik, L.: Computational Intelligence assisted Theory Building in Economics. In: Proceedings of the 10th International Symposium of Hungarian Researchers on Computational Intelligence and Informatics, November 2009, p. 123 (2009)
[4] Andrášik, L.: The Theory of Computer-aided Experimentation in an Artificial Economy - Some Unconventional Approaches to Simulation of Models of Economical Evolution and to Experimentation in Successive Environment. Economic Journal/Ekonomický časopis 52(8), 996 (2004)

[5] Andrášik, L.: Digital Stories in Non-Linear Dynamical Economies in Discrete Time. Economic Journal/Ekonomický časopis 56(3), 239 (2008)

[6] Andrasik, L.: Virtual Life and Perpetualogics (Self-Preservation of Virtual Entities in Computational Intelligent Technology). Philosophy/Filozofia 53(1), 15–26 (1998)

[7] Andrasik, L.: Learning by Evolution - in an Artificial Economy. Economic Journal/Ekonomický časopis 46(1), 72–98 (1998)

[8] Cathala, J.C.: On the Boundaries of Absorbing and Chaotic Areas in Second-Order Endomorphism. Nonlinear Analysis, Theory, Methods & Applications 29(1), 77–119 (1997)

[9] Chiarella, C., Dieci, R., Gardini, L.: Speculative Behaviour and Complex Asset Price Dynamics: a Global Analysis. Journal of Economic Behaviour & Organization 49, 173–197 (2002), The paper is available on Internet:
http://www.elsevier.com/locate/econbase

[10] Dieci, R.: Critical Curves and Bifurcations of Absorbing Areas in a Financial Model. Nonlinear Analysis 47, 5265–5276 (2001),
http://www.elsevier.nl/locate/na

[11] Guckenhaimer, J., Oster, G.F., Ipaktchi, A.: The Dynamics of Density Dependent Population Models. Journal of Mathematical Biology 4, 101–147 (1977)

[12] Gumowski, I., Mira, C.: FSE 1993. Lecture Notes in Mathematics, vol. 809. Springer, Berlin (1980)

[13] Lines, M., Medio, A.: iDmc (interactive Dynamical Model Calculator), user's guide (2005), http://www.dss.uniud.it/nonlinear

[14] Lorenz, H.-W.: Nonlinear Dynamical Economics and Chaotic Motion. Springer, Heidelberg (1993)

[15] Puu, T.: Nonlinear Economic Dynamics. Springer, Heidelberg (1997)

[16] Smale, S.: Differentiable Dynamical Systems. Bulletin of American Mathematical Society 73, 747–817 (1967)

[17] Zeeman, E.C.: On the Unstable Behaviour of Stock Exchanges. Journal of Mathematical Economics 1, 39–49 (1974)

A Spectral Projected Gradient Optimization for Binary Tomography

Tibor Lukić[1] and Anikó Lukity[2]

[1] Faculty of Engineering, University of Novi Sad
Novi Sad, Serbia
tibor@uns.ac.rs
[2] Budapest University of Technology and Economics
Budapest, Hungary
lukity@math.bme.hu

Abstract. In this paper we present a deterministic binary tomography reconstruction method based on the Spectral Projected Gradient (SPG) optimization approach. We consider a reconstruction problem with added smoothness convex prior. Using a convex-concave regularization we reformulate this problem to a non-integer and box constrained optimization problem which is suitable to solve by SPG method. The flexibility of the proposed method allows application of other reconstruction priors too. Performance of the proposed method is evaluated by experiments on the limited set of artificial data and also by comparing the obtained results with the ones provided by the often used non-deterministic Simulated Annealing method. The comparison shows its competence regarding to the quality of reconstructions.

Keywords and phrases: Binary Tomography, Spectral Projected Gradient method, non-monotone line search, convex-concave regularization, smooth regularization.

1 Introduction

Tomography is imaging by sections. It deals with recovering images from a number of projections. Since it is able to explore inside of object without touching it at all, tomography has a various application areas, for example in medicine, archaeology, geophysics and astrophysics. From the mathematical point of view, the object corresponds to a function and the problem posed is to reconstruct this function from its integrals or sums over subsets of its domain. In general, the tomographic reconstruction problem may be continuous or discrete. In *Discrete Tomography* (DT) the range of the function is a finite set. More details about DT and its applications you can find in [12, 13]. In addition to other, it has a wide range of application in medical imaging, for example within Computer Tomography (CT), Positron Emission Tomography (PET) and Electron Tomography (ET). A special case of DT, which is called *Binary Tomography* (BT), deals with the problem of the reconstruction of a binary image.

I.J. Rudas et al. (Eds.): Computational Intelligence in Engineering, SCI 313, pp. 263–272.
springerlink.com   © Springer-Verlag Berlin Heidelberg 2010

DT reconstruction problem usually leads to solving a large-scale and ill-posed optimization problem. Performance of this optimization is very important for the applicability of the DT method. This issue is topical and several different optimization methods are proposed in literature for this purpose. *Simulated Annealing* (SA) [19] is an often used method. Although its applicability is very wide, it has drawbacks, inter alia slow convergence and non-deterministic nature. Schüle et al. [24, 22, 23, 25] introduced a powerful method based on D.C. Programming [20]. The objective function of D.C. based method has to be expressed as a difference of convex functions, what can limit its flexibility for including new priors and constrains. Balázs et al. [2] proposed an interesting approach based on the Genetic optimization. This method is applicable in limited cases of reconstructions when images representing disks. Also, efforts have been made for adaptation of Branch and Bound method too [1]. In this paper we propose a new optimization approach for BT reconstruction problem based on the SPG optimization method in combination with *convex-concave* regularization [22]. SPG is introduced by (Birgin et al. 2000) in [6] and it is further analyzed and improved in [5, 8, 9]. The main motivation for application of SPG lies in the fact that SPG is a very efficient for solving large-scale and convex-constrained problems, especially when the projection onto the feasible set is easy to compute, see [18]. Numerous numerical tests in [6] and [7] show the superiority of SPG method in compare with other ones. Our problem is obviously a large-scale (regarding to the image resolution) and we reformulate it as a convex and box-constrained optimization problem where the projection is trivial to compute. Therefore the application of the SPG method became a suitable choice.

The paper is organized as follows. In Section 2 we describe the reconstruction problem. Section 3 describes the general SPG method for convex-constrained minimization. In Section 4 we define and analyze the proposed reconstruction method. Section 5 contains experimental results and finally, Section 6 is for conclusion and further work.

2 Reconstruction Problem

A main problem in connection with DT refers to the image reconstruction. We consider a BT image reconstruction problem where the imaging process is represented by the following linear system of equations

$$Ax = b, \quad A \in R^{m \times n}, \quad x \in \{0,1\}^n, \quad b \in R^m. \tag{1}$$

The matrix A is a so called projection matrix, whose each row corresponds to one projection ray, the corresponding components of vector b contain the detected projection values, while binary-vector x represents the unknown image to be reconstructed. The row entries a_i of A represent the length of the intersection of pixels of the discretized volume and the corresponding projection ray, see Figure 1. Components of the vector x are binary variables indicating the membership of the corresponding pixel to the object: for $x_i = 1$ pixel belongs to the object, while for $x_i = 0$ it does not. In a general case the system (1) is under-determined ($m < n$) and has no unique solution. Therefore the minimization of the squared projection error

$$\min_{x \in \{0,1\}^n} \|Ax - b\|^2$$

can not lead to the satisfactory result. To avoid this problem an appropriate prior regularization is need. We consider a well know smoothness prior defined by

$$\sum_i \sum_{j \in N(i)} (x_i - x_j)^2, \tag{2}$$

where $N(i)$ represents a set of indices of image neighbour pixels right and below from x_i. This prior is quadratic and convex and its role is to enforce the spatial coherency of the solution. In this paper we focus on the binary tomography problem given by

$$\min_{x \in \{0,1\}^n} \Phi_\alpha(x), \tag{3}$$

where the objective function is defined by

$$\Phi_\alpha(x) = \frac{1}{2} \left(\|Ax - b\|^2 + \alpha \sum_i \sum_{j \in N(i)} (x_i - x_j)^2 \right), \tag{4}$$

parameter $\alpha > 0$ is the balancing parameter between projection error and smoothing term. First term in (4) measures the accordance of a solution with a projection data while a rule of the last term is to enforce the coherency of the solution.

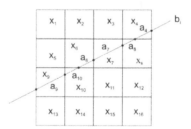

Fig. 1. The discretization model. The corresponding reconstruction problem is presented in a form of linear system of equations, see (1).

2.1 SPG Optimization Algorithm

In this section we give a short description of the SPG optimization algorithm. It is a deterministic, iterative algorithm, introduced by (Birgin et al. 2000) in [6] for solving a convex-constrained optimization problem

$$\min_{x \in \Omega} f(x),$$

where here the feasible region Ω is a closed convex set in R^n. We can roughly say that the method is a combination of a *Projected Gradient method* [4] with Grippo type *nonmonotone line search* [11] and *Spectral steplength* [3] schemes. The requirements for application of SPG algorithm are: *i*) the objective function, f is defined and has continuous partial derivatives on an open set that contains Ω; *ii*) the projection P_Ω of an arbitrary point $x \in R^n$ onto a set Ω is defined. Global convergence of this method is proved in [6]. The parameters of the algorithm are as follows. Integer $m \geq 1$ is a number of memorized previous objective function values used in line search procedure in each iteration. Especially, for $m = 1$ we got monotone line search. Parameters $0 < \alpha_{min} < \alpha_{max}$ and $0 < \sigma_1 < \sigma_2 < 1$ have safeguarding function: they keep α_k and λ_{temp} inside given limits. Parameter $\gamma \in (0,1)$ controls the non-monotone objective function decrease condition (5) [11]. Further details about parameters you can find in [7]. Starting from an arbitrary initial iteration $x^0 \in \Omega$, the below computation is iterated until convergence.

SPG iterative step [7]e

Given x^k and α_k, the values x^{k+1} and α_{k+1} are computed as follows:
$d^k = P_\Omega(x^k - \alpha_k \nabla f(x^k)) - x^k$;
$f_{max} = \max\{f(x^{k-j}) \mid 0 \leq j \leq \min\{k, m-1\}\}$;
$x^{k+1} = x^k + d^k$; $\delta = \langle \nabla f(x^k), d^k \rangle$; $\lambda_k = 1$;

%%% Line Search Procedure
while
$$f(x^{k+1}) > (f_{max} + \gamma \lambda_k \delta), \tag{5}$$
$\lambda_{temp} = -\frac{1}{2} \lambda_k^2 / (f(x^{k+1}) - f(x^k) - \lambda_k \delta)$;
if $(\lambda_{temp} \geq \sigma_1 \wedge \lambda_{temp} \leq \sigma_2 \lambda)$ **then** $\lambda_k = \lambda_{temp}$;
else $\lambda_k = \lambda/2$;
end if
$x^{k+1} = x^k + \lambda_k d^k$;
end while;
%%%
$s^k = x^{k+1} - x^k$; $y^k = \nabla f(x^{k+1}) - \nabla f(x^k)$; $\beta_k = \langle s^k, y^k \rangle$;

if $\beta_k \leq 0$ **then** $\alpha_{k+1} = \alpha_{max}$;
else $\alpha_{k+1} = \min\{\alpha_{max}, \max\{\alpha_{min}, \langle s^k, s^k \rangle \beta_k\}\}$;

end if

 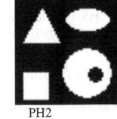 

PH1   PH2   PH3

Fig. 2. Phantom images used in our experiments. All images have the same resolution 64×64.

The SPG algorithm is particularly suited for the situations when the projection calculation is inexpensive, as in box-constrained problems, and its performance is shown to be very good in large-scale problems (see [7]).

3 Proposed Method

We transform the binary tomography problem (3) to the convex-constrained problem defined by

$$\min_{x \in [0,1]^n} \Phi_\alpha(x) + \mu \cdot x^T (e - x), \quad \mu > 0 \tag{6}$$

where $e = [1,1,1,\ldots,1]^n$. In (6) we relax the feasible set of the optimization to the convex set, $[0, 1]^n$ and add a concave regularization term $x^T(e-x)$ with aim to enforce binary solution. Parameter μ regulates the influence of this term. Due to the convex smoothness regularization (2) and the concave binary enforcing regularization the problem (6) belongs to the class of *convex-concave* regularized methods [27, 22]. Soundness of the problem (6) is ensured by the following theorem which establishes an equivalence between (3) and (6).

Theorem 1. *[10, 14] Let E be Lipschitzian on an open set $A \supset [0,1]^n$ and twice continuously differentiable on $[0,1]^n$. Then there exist a $\mu_* \in R$ such that for all $\mu > \mu_*$:*
(i) the integer (binary) programming problem

$$\min_{x \in \{0,1\}^n} E(x)$$

is equivalent with the concave minimization problem

$$\min_{x \in [0,1]^n} E(x) + \frac{1}{2}\mu \langle x, e - x \rangle,$$

(ii) the function $E(x) + \frac{1}{2}\mu \langle x, e - x \rangle$ is concave on $[0,1]^n$.

Requirements for application of the SPG algorithm for solving the problem (6) are satisfied. Indeed, it is obvious that the objective function is differentiable and the projection onto a feasible set P_r is given by

$$[P_r(x)]_i = \begin{cases} 0, & x_i \leq 1 \\ 1, & x_i \geq 1 \\ x_i & \text{elsewhere} \end{cases}, \quad \text{where } i = 1,\ldots,n. \tag{7}$$

where $x \in R^n$. P_r is a projection with respect to the Euclidean distance and its calculation is inexpensive. Therefore the SPG algorithm is suitable choice for solving (6) for every fixed $\mu > 0$.

Our strategy is to solve a sequence of optimization problems (6), with gradually increasing μ, which will lead to a solution of the binary solution. More precisely, we suggest the following optimization algorithm.

SPG Algorithm for Binary Tomography

Parameters: $\varepsilon_{in} > 0$; $\varepsilon_{out} > 0$; $\delta > 1$; μ_0; *maxit.*
$x^0 = [0.5, 0.5, \ldots, 0.5]^T$; $\mu = \mu_0$; $k = 0$;
do
    do
            x^{k+1} from x^k by SPG iterative step; $k = k + 1$;
      until $\left\| P_\Omega(x^k - \nabla\Phi(x^k)) - x^k \right\| > \varepsilon_{in}$ and $k < maxit$
    $\mu = \delta * \mu$;
  until $\max_i \{ \min\{x_i^k, 1 - x_i^k\} \} > \varepsilon_{out}$.

The initial configuration is the image with all pixel values equally to 0.5. In each iteration in the outer loop we solve an optimization problem (6) for a fixed binary factor $\mu > 0$ by using the SPG method. By iteratively increasing the value of μ in the outer loop the binary solutions are enforced. The termination criterion for the outer loop, ε_{out}, regulates the tolerance for the finally accepted (almost) binary solution.

It is easy to show that the function Φ_α is quadratic and convex, see for example [22]. However, by increasing the μ factor during the optimization process the influence of the concave regularization term becomes larger which leads to the non-convex objective function. Therefore, we cannot guaranty that this approach always end up in a global minimum. However, experimental results in Section 4 show its very good performance.

An important advantage of this method is its flexibility regarding the inclusion of other regularization terms (priors), for example, inter alia, the *Total Variation* [21] smoothness prior [26]. The only requirement is the differentiability of the used term. However, in this paper we are focused on the reconstruction problem defined by (3).

4   Evaluation

In this section we evaluate the performance of the proposed method. We compare its performance with the performance of SA. SA algorithm in [24] is compared with a powerful D.C. based reconstruction algorithm. The main conclusion of this Benchmark evaluation was that "there is no huge difference between the qualities of the reconstructed images of the two methods". This fact gives more validity for our evaluation.

We performed experiments on the binary test images (phantoms) presented in Figure 2. Reconstruction problems are composed by taking parallel projections from different directions. We take 64 parallel projections for each direction. Regarding to direction we distinguish reconstructions with 2, 3, 5 and 6 projections. For 2, 3 and 5 projections directions are uniformly chosen within [0°,90°] and for 6 projections within [0°,150°].

The quality of reconstruction (solution) is measured by the following two error measure functions

$$E_1(x^r)=\|Ax^r- b\|,$$
$$E_2(x^r)=\|x^r- x^*\|_1,$$

where x^r is the reconstructed image. Function E_1 measures the accordance with the projection data, while E_2 gives the number of failed pixels in compare with the original image x^*.

SA is a stochastic optimization algorithm based on the simulation of physical process of slow cooling of the material in a heat bath. Based on the ideas from a paper published by (Metropolis et al. 1953) [19] the SA algorithm is introduced by (Kirkpatrick et al.1983) [15]. In our experiments we use the following SA algorithm adapted for BT problem by (Weber et al. 2006) in [24].

SA Algorithm

Parameters:
$\alpha > 0$; $T_{start} > 0$; $T_{min} > 0$; $T_{factor} \in (0,1)$; $S > 0$.
Initial variable setting:
$x = [0, 0,..., 0]^T$; $T = T_{start}$; $S_{nr} = 0$; $E_{old} = \Phi_\alpha(x^0)$.
while ($T \geq T_{min}$) $\wedge$ ($S_{nr} < S$)
   for $i = 1$ to sizeof(x),
        choose a random position j in the vector x;
        $\tilde{x} = x$;
        $\tilde{x}[j] = 1 - x[j]$;
        $E_{new} = \Phi_\alpha(\tilde{x})$;
        $z = rand()$;
        $\Delta E = E_{new} - E_{old}$;
        if $\Delta E < 0 \vee \exp(-\Delta E/T) > z$, **then**
        $x = \tilde{x}$ {accept changes}
        $E_{old} = E_{new}$; $S_{nr} = 0$;
        end if
   end for
   $T = T * T_{factor}$; $S_{nr} = S_{nr} + N$;
end while.

For SA algorithm we use the following parameter settings: $\alpha = 5$, $T_{start} = 4$, $T_{min} = 10^{-14}$, $T_{factor} = 0.97$, $S = 10*$sizeof(x). Parameter settings for SPG based algorithm are following: $m = 10$, $\gamma = 10^{-4}$, $\sigma_1 = 0.1$, $\sigma_2 = 0.9$, $\theta_{min} = 10^{-3}$, $\theta_{max} = 10^3$, $\alpha = 3$, $\delta = 1.2$, $E_{in} = 0.1$, $E_{out} = 0.001$, $maxit = 100$.

Regarding to elapsed CPU time the performance of SPG is significantly better than the performance of SA. Due to the non-deterministic nature of SA the running time, but also the obtained solution, can vary in different runs. Therefore we decide to omit reporting about CPU time. Visual look of the reconstructed images from 2, 5 and 6 projections are presented in Figure 3. The results are similar or exactly the same. Error measure values of the reconstructions are reported in Table 1. Out of the 12 tests, SPG has better reconstruction quality in 5 cases, while in 4 cases its quality is exactly

Table 1. The measured values E_1 and E_2 of the reconstructed images

Proj.	Alg.	PH1		PH2		PH3	
		$E_1(x')$	$E_2(x')$	$E_1(x')$	$E_2(x')$	$E_1(x')$	$E_2(x')$
2	SA	7.874	806	6.782	950	7.615	1177
	SPG	6.324	849	5.831	499	11.225	1432
3	SA	2.554	3	9.587	580	11.959	858
	SPG	2.554	3	3.182	3	11.769	1198
5	SA	0	0	4.030	3	17.015	589
	SPG	0	0	3.070	2	12.321	538
6	SA	0	0	3.444	2	0	0
	SPG	0	0	2.205	1	0	0

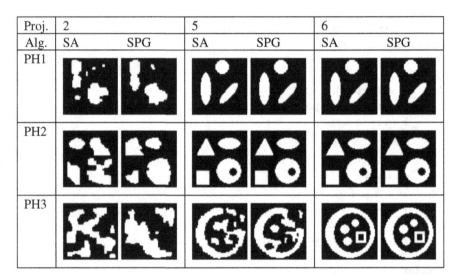

Fig. 3. Reconstruction of the phantom images presented in Figure 2. They are obtained from 2, 5 and 6 projections.

the same as quality of the reconstructions obtained by SA algorithm. Limited experiments on noisy data were performed in our recent work [17]. The obtained reconstructions by SPG and SA are very similar.

5 Concluding Remarks

We have successfully developed a deterministic optimization method based on the SPG optimization for binary tomography reconstruction problem. Performance evaluation based on the comparison with often used SA method shows its competence regarding to the quality of reconstructions. Flexibility of the proposed method allows to use different regularization (priors) in reconstruction which gives to this method more freedom for further application.

Regarding to further work we suggest adaptation of Spectral Conjugate Gradient method for BT reconstruction. Our recent studies have confirmed its excellent performance for similar, denoising application [16].

Acknowledgments. Tibor Lukić acknowledges to the Hungarian Academy of Sciences for support through the Project ON2009B00056E of HTMT.

References

[1] Balázs, P.: Discrete Tomography Reconstruction of Binary Images with Disjoint Components using Shape Information. Int. J. of Shape Modeling 14, 189–207 (2008)

[2] Balázs, P., Gara, M.: An Evolutionary Approach for Object-based Image Reconstruction using Learnt Priors. In: Salberg, A.-B., Hardeberg, J.Y., Jenssen, R. (eds.) SCIA 2009. LNCS, vol. 5575, pp. 520–529. Springer, Heidelberg (2009)

[3] Barzilai, J., Borwein, J.M.: Two Point Step Size Gradient Methods. IMA Journal of Numerical Analysis 8, 141–148 (1988)

[4] Bertsekas, D.P.: On the Goldstein-Levitin-Polyak Gradient Projection Method. IEEE Transactions on Automatic Control 21, 174–184 (1976)

[5] Birgin, E., Martínez, J.: A Box-Constrained Optimization Algorithm with Negative Curvature Directions and Spectral Projected Gradients. Computing 15, 49–60 (2001)

[6] Birgin, E., Martínez, J., Raydan, M.: Nonmonotone Spectral Projected Gradient Methods on Convex Sets. SIAM J. on Optimization 10, 1196–1211 (2000)

[7] Birgin, E., Martínez, J., Raydan, M.: Algorithm: 813: Spg – Software for Convex-Constrained Optimization. ACM Transactions on Mathematical Software 27, 340–349 (2001)

[8] Birgin, E., Martínez, J., Raydan, M.: Inexact Spectral Projected Gradient Methods on Convex Sets. IMA Journal of Numerical Analysis 23, 539–559 (2003)

[9] Birgin, E., Martínez, J., Raydan, M.: Spectral Projected Gradient Methods. Encyclopedia of Optimization, 3652–3659 (2009)

[10] Giannessi, F., Niccolucci, F.: Connections between Nonlinear and Integer Programming Problems. Symposia Mathematica, 161–176 (1976)

[11] Grippo, L., Lampariello, F., Lucidi, S.: A Nonmonotone Line Search Technique for Newton's Method. SIAM J. Numer. Anal. 23, 707–716 (1986)

[12] Herman, G.T., Kuba, A.: Discrete Tomography: Foundations, Algorithms and Applications. Birkhäuser, Basel (1999)

[13] Herman, G.T., Kuba, A.: Advances in Discrete Tomography and Its Applications. Birkhäuser, Basel (2006)

[14] Horst, R., Tuy, H.: Global Optimization: Deterministic Approaches, 3rd edn. Springer, Berlin (1996)

[15] Kirkpatrick, S., Gelatt, C., Vecchi, M.: Optimization by Simulated Annealing. Science 220(4598), 671–681 (1983)

[16] Lukić, T., Lindblad, J., Sladoje, N.: Image Regularization based on Spectral Conjugate Gradient Optimization (submitted for publication)

[17] Lukić, T., Lukity, A.: Binary Tomography Reconstruction Algorithm based on the Spectral Projected Gradient Optimization. In: Proc. of the 10th International Symposium of Hungarian Researches on Computational Intelligence and Informatics, Budapest, pp. 253–263 (2009)

[18] Lukić, T., Sladoje, N., Lindblad, J.: Deterministic Defuzzification Based on Spectral Projected Gradient Optimization. In: Rigoll, G. (ed.) DAGM 2008. LNCS, vol. 5096, pp. 476–485. Springer, Heidelberg (2008)

[19] Metropolis, N., Rosenbluth, A.W., Rosenbluth, M.N., Teller, A.H., Teller, E.: Equation of State Calculations by Fast Computing Machines. Journal of Chemical Physics 21(6), 1087–1092 (1953)

[20] Pham Dinh, T., Elbeirnoussi, S.: Duality in d.c (difference of convex functions) Optimization, Subgradient Methods. Trends in Math. Opt. 84, 276–294 (1988)

[21] Rudin, L., Osher, S., Fatemi, E.: Nonlinear Total Variation-based Noise Removal Algorithms. Physica D 60, 259–268 (1992)

[22] Schüle, T., Schnörr, C., Weber, S., Hornegger, J.: Discrete Tomography by Convex-Concave Regularization and D. C. Programming. Discrete Appl. Math. 151, 229–243 (2005)

[23] Schüle, T., Weber, S., Schnörr, C.: Adaptive Reconstruction of Discrete-valued Objects from Few Projections. In: Proceedings of the Workshop on Discrete Tomography and its Applications. Electronic Notes in Discrete Mathematics, vol. 20, pp. 365–384. Elsevier, Amsterdam (2005)

[24] Weber, S., Nagy, A., Schüle, T., Schnörr, C., Kuba, A.: A Benchmark Evaluation of Large-Scale Optimization Approaches to Binary Tomography. In: Kuba, A., Nyúl, L.G., Palágyi, K. (eds.) DGCI 2006. LNCS, vol. 4245, pp. 146–156. Springer, Heidelberg (2006)

[25] Weber, S., Schnörr, C., Schüle, T., Hornegger, J.: Binary Tomography by Iterating Linear Programs from Noisy Projections. In: Klette, R., Žunić, J. (eds.) IWCIA 2004. LNCS, vol. 3322, pp. 38–51. Springer, Heidelberg (2004)

[26] Weber, S., Schüle, T., Kuba, A., Schnörr, C.: Binary Tomography with Deblurring. In: Reulke, R., Eckardt, U., Flach, B., Knauer, U., Polthier, K. (eds.) IWCIA 2006. LNCS, vol. 4040, pp. 375–388. Springer, Heidelberg (2006)

[27] Weber, S., Schüle, T., Schnörr, C.: Prior Learning and Convex-Concave Regularization of Binary Tomography. In: Proceedings of the Workshop on Discrete Tomography and its Applications. Electronic Notes in Discrete Mathematics, vol. 20, pp. 313–327. Elsevier, Amsterdam (2005)

Product Definition Using a New Knowledge Representation Method

László Horváth and Imre J. Rudas

Institute of Intelligent Engineering Systems, John von Neumann Faculty of Informatics,
Óbuda University, Bécsi út 96/b, H-1034 Budapest, Hungary
horvath.laszlo@nik.uni-obuda.hu, rudas@uni-obuda.hu

Abstract. A new style of engineering was introduced in leading industries during the past decade. Based on the achievements in integrated product modeling, a paradigm was developed. This new paradigm is management of product data for lifecycle where fully integrated product description is supported by fully integrated model creation functionality and serves all product related activities. This paper shows a redefined application of knowledge at description the background of decisions for more efficient application of product model at its development and evaluation. The recently prevailing product model is characterized as classical one. Influences on engineering object definition were analyzed in case of this classical product modeling. The authors of this paper proposed to replace the current direct application of knowledge by an indirect method where knowledge is mapped to the contextual chain of human intent, engineering objective, contextual connection, and status driven decision. This contextual chain is called as information content. In this way, knowledge controls engineering objects indirectly through information content. The paper discusses background of human decision in product model, origin and place of information content, human-modeling procedure communication, and information content extension to classical product model.

1 Introduction

Engineering has changed to an information technology intensive style. Routine activities such as development of engineering objects using engineer defined parameters, placing variant information in engineering objects, retrieving existing engineering objectives as a basis of new variants or types, have been taken by modeling procedures. Engineer can really concentrate on decisions. The most frequent task is contribution to earlier established solutions for engineering objects. Importance of knowledge about earlier decisions for entities in existing models was grove by wide application of the new style of engineering.

Knowledge for background of a decision is critical prerequisite for effective assistance of understanding, evaluation, change, and for the connection of new decisions to product model. When an engineer demands results about development of earlier products, the background must be extended beyond the product under development.

I.J. Rudas et al. (Eds.): Computational Intelligence in Engineering, SCI 313, pp. 273–285.
springerlink.com

The authors proposed a new product model as an extension to currently applied product model. Current product modeling reached a high level in the description of parameters and parameter relationships for engineering objects. In this context, engineering object is any object that is demanded to define by any engineering activity during lifecycle of a product. Authors characterized current product models as classical ones for which they analyzed human influences in case of engineering object definition. One of the most problematic features of this classical product model that knowledge is mapped directly to engineering objects. The authors of this paper proposed to replace this direct application of knowledge by an indirect method where knowledge is mapped to contextual chain of human intent, engineering objective, contextual connection, and status driven decision. This contextual chain is called as information content. In this way, knowledge controls engineering objects indirectly through information content. The paper discusses background of human decision in product model, origin and place of information content, human-modeling procedure communication, and information content extension to classical product model.

2 Influences on Engineering Object Definition

The current way of product modeling handles influences by the control of access to modeling procedures and models. However it does not handle connection of various influences to engineering objects.

Classical applications of knowledge are based on calculation or inference by using of well-defined rules, etc. In modeling for engineering activities at development and application of products, influence by different humans on the same decision should be considered (Fig. 1). Direct acting humans are authorized and responsible for well-defined engineering tasks. They work in groups within extended companies where

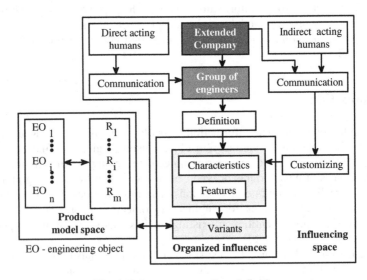

Fig. 1. Influences on product definition

extension changes according to actual task and may also be different at different parallel tasks. The authors proposed handling of organized influence of engineers within a group working on an actual project. They defined influences for product and its environment. However, these definitions do not act directly on engineering objects; they are organized for the definition of characteristics, features, and variants of product. Indirect acting humans are also considered mainly to customize the product. These influences act through instructions, specifications, standards, and customer demands. Influences control engineering objects (EO) and their relationships (R) in a product model space.

3 Direct and Indirect Control by Knowledge

The next step towards modeling of decision background is definition of knowledge in product model. The knowledge that is appropriate for this purpose must be both easy to define and understand by engineers in their every day work. This is why recent industrial practice prefers rules, checks, controls, and constraints (Fig. 2/a). Rules and checks can be defined in sets while constraints can be analyzed for consistence. Rules

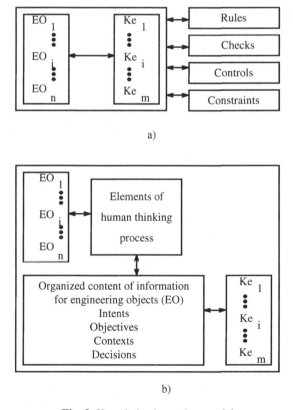

a)

b)

Fig. 2. Knowledge in product model

assist engineer activities for the definition of object parameters. Checks contain restrictions for the work of engineers and will be active when something alters from actual threshold knowledge or earlier decided results. Controls can be used when a given result needs a definite measure.

In the case of classical product model, direct mapping of knowledge entities (Ke) to engineering objects (EO) records means of the creation applied at the definition of that object. However, origin and purpose of knowledge are unknown. Engineer who apply the model can track only the history of engineering object definition instead of tracking of intents of the influencing humans and fulfilling of engineering objectives. At the same time, it is impossible to track the status of decisions from the proposal to the accepted. Moreover, after an engineer proposes a decision, any possible consequence must be analyzed. This is impossible in case of direct knowledge mapping.

The authors of this paper proposed an indirect application of knowledge in order to establish a possible solution for the above problem (Fig. 2/b). Their method uses contextual chain of human intent, engineering objective, contextual connection, and status driven decision definitions. Knowledge is mapped to these entities called as information content. Knowledge controls engineering objects indirectly through information content.

4 Background of Human Decision in Product Model

Because experience, expertise, and intelligence should be involved in product model as human resource in order to better assistance of product model application, new ways are investigated for their representation as knowledge. Authors of [1] propose method for handling personal and organizational nature of knowledge. In [2], personalization and codification are shown in the development of a multidisciplinary framework. The authors of [3] propose tools and models for knowledge capitalization. An approach to definition and mapping of knowledge, based on the point of view of an expert in manufacturing is discussed. Most of traditional methods assume that knowledge comes from a single expert. In [4], multiple expert sources are proposed. This approach is considered as a more feasible alternative. Research that is reported in [5], introduces the product lifecycle aspect and the feature principle in a new distributed database and knowledge base modeling approach for concurrent design. The knowledge for product modeling is described by collections of rules in rule-bases. Databases and knowledge bases are placed at different locations for modeling according to different life-cycle aspects of the same product. In paper [6], interfacing knowledge oriented tools and computer aided design application is identified as a technical gap for intelligent product development. The authors of [6] consider definition of associative features in the form of self-contained and well-defined design objects as essential for high-level reasoning and execution of decisions.

Authorized and responsible engineers make decision on engineering object. In current product modeling, final result of human thinking process is recorded in the form of directly defined parameters of engineering objects or knowledge for the calculation of those parameters. The problem is that application of product model would require at least the critical elements of this thinking process at evaluation for criticism or additional contribution. The authors of this paper analyzed human thinking process for engineering object definition and concluded that key thinking process elements and partial decision points would be most appropriate to record human intent. It must be emphasized that in the current stage of the reported research not a deep analysis of human thinking process is aimed. Whereas a minimum background information is extracted and recorded for each decision as an extension to currently prevailing product models.

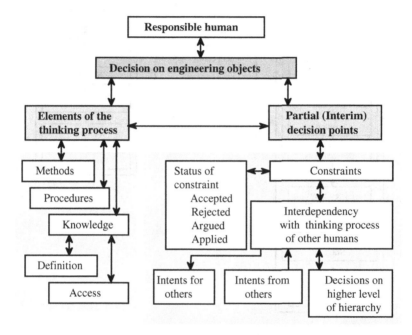

Fig. 3. Human thinking for the definition of engineering objects

The concept by the authors can be outlined in Fig. 3. Methods, procedures, knowledge entities are considered as elements of thinking process for the definition of engineering object parameters. Partial decision points can be characterized by constraints. Status of a constraint and its interdependence with thinking process of other influencing humans are main characteristics of partial decision points. Interdependency also includes intent for other influencing humans and decisions on higher levels of hierarchy. Sequence of elements in a human thinking process and partial decision points represent logical routes to the solution for parameters of engineering objects including all of the relationships.

Elements in a human thinking process and partial decision points are explained on Fig. 4 by using of representative examples. In the course of human thinking process, engineer observes related physical and logical phenomena, perceives relevant real world information, and retrieves product model and other stored information and information content from the computer system. Engineer also retrieves related information and information content stored in mind as knowledge and fact-like experience. This is the stage A of the thinking process on an engineering object related decision.

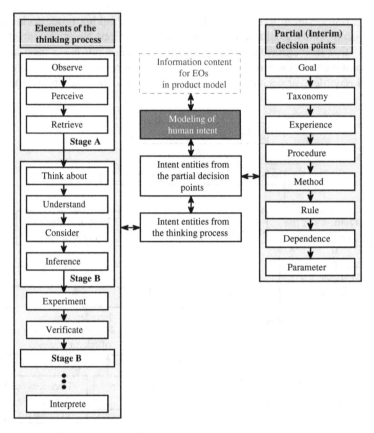

Fig. 4. Examples for thinking process elements to record

Having all initial information, human understands the problem, thinks about it, considers the circumstances, and inferences. This is the stage B of the thinking process on an engineering object related decision. In the meantime, human produces results at interim decision points. Human conducts experiments, then evaluates and verifies results then interprets results or goes back to the stage B. Any result is mapped to the actual engineering objects as intent. Structure of elements of thinking process may be represented in tree or network according to the task. One of the knowledge or fact carriers accepted in the product development and management organization is produced as result at a partial or interim decision point. Intent entities are involved in model of the human intent. Partial decision points are defined by the human at points where an interim result is considered to record for later use. Sometimes interim decision points depend on elements of the thinking process, sometimes not. Fig. 4 shows a typical example. A goal is produced, and then concepts are placed in taxonomy. Following this, an experience is thought then recorded. Engineer conceptualizes a procedure and a method to be applied in it. Finally, partial results are recorded for rule, dependence, and parameters. Partial decision point information also may be represented in tree or network according to the task.

5 Origin and Place of Information Content

The authors stated above that product information has been extended to its content in the extended classical product model they proposed. Information content is a set of new features of information in order to define its content. The information based nature of classical product model was developed on the basis of classification by Russell Ackoff by the authors. Ackoff classified content of the human mind into five categories [7]. These categories are data, information, knowledge, understanding, and wisdom. Ackoff characterizes data, as that does not have meaning in itself. In product model space, data for existence of an engineering object and collection of data for its description are available. According to Ackoff, information is data that has been given meaning by the way of relational connections. Information provides answers to questions who, what, where, and when. Data becomes information where it is connected to its environment in the product model. However, according to its above definition, information is no more than connected data. Attributes are connected within an engineering object, among different engineering objects, and between the product model space and the world outside of it. Information is suitable to assure consistency of product model.

The term classical product model (CPM) was introduced by the authors for the product model space that includes information representations of engineering objects for a product. The classical attribute refers to an age of product modeling achievements of that has become classical. It is also a tribute to people involved in significant research and development in this area. CPM is theoretically and methodologically complete and well-established system. It is appropriate for accommodation of information structures that are produced by classical product model generation procedures under control of information content based modeling procedures.

Fig. 5 illustrates extension to information based product model by information content based product model where engineer control is repositioned. Classical product modeling can handle concurrency of engineering activities. However, it can not handle different influences of the same decision. An influencing space is necessary to establish where all influences can be mapped to engineering object parameters influenced (upper part of Fig. 5). The next step is repositioning of influencing space from the direct connection with engineering object parameters to the indirect control through information content (lower part of Fig. 5)

As it was stated above, background of a decision should be represented in the product model because engineer who receives that decision can not find answer for several questions in the classical product model. The supposed questions are involved in Fig. 6 together with answers and the modeled information content. Question about origin of the decision on engineering objects can be easily answered, because it is required by influencing humans (IHs). Intent of IHs should be recorded together with their agreed hierarchy. Definition of intents includes new or modified concepts. Meaning of these concepts is the next element of the modeled information content. Engineering objects should match with engineering objectives. These objectives are specified directly by IHs or they come from human intent definitions. An engineering object is defined in the knowledge of information about other engineering objects that are in some relationship with it. For this purpose, contextual and non-contextual

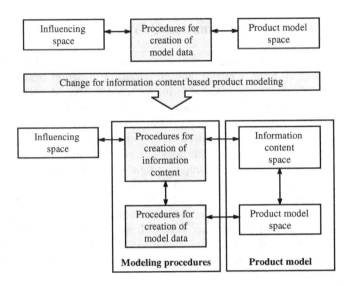

Fig. 5. Repositioned human control in extended classical product modeling

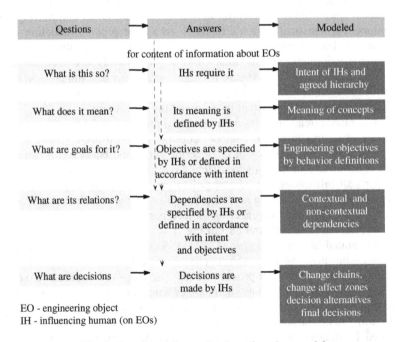

Fig. 6. Question at the application of product model

dependencies are to be modeled. Utmost purpose of information content is supporting decision making on engineering objects. Some content is necessary to know about decisions as the basis of control of engineering objects in the product model space. A decision establishes or changes engineering objects. Consequences of these changes are change of other engineering objects that are in direct or indirect dependence connection with the originally changed engineering objects.

6 Human-Modeling Procedure Communication

Information based communication between two cooperating humans in a product development in Fig. 7 shows that indirect communication through product model is not enough because product model is not capable of transfer content behind model information. An auxiliary "bypass" or out of model communication is necessary. This communication is done by using of one of the conventional methods and its result can not be used in model and modeling procedures. Human is forced to convert content of mind into information. Human defines information in order to communicate it with information based or classical product modeling procedures. A choice of model entities is available and high number of purposeful procedures is applied to generate model information so that consistency of model highly depends on human attention. In this way, human controls procedures at definition of characteristics of engineering objects. Human who receive product model, must understand earlier object definitions by using of information based product model and auxiliary or outside of model communication. Human "B" perceives information for own mind and converts it into content of own mind. This communication between humans is two directional and there may be also other participants.

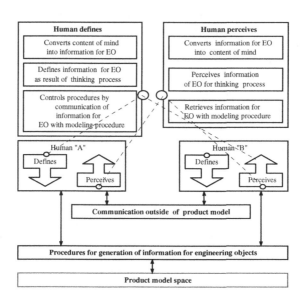

Fig. 7. Communication between human and classical product modeling

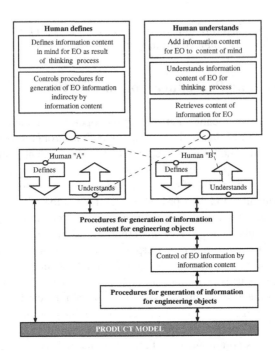

Fig. 8. Communication between human and information content handling

Fig. 8 explains the above communication in case of information content based modeling. Human "A" defines content of mind for definition of engineering object then communicates it with information content based modeling procedures. Content controls procedures for generation of engineering object information indirectly. Human "B" retrieves information content for engineering object on which decides. Function of the auxiliary "bypass" out of model communication is taken by information content based product model. Definition process for engineering object must record information content enough for application. If not enough, feedback may be generated in order to notice at the original decision maker. The process works also in the opposite direction as it is illustrated in Fig. 8.

7 Extension to Classical Product Model

The authors developed a method for the information content based extension to the classical product model. This work was based on their earlier results. They analyzed modeling methods and models in order to reveal functions, relationships, and human-computer interactivity in current product modeling practice [8]. Modeling of human intent was researched in close connection with grouping of model entities according to associative connections in environment adaptive product model objects [9]. Additional issue for information content based modeling is modeling of products behaviors. An organized description of engineering object behavior definitions is proposed in [10]. Management of product model changes [11] and its human intent and knowledge management aspects [12] were also important research issues.

In the model proposed by the authors information content is organized in a multi-level structure where levels are placed in a hierarchical sequence. Development of information content model can be followed in Fig. 9 where definition, essential elements, and application of information content are shown. Influencing humans define intent on the basis content of their mind. Intent related content elements among others are statements, intent attributes, humans, relationships, and citations. Intents are organized in intents space where intents are mapped to engineering objects. Intent space is the space of highest hierarchy in the information content model.

Meaning of a concept as information content is corporate or personal owned and it may include definitions, attributes, application, taxonomy, relationships, and citations. Humans may define the same concept in different manner.

Engineering objectives are represented as behaviors by using of situations and circumstances for situations. Behaviors are organized in behavior space where intents are mapped to engineering objects. Behavior space is the second highest hierarchy in the information content model.

Engineering is inherently contextual. Any engineering object is to be defined in the context of other engineering objects. Only existing engineering objects can be applied for the definition of engineering objects. At the same time, any engineering objective is contextual with appropriate intents and context is contextual with engineering objectives.

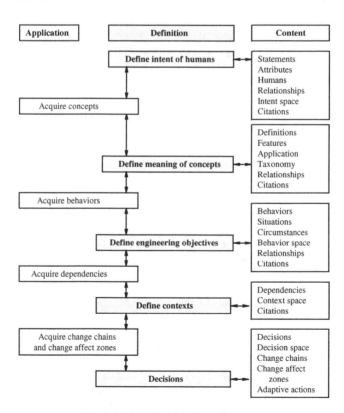

Fig. 9. Definition of information content

Decisions are represented as adaptive actions. An adaptive action controls relevant information based model creating procedures and takes control from human. Decisions are supported by tools for tracking of engineering object change consequences. These tools are change affect zone within which consequence changed objects are found, and change chains along which consequences are tracked.

8 Conclusions

The paper emphasizes some essential problems at current product modeling. The currently prevailing modeling has not capability for the representation of background of decisions. Instead, engineers record results of their thinking processes in product model in the form of object parameters and relationships of these parameters. Moreover, knowledge definitions serving definitions for engineering object relationships are mapped to engineering object directly. Some more information is necessary about background of decisions on engineering objects in order to assist application of product model. The authors proposed one of the possible solutions in the form of indirect knowledge application. In this case, knowledge is attached to a contextual chain of human intent, engineering objective, contextual connection, and decision for engineering object parameter. An indirect knowledge definition is realized by the application of the above contextual chain called as information content at the control of engineering object definition in classical product modeling. Human can record the required amount of decision background as information content.

Acknowledgments. The authors gratefully acknowledge the grant provided by the OTKA Fund for Research of the Hungarian Government. Project number is NKTH-OTKA K 68029.

References

[1] Kitamura, Y., Kashiwase, M., Fuse, M., Mizoguchi, R.: Deployment of an Ontological Framework of Functional Design Knowledge. Advanced Engineering Informatics 18(2), 115–127 (2004)

[2] McMahon, C., Lowe, A., Culley, S.: Knowledge Management in Engineering Design: Personalization and Codification. Journal of Engineering Design 15(4), 307–325 (2004)

[3] Renaud, J.: Improvement of the Design Process through Knowledge Capitalization: an Approach by Know-how Mapping. Concurrent Engineering 12, 25–37 (2004)

[4] Richardson, M., Domingos, P.: Learning with Knowledge from Multiple Experts. In: proc. of the Twentieth International Conference on Machine Learning, Washington, DC, pp. 624–631. Morgan Kaufmann, San Francisco (2003)

[5] Zhang, F., Xue, D.: Distributed Database and Knowledge Base Modeling for Concurrent Design. Computer-Aided Design 34(1), 27–40 (2002)

[6] Ma, Y.-S., Tong, T.: Associative Feature Modeling for Concurrent Engineering Integration in Computers in Industry, vol. 51(1), pp. 51–71 (2003)

[7] Ackoff, R.L.: From Data to Wisdom. Journal of Applied Systems Analysis 16(1), 3–9 (1989)

[8] Horváth, L., Rudas, I.J.: Modeling and Problem Solving Methods for Engineers, p. 330. Elsevier, Academic Press (2004)

[9] Horváth, L., Rudas, I.J.: Human Intent Description in Environment Adaptive Product Model Objects. Journal of Advanced Computational Intelligene and Intelligent Informatics 9(4), 415–422 (2005)

[10] Horváth, L.: Supporting Lifecycle Management of Product Data by Organized Descriptions and Behavior Definitions of Engineering Objects. Journal of Advanced Computational Intelligence and Intelligent Informatics 11(9), 1–7 (2007)

[11] Horváth, L., Rudas, I.J.: An Approach to Processing Product Changes During Product Model Based Engineering. In: Proc. of the 2007 IEEE International Conference on System of Systems Engineering, San Antonio, Texas, USA, pp. 104–109 (2007)

[12] Rudas, I.J., Horváth, L.: Emphases on Human Intent and Knowledge in Management of Changes at Modeling of Products. WSEAS Transactions on Information Science and Applications 3(9), 1731–1738 (2006)

Incremental Encoder in Electrical Drives: Modeling and Simulation

Ioan Iov Incze, Csaba Szabó, and Mária Imecs

Technical University of Cluj-Napoca
15 C. Daicoviciu Str., 400020 Cluj-Napoca
PO 1, Box 159, RO-400750 Cluj-Napoca, Romania
ioan.incze@edr.utcluj.ro, csaba.szabo@edr.utcluj.ro,
imecs@edr.utcluj.ro

Abstract. Incremental encoders are electro-mechanical devices used as position sensors in electrical drives. They provide electrical pulses when their shaft is rotating. The number of pulses is proportional to the angular position of the shaft. The paper focuses on the modeling and simulation of an incremental encoder, and associated units serving for direction identification and position computing. Matlab-Simulink® structure was realized and tested. The proposed structure identifies the direction of the rotation in an angular interval equal to a quarter of the angular step of encoder graduation. The incremental encoder was integrated into the simulation structure of an induction motor based drive control system in order to provide the position of the motor shaft. Experimental results are also presented.

1 Introduction

One of the most frequently used position transducer is the incremental encoder. It is a device which provides electrical pulses if its shaft is rotating [1, 4, 5, 7]. The number of generated pulses is proportional to the angular position of the shaft. Incremental encoders are realized using different technologies: optical, electromechanical, magnetic etc. Optical encoders are the most wide-spread ones.

On the shaft of the optical encoder there is fixed a transparent (usually glass) rotor disc with a circular graduation-track realized as a periodic sequence of transparent and non-transparent radial zones, as is shown on Fig. 1.

On one side of the disk – mounted on the fixed part of the encoder (stator) –there are light sources. The emitted light beam traverses the graduation zone of the disk and then it is sensed by a group of optical sensors. If the shaft is rotating, the graduation modulates the light beam. Electronic circuits process the light sensors outputs and generate electrical pulses at the output of the encoder. The period of the pulses corresponds to a rotation angle of the shaft equal to the angular step of graduation θ_p, i.e. the angle corresponding to one consecutive transparent and non-transparent zone. The number of pulses (counted by external electronic counters) is proportional to the angular position of the shaft [3, 5]. In order to provide information regarding the direction of the rotation, there is a second light-beam (with associated sensors and processing circuits) placed with an angular shift equal to a quarter angular step $\theta_p/4$ of

I.J. Rudas et al. (Eds.): Computational Intelligence in Engineering, SCI 313, pp. 287–300.
springerlink.com

graduation compared to first beam, which is generating pulses at a second output of the encoder. The two output pulses (usually named *A* and *B*) have the same period; their mutual phase shift provides the information regarding the direction of the rotation. For positive sense, i.e. counter-clockwise (CCW) rotation, the *B* signal is in phase-lag compared to *A*; for negative one – clockwise (CW) rotation – *A* is lagging behind to *B* [7].

In order to have a reference angular position of the shaft, the disc of the encoder is provided with a second track and the corresponding light-source, light-sensor, processing circuits and a third output (named usually *Z* or "marker"). This second track has a single graduation, therefore produces a single pulse (of width equal to a quarter angular step of graduation $\theta_p/4$) in the course of a complete (2π rad) rotation of the shaft. The shaft position corresponding to this marker pulse is considered as the reference position.

The diagram of the generated output signals versus rotation angle for CCW and CW rotation sense is presented in Fig. 2.

The control algorithm of the modern adjustable-speed electrical drives needs the knowledge of position and/or speed of the rotor [1, 4]. This information in the case of sensored drives is provided by position or speed transducers. The use of the incremental encoders on this purpose is very popular.

Nowadays the modern drive systems are controlled digitally. Usually in these systems the processing is accomplished at a fixed sampling rate. However, the position information (and also the speed information computed using encoder signals) furnished by incremental encoder is inherently digital, with a rotation-speed-dependent sampling rate. Accordingly, the whole drive system may be considered as a complex system composed of some parts working with fixed sampling rate, other parts with a

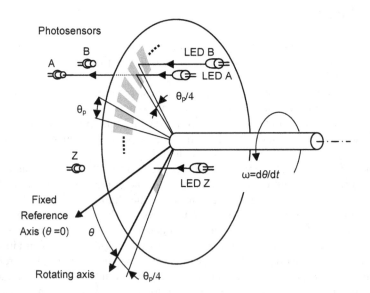

Fig. 1. Construction principle of the incremental encoder: the gray surfaces are optically transparent

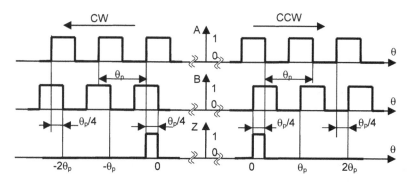

Fig. 2. Diagram of the output signals generated by the incremental encoder for counter-clockwise (CCW) and clockwise (CW) rotation

variable one. Frequently, the sampled character of the position or speed information provided by the encoder is neglected and they are treated as analogue quantities.

2 Modeling of the Incremental Encoder and Associated Units

In the course of research and early developments of the drive systems extensive simulations are performed. In these stages it is important to take already into account the real, sampled character of the information provided by the encoder. Therefore, a mathematical model and the corresponding simulation structure of the encoder and its associated units for identification of the direction and position computation were created [1, 3, 6].

2.1 Encoder Modeling

The input signal of the incremental encoder is the angular position θ of its shaft with respect to a fixed reference axis. The output signals are the two pulses shifted by a quarter angular step $A(\theta)$ and $B(\theta)$, respectively the marker signal $Z(\theta)$. If θ_p is the angular step of the encoder, the outputs may be described by the following equations:

$$A(\theta) = \begin{cases} 1 & \text{if} \quad 0 \leq (\theta \bmod \theta_p) \leq \theta_p/2; \\ 0 & \text{if} \quad \theta_p/2 < (\theta \bmod \theta_p) \leq \theta_p; \end{cases}$$

$$B(\theta) = \begin{cases} 1 & \text{if} \quad 0 \leq ((\theta - \theta_p/4) \bmod \theta_p) \leq \theta_p/2; \\ 0 & \text{if} \quad \theta_p/2 < ((\theta - \theta_p/4) \bmod \theta_p) \leq \theta_p; \end{cases} \quad (1)$$

$$Z(\theta) = \begin{cases} 1 & \text{if} \quad \theta \bmod (2\pi) = 0; \\ 0 & \text{if} \quad \theta \bmod (2\pi) \neq 0. \end{cases}$$

It is important to note, that during a rotation angle of the shaft equal to the angular step of graduation θ_p there are four switching events in the shape of the output pulses. Therefore, the minimal rotation-angle-increment, detectable by the encoder, is $\theta_p/4$. It is obvious that the number of pulses during a rotation, generated by the encoder, will be

$$N_r = \frac{2\pi}{\theta_p},\tag{2}$$

which is equal to the number of angular steps of the graduation on the circular track on the rotor. The number of pulses per rotation together with the angular speed of the shaft will determine the frequency of the encoder output signals:

$$f_A = f_B = \frac{\omega}{2\pi} N_r.\tag{3}$$

From point of view of the simulation the above equation is important because it offers a starting point in choosing of the simulation step. For usual values of N_r and ω one concludes that the encoder signal is far from the highest frequency quantity in a usual drive system. A numerical example underlines this fact: the $N_r = 1000$ pulses/rotation value is very common for low-cost incremental encoders, angular speed $\omega = 314$ rad/s is typical for the wide-spread 1 pole-pair/50 Hz motors; this combination yields an encoder signal frequency of 50 kHz (accordingly, the period is $2*10^{-5}$ s); it is about an order of magnitude higher than the highest frequency in the drive system, supposing a 5 kHz PWM frequency.

The environment used for simulation is the Matlab/Simulink®, because it is the most used software in simulation of electrical drives. The simulation structure is shown in Fig. 3. It is created mainly based on equations (1) with some modifications in order to make easy the building of the structure. The "Modulo 2π" block reduces the input angle to a 2π interval. This is further reduced by the block "Modulo $\theta_p/2$" to the angle θ_p. The output of the block is compared to the upper-limit U and the lower one L. Based on the four possible outputs of the "Relational Operators" block and the *Sense* signal (which indicates that the input angle θ is increasing or decreasing), the "AB Logic" block generates the A_t or B_t switching-moment signals. The output signals of the encoder are obtained using two T-type flip-flops, one "T-FF A" triggered by A_t for outputs A, A_N, and the other "T-FF B" triggered by B_t for B, B_N. The flip-flops hold their actual state until a new switching moment occurs due to the rotation of the shaft.

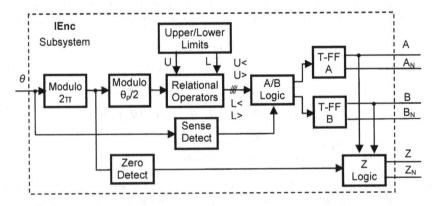

Fig. 3. Simulation structure of the incremental encoder (the subscript "N" of the output signals denotes the negated logic variable)

By every complete (i.e. 2π rad) shaft rotations – identified by "Modulo 2π" block – the "Zero Detect" block produces an output signal, which after synchronizing in "Z Logic" block with A and B signals yields the marker outputs Z and Z_N. The above blocks are integrated in a Simulink® subsystem named "IEnc".

2.2 Identification of the Rotation Direction

The direction of rotation is identified by comparing the relative position of the outputs A to B. This may be accomplished on different ways.

A trivial solution of the problem may be a sampling with a D type flip-flop at every rising front of the B pulses the logic value of A output. For CCW rotation the output of flip-flop will be at logic value 1, for the CW rotation at value 0. The drawback of the method is that the direction changing is detected only after a time interval according to a rotation of $3\theta_p/4 - 5\theta_p/4$; it is depending on the moment of direction changing with respect to signals A and B.

A more precise approach takes into account all four possible combinations of A and B signals for the both situations: when the actual direction is CCW and is changing to CW, or the CW is changing to CCW. Table 1 summarizes all combinations of signals, which detect the reversal of the rotation direction.

Table 1. Possible combinations of A and B signals during the reversal process

From CCW to CW $S=1$ to $S=0$			From CW to CCW $S=0$ to $S=1$		
Occurs if		Triggered by	Occurs if		Triggered by
A	B	Res	A	B	Set
0	0	A_N & B↑	0	0	B_N & A↑
0	1	B & A↑	0	1	A_N & B↓
1	0	B_N & A↓	1	0	A & B↑
1	1	A & B↓	1	1	B & A↓

Note: The subscript "N" denotes the negated logical variable; the ↑ and ↓ symbols denote the raising- and falling-edge of associated logic variable, respectively; symbol & denotes the AND logic function.

This strategy permits the detection of the direction changing in all cases during a rotation of $\theta_p/4$, namely in minimal detectable rotation-angle-increment.

The realized direction-detector block is based on the above idea. The block scheme of the simulation structure is presented in Fig. 4.

Corresponding to the two possible direction changes, the logic blocks detect all combinations of input signals shown in Table 1. The actual combination triggers a flip-flop "FF" of which output retains the direction of rotation until another changing occurs. The Simulink® subsystem named "DiD" contains the above logic functions.

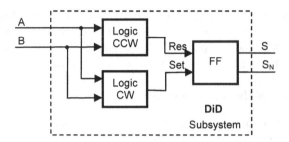

Fig. 4. The block diagram of the simulation structure of the direction-detector block

2.3 Encoder-Based Position Computation

In incremental encoder based system the angular position θ is referred to a fixed reference axis (corresponding to $\theta=0$ rad, as is shown on Fig. 1). The angular position is obtained by algebraic counting the number of generated encoder pulses according to the two direction of rotation and multiplying it with the angular step θ_p of the encoder. Mathematically:

$$\theta = \theta_p \left(\sum_{CCW} N_i - \sum_{CW} N_j \right) = \theta_p N , \qquad (4)$$

where ΣN_i represents the counted number of pulses in CCW direction and ΣN_j the number of pulses in CW direction. The counting of the pulses is made by electronic counters, and usual it is not contained in the encoder.

The simulation structure realized as a Simulink® subsystem named "Poz" is presented on Fig. 5. The encoder-output-signal A is the common input signal for the two counter blocks: "Counter CCW" and "Counter CW". The counters are enabled to count on their "En" input by signals provided by direction-detector bloc "DiD", i.e. "Counter CCW" is enabled by signal S and it will contain the sum of N_i, and "Counter CW" by negated signal S_N and it contains the sum of N_j. The difference between the contents of the two counters N will indicate the number of pulses proportional to the position of the shaft. This multiplied by the angular step θ_p yields the angular position θ.

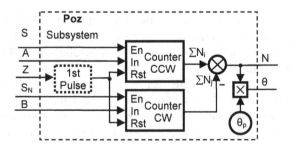

Fig. 5. The simulation structure of the position computing block

The Z marker pulse at every complete rotation, when the shaft passes the reference position, resets the counters. Therefore the position information is obtained in every $0–2\pi$ rad interval. In order to extend the position measurement on more rotations the structure has to be provided with "1st Pulse" block (dotted-line block on Fig. 5), which resets the counters only at the first pass through the reference position after the switch on.

3 Modeling of the Drive System

Encoders are frequently used as position transducers in drive systems. Usually a drive system is composed of a motor coupled on shaft with a load machine and the associated control equipment. The dynamic behavior of the system (with constant moment of inertia J) is described by the equation of motion [2]:

$$m_e - m_r = \frac{J}{z_p} \frac{d\omega}{dt} , \qquad (5)$$

where m_e is the electromagnetic torque developed by the electrical motor with z_p pole-pair number and m_r is the load torque. If θ represents the electrical angular position of the shaft, then it will rotate with $\omega = d\theta/dt$ electrical angular speed.

In present application, used to verify the encoder model, a squirrel-cage induction motor was considered, which drives a constant torque load machine.

The mathematical model of the induction motor is based on the so-called general equations written with space-phasors [2]. Choosing the stator- and rotor-currents and also the electrical angular speed as state-variables, the motor state-equations are deduced.

In this way the motor current model was conceived in form of a Matlab®Simulink *s-function*. The input quantities are the stator- and rotor-voltages (in this application the last is considered equal to zero, supposing a squirrel-cage motor) and the load torque. The model outputs the currents, fluxes, electromagnetic torque, rotor electrical angular speed and its electrical angular position. The *s-function* is interfaced to input and output quantities by the "Mux" and "Demux" blocks and they form a Simulink® subsystem. The motor parameters are entered using a subsystem mask. The structure of the induction motor simulation model is given on Fig. 6.

In simulations the speed of the induction motor was controlled in open-loop, according to the constant Volt/Hertz control principle, i.e. the amplitude of the voltage applied to the motor is proportional to the stator frequency. The speed of the shaft is determined by the load machine characteristics; this imposes the necessary stator frequency f_s. The applied stator voltage U_s is computed using the name-plate voltage U_{sN} and frequency f_{sN} in accordance to the following relation:

$$U_s^{Ref} = f_s \frac{U_{sN}}{f_{sN}} \qquad (6)$$

The procedure excels in its simplicity. Also, it is computational non-intensive. On the other hand, the dynamic behavior of the drive is weak.

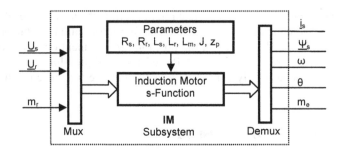

Fig. 6. The simulation structure of the induction motor

4   Simulation Results

The conceived simulation structure was investigated in two stages:

-     the functionality of the encoder, direction-detector and position computing structure was first verified independently;
-     the encoder and related blocks are integrated in an induction-motor drive structure controlled according the constant V/Hz principle.

Below are presented the procedures and results of the investigations.

4.1   Simulation of the Encoder

Using the functional units described in Chapter 2 the simulation structure presented on Fig. 7 was build.

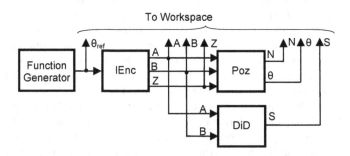

Fig. 7. The simulation structure of the encoder and the interconnected functional units. IEnc – incremental encoder; Poz - position computing block; DiD - direction-detector bloc

The "Function Generator" block, realized by using the "Repeating Sequence" Simulink® block, produces the input signal θ_{ref} for the encoder (it is equivalent to the angular position of the shaft). Its magnitude and the time-profile of the variation are editable by the user. The "IEnc" encoder block generates the A, B and Z output signals. Based on these signals the "DiD" block determines the direction of the rotation, and the "Poz" block counts the number N of the impulses and computes the angular position θ of the shaft. All signals are saved in Matlab's "Workspace" for further analysis. The generator was programmed for a ramp-shaped reference signal, which is considered the encoder shaft position. The encoder used in simulations has an angular step of graduation $\theta_p = 2\pi/500$ rad. The reference angle starts in CW direction with a ramp of 80 rad/s, at 0.05 s occurs the reversal in CCW direction, with the same ramp. At 0.015 s occurs another reversal. Fig. 8a) presents the computed position angle in comparison with the reference angle. The figure shows, that the computed angular position follows the reference very well, and the marker pulses Z occurs every time when the angle passes through zero value. Fig. 8b) presents in enlarged detail the marker pulse referred to the A and B output signals, its width obviously is a quarter angular step.

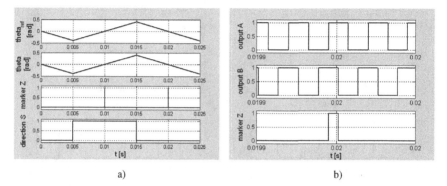

a)   b)

Fig. 8. The simulated reference angle and computed angle versus time during reversal;
a) Reference angle θ_{ref}, computed angle θ, marker output Z, direction signal S;
b) Detail showing the marker signal Z referred to outputs A and B

In the next step the reversal process was analyzed. The reference-angle generated by function generator has a saw-tooth-like variation versus time. Its parameters were selected in such a manner, that all possible combinations of reversal were captured and presented in Fig. 9 a)-d). The captures cover all situations enumerated from Table 1 in Chapter 2.2. The left column of Fig. 9 corresponds to CCW to CW reversal, the right one to CW to CCW changing. The pictures situated in the same row of the columns present the same combinations of A and B signals for the two reversals. In all cases the sensing of the reversal is done in a quarter of the angular step. This fact is underlined in the diagrams also for the case of CCW to CW reversal (left column): the S signal starts with 0 default value (corresponding to CW rotation), but after a quarter angular step the "DiD" block observes the CCW rotation and rises its S output up to 1, corresponding to CCW rotation direction.

The simulation results correspond with those expected.

4.2 Simulation of the Induction Motor Drive Control

The structure presented in Fig. 7 was integrated in the simulation structure of an induction motor drive. In this case the angular position – the input signal of the

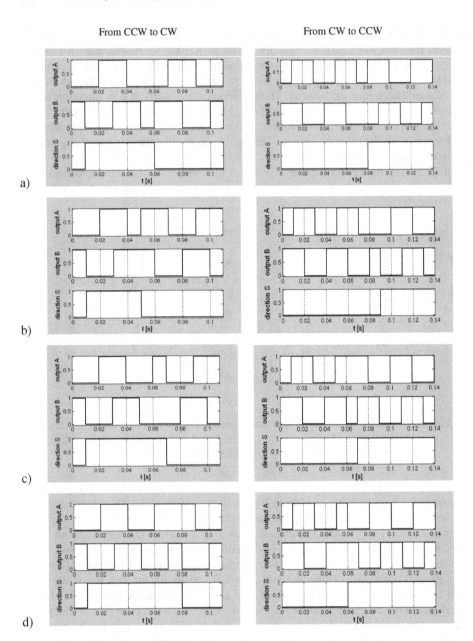

Fig. 9. The simulation results representing all possible combinations of *A* and *B* versus time during the reversal process. Left column: From CCW to CW, Right column: from CW to CCW, *Top trace*: signal A, *Middle trace*: signal B, *Bottom trace*: direction signal S.
Reversal occurs at: a) *A*=0, *B*=0; b) *A*=0, *B*=1; c) *A*=1, *B*=0; d) *A*=1, *B*=1.

encoder – will be provided by the drive structure; it is obtained from the model of the electrical machine. The control scheme of the drive system will use as position feedback value that determined by using the encoder signals from the "Poz" block.

In the simulation structure presented in Fig. 10 the shaft of the encoder is coupled virtually to the shaft of the induction motor. The input variable of the drive is the frequency f_s^{Ref}. It is provided by the user-programmable "Repeating sequence" Simulink® block. The "V/Hz=ct" block computes the amplitude of the stator voltage U_s^{Ref} according to the equation (6).

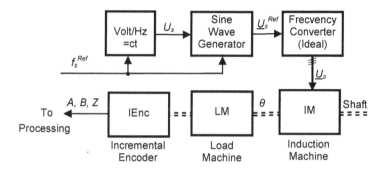

Fig. 10. The simulation structure of the encoder equipped induction motor drive system

Based on these quantities the "Sine Wave Generator" bloc generates the three-phase-voltage system, which feeds directly the induction motor (the power electronic frequency converter was considered an ideal one). The angular position of the shaft θ is provided by the motor model (obtained by integration of the rotor angular-speed) and it is the input quantity for the encoder model. The encoder generates the A, B, and Z pulses. After that they are processed by blocks "DiD" and "Poz", resulting the measured angular position of the shaft θ_m. It is compared with the position of the shaft.

The name-plate data of the simulated 1LA7073 series induction motor are: $U_{sN} = 230$ V, $I_{sN} = 1.35$ A, $P_N = 0.55$ kW, $n_N = 2800$ rpm, $\cos\varphi_N = 0.81$, $\eta = 0.71$, $M_N = 1.9$ Nm, (identical to the name-plate data of motor used in the experimental rig). The load torque is a constant one, equal to the no-load torque of the induction motor-load machine assemble (0.2 Nm). The motor is started at t = 0 s in positive direction, with a ramp of 100 Hz/s. At about 0.1 s achieves the desired 10 Hz. After a 0.3 s run at 10 Hz a reversal is commanded. At t = 0.5 s the rotation sense is changed, at t = 0.6 s the new steady state is reached at CW direction.

Fig. 11 a) and b) present the simulation results of the drive system. The top trace on Fig. 11 a) shows the prescribed angular speed (w_{ref}) and the motor speed (w) versus time profile. The shape of the motor speed diagram is typical for the open-loop controlled induction motor in low-speed region, characterized by a weak transient response. The middle trace shows the shaft angular position resulted from the motor model; the trace below presents the encoder signal based, computed shaft position. They have an identical shape versus time in CCW and also in CW direction.

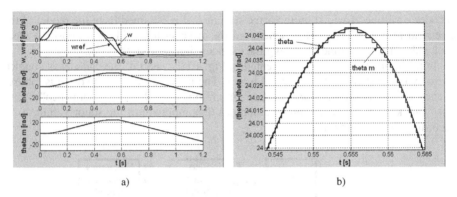

a)   b)

Fig. 11. Simulation results of the drive system.
   a) *Top trace*: Speed reference (*wref*) and motor speed (*w*) versus time;
   Middle trace: the motor shaft-angular-position (*theta*) versus time;
   Bottom trace: the computed shaft-angular-position (*theta m*) versus time
   b) Detail of the motor shaft angular position (theta) superposed to the computed shaft
   angular position (theta m) versus time

Fig. 11 b) shows the zoomed detail of the rotor angular position superposed with the computed one. The computed position overlaps very well the motor shaft position.

The incremental character of the computed position is obvious. The simulation results confirm the validity of the created structures.

6   Experimental Measurements

Experimental measurements were performed using a 1XP8001-1 (*Siemens*) incremental encoder. The encoder has 1024 divisions per 360°, and provides the output signals A, B, Z and their negated values A_N, B_N, Z_N compatible with HTL logic. The encoder was mounted on the induction motor described before for simulations. The rotational speed of the motor was controlled by a *Micromaster 440* static frequency converter configured in U/f mode. The converter uses the encoder signals as speed-feedback.

The scheme of the experimental set-up is presented on Fig. 12. Figure 13 a) presents the three encoder signals, i.e. A, B and Z captured, by a storage scope at a CW running. It is obvious that the width of the Z marker pulse is a quarter of the A (or B) signal width. Fig. 13 b) shows the A and B signals at CCW to CW rotation sense reversal according to the situation when the reversal occurs at A=1, B=1. The comparison of the experimental captured figures with the simulation results, i.e. Fig. 13 a) with Fig. 8 b), and Fig. 13 b) with Fig. 9 d) confirms the correctness of the models. The captured figures correspond with those simulated.

The experimental board (under development) is built around of the *TMS320F2812 TI* DSP based *eZdsp kit* from *Spectrum Digital*. The used software environment is the Matlab/Simulink[®] and Code Composer[®]. The verified simulation structure is completed with the *Real Time Interface* blocks in order to link to the hardware. Based on

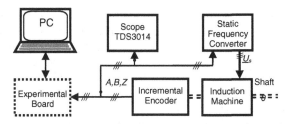

Fig. 12. The block-scheme of the experimental set-up

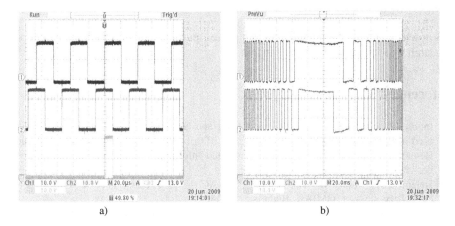

a)   b)

Fig. 13. Experimentally captured encoder signals:
a) *Top trace*: *A* output signal, *Middle trace*: *B* output signal; *Bottom trace*: *Z* marker
signal in case of CW rotation;
b) *Top trace*: *A* output signal, *Bottom trace*: *B* output signal at CCW to CW reversal,
occurred at *A*=1 and *B*=1

this structure automated DSP code generation is performed. The Code Composer®
permits the downloading of the code to the target-DSP memory, its real-time execution and data acquisition.

The "DiD" direction-detector unit and the "Poz" position computing block (together
with other functions, i.e. speed computing) may be implemented also hardware in
FPGA or other programmable logic. This approach allows parallel data processing enhancing the overall computing speed of the system. The information exchange with the
main processing unit of the drive may be managed by parallel data-bus architecture.

7 Conclusions

The information provided by the incremental encoders is inherently digital. The generated pulses are counted by an electronic counter; their number is proportional to the
angular position of the encoder shaft.

The direction of the rotation may be identified by a digital decoding scheme using the two quadrature signals provided by the encoder. The changes of the direction are detected in an angular interval equal to a quarter of the angular step of the encoder disc.

The presented simulation structure of the incremental encoder may be successfully integrated in other Simulink® simulation structures as is confirmed by the described application with induction motor drive.

Due to the fact that the encoder signals are the highest frequency quantities in an electrical drive system, consequently the simulation step has to be determined accordingly.

The simulation results correspond to those expected and are very close to the experimental one.

Acknowledgments. This work was supported partially by the National University Research Council (CNCSIS) of Romania.

References

[1] Incze, J.J., Szabó, C., Imecs, M.: Modeling and Simulation of an Incremental Encoder used in Electrical Drives. In: Proceedings of 10th International Symposium of Hungarian Researchers on Computational Intelligence and Informatics, Budapest, Hungary, November 12-14, pp. 97–109 (2009)

[2] Kelemen, Á., Imecs, M.: Vector Control of AC drives Vector Control of Induction Machine Drives, vol. 1. OMIKK, Budapest (1991)

[3] Koci, P., Tuma, J.: Incremental Rotary Encoders Accuracy. In: Proceedings of International Carpathian Control Conference ICCC 2006, Roznov pod Radhostem, Czech Republic, pp. 257–260 (2006)

[4] Lehoczky, J., Márkus, M., Mucsi, S.: Szervorendszerek, követő szabályozások. Műszaki, Budapest (1977)

[5] Miyashita, I., Ohmori, Y.: A New Speed Observer for an Induction Motor using the Speed Estimation Technique. In: 5th European Conference on Power Electronics and Applications EPE 1993, Brighton, UK, September 14-17, vol. 5, pp. 349–353 (1993)

[6] Petrella, R., Tursini, M., Peretti, L., Zigliotto, M.: Speed Measurement Algorithms for Low Resolution Incremental Encoder Equipped Drives: Comparative Analysis. In: Proceedings of International Aegean Conference on Electrical Machines and Power Electronics, ACEMP 2007 - ELECTROMOTION 2007 Joint Meeting, Bodrum, Turkey, September 10-12, pp. 780–787 (2007)

[7] *** Encoder vs. Resolver-based Servo Systems. Application Note. ORMEC, http://www.ormec.com/Services/ApplicationsEngineering/ ApplicationNotes.aspx (accessed September 29, 2009)

Real-Time Modeling of an Electro-hydraulic Servo System

Attila Kővári

Collage of Dunaújváros
Táncsics M. út 1/A, H-2400 Dunaújváros, Hungary
kovari@mail.duf.hu

Abstract. Real-Time Hardware-in-the-loop (HIL) test is a good ground for examine the dynamic model of controlled equipments, for example electro-hydraulic servo systems, to observe the main parameters and monitor the dynamic behavior of the system. Electro-hydraulic servo actuators are widely used in industrial applications because it has high moving force, power/volume ratio and they proof against environmental impacts so it can be used as built-in element at the acting location directly. Using HIL test, it has possibility to examine the dynamic phenomena inside hydraulic system using trials to test how system behavior changes when system's parameters are varying. In this paper a HIL real-time advanced more realistic nonlinear model was developed, realized and step response analysis is used to test the dynamic behavior of the complete electro-hydraulic servo system which enables us to observe the system's parameter variations for example when seals are worn out in the hydraulic actuator.

Keywords: electro-hydraulic servo system, hardware-in-the-loop simulation, real-time control.

1 Introduction

Electro-hydraulic servos derive their flexibility from the electronic portion and their power handling ability from hydraulic portion. The range of applications for electro-hydraulic servo systems is diverse, and includes manufacturing systems, materials test machines, active suspension systems, mining machinery, fatigue testing, flight simulation, paper machines, ships and electromagnetic marine engineering, injection moulding machines, robotics, and steel and aluminium mill equipment.

Hardware-in-the-Loop (HIL) simulation is a technique that is used increasingly in the development and test of complex real-time systems so the purpose of HIL simulation is to provide an effective platform for developing and testing real-time systems. Hardware-In-the-Loop differs from pure real-time simulation by the addition of a real component in the loop and simulation is achieving a highly realistic simulation of equipment in an operational virtual environment.

In this paper an electro-hydraulic servo system is examined and HIL tested. The HIL electro-hydraulic plant is driven by a Texas Instruments DSP control unit.

I.J. Rudas et al. (Eds.): Computational Intelligence in Engineering, SCI 313, pp. 301–311.
springerlink.com © Springer-Verlag Berlin Heidelberg 2010

2 Hardware-in-the-Loop Test

The current industry definition of a Hardware-In-the-Loop system is shown in Figure 2.1. It shows that the plant is simulated and the control unit is real. The purpose of a Hardware-In-the-Loop system is to provide all of the electrical stimuli needed and a typical HIL system includes sensors, actuators to receive data from the control system, actuators to send data, a controller to process data, a human-machine interface (HMI) and a development post-simulation analysis platform. The value of each electrically emulated sensor is controlled by the plant simulation and is read by the embedded system under test. Likewise, the embedded system under test implements its control algorithms by outputting actuator control signals. Changes in the control signals result in changes to variable values in the plant simulation.

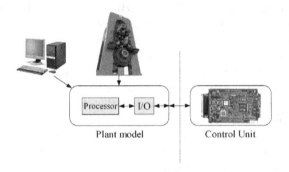

Fig. 2.1. Hardware-in-the-Loop System

Advantages of HIL systems:

- enable testing the hardware without building a "plant prototype"
- supports reproducible test runs that can assist in uncovering and tracking down hard to find problems
- enables testing less risk of destroying system
- provides cost savings by shortened development time
- complete, consistent test coverage
- supports automated testing
- simulator performs test outside the normal range of operation

The most evident advantage of HIL simulation is that real-world conditions are achieved without the actual risks involved. HIL simulation is achieving a highly realistic simulation of equipment in an operational virtual environment. With HIL, you can test the control units with extreme conditions that might not be feasible in the real world. HIL enables you to isolate deficiencies in the control unit even if they occur only under certain circumstances. Robust, high-fidelity real-time HIL simulations not only enable shorter time to market by reducing the development period, but also reduce cost by eliminating the need for actual hardware during testing, as well as associated maintenance costs. With the power and flexibility of today's computers, engineers and scientists are increasingly using PC-based systems for HIL simulation applications. A key element of the development of such a system is the integration of signal generation/acquisition I/O functions with the software used to simulate the system. A normal desktop PC was used as hardware of the HIL simulator with a National Instruments PCI-6251analog-digital data acquisition card. The real-time operating system solution for the plant model was xPC Target real-time kernel.

3 Plant and Control Unit

Plant is an electro-hydraulic servo system shown in Figs. 3.1 and 3.2. Electro-hydraulic servos are capable of performance superior to that of any other type of servo. Large inertia and torque loads can be handled with high accuracy and very rapid response. A typical position controlled hydraulic system consists of a hydraulic power supply, flow control valve, linear actuator, displacement transducer, and electronic servo-controller [2].

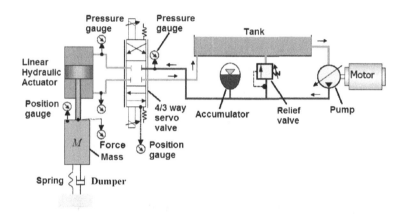

Fig. 3.1. Electro-hydraulic Servo System

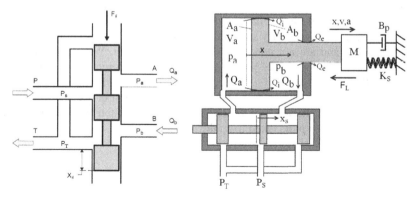

Fig. 3.2. 4/3 Flow control servo valve (Three Land, Four-way) and Single-acting, Linear Hydraulic Actuator with Load and Valve

Fig. 3.2 shows the spool configuration of a typical "4/3" way flow control valve. The ports are labeled P (pressure), T (tank), A and B (load control ports). The spool is shown displaced a small distance (x_s) as a result of a command force applied to one end, and arrows at each port indicate the direction of fluid flow which results. With no command force applied (F_s=0), the spool is centralized and all ports are closed off by the lands resulting in no load flow [2].

The control unit F2812 eZdsp™ card is a stand alone module with a single chip parallel port to JTAG scan controller shown in Fig. 3.3. The eZdsp F2812 allows developers to get started using the F2812 DSP. The C28x core is the world highest performance DSP core for digital control applications. The 32-bit F2812 has on board flash, 64K words on board RAM and runs at 150MHz, making it capable of numerous sophisticated control algorithms in real-time. The module can be operated without additional development tools such as an emulator. The combination of a bundled version of Code Composer and an on-board JTAG interface means that the eZdsp can be used to develop and debug code without the requirement for an external emulator or debugger [3].

Fig. 3.3. eZdsp F2812 Development Package

4 Mathematical Model of the Electro-hydraulic Servo System

A servo-valve is a complex device which exhibits a high-order non-linear response, and knowledge of a large number of internal valve parameters is required to formulate an accurate mathematical model. Indeed, many parameters such as nozzle and orifice sizes, spring rates, spool geometry and so on, are adjusted by the manufacturer to tune the valve response and are not normally available to the user. When modeling complex servo-valves, it is sometimes possible to ignore any inherent non-linearities and employ a small perturbation analysis to derive a linear model which approximates the physical system. Such models are often based on classical first or second order differential equations, the coefficients of which are chosen to match the response of the valve based on frequency plots taken from the data sheet [2]:

$$\frac{d^2 x_s}{dt^2} + 2 \cdot \zeta_s \cdot \omega_s \cdot \frac{dx_s}{dt} + \omega_s^2 \cdot x_s = \omega_s^2 \cdot k_t \cdot i \tag{1}$$

where ω_s is the natural frequency and ζ_s damping ratio of the spool, k_t proportionality coefficient between the control current and valve spool displacement x_s, i is the servo-valve control current which produce F_s electromagnetic spool moving force (Fig 3.2) [2].

The servo-valve delivers a control flow proportional to the spool displacement for a constant load. For varying loads, fluid flow is also proportional to the square root of the pressure drop across the valve. Control flow, input current, and valve pressure drop are related by the following equations [2]:

$$Q_a = c_s \cdot w \cdot x_s \cdot \sqrt{\frac{2}{\rho} \cdot (P_S - P_a)} \tag{2}$$

$$Q_b = c_s \cdot w \cdot x_s \cdot \sqrt{\frac{2}{\rho} \cdot (P_b - P_T)} \tag{3}$$

where c_s is the volumetric flow coefficient and w the valve-port width – area gradient, ρ volumetric density of the oil, P_S, P_T are system pressure and tank pressure, P_a and P_b are the load and return pressure, Q_a and Q_b the load and return flow of the valve according to Fig. 3.2. P_S can be noted as constant because of the hydraulic power supply system consists an accumulator.

The relationship between valve control flow and actuator chamber pressure is important because the compressibility of the oil creates a "spring" effect in the cylinder chambers which interacts with the piston mass to give a low frequency resonance. This is present in all hydraulic systems and in many cases this abruptly limits the usable bandwidth. The effect can be modeled using the flow continuity equation from fluid mechanics [4]:

$$Q_a - Q_b = \frac{dV}{dt} + \frac{V}{\beta} \cdot \frac{dP}{dt} \tag{4}$$

where V is the internal fluid volume (in pipe and cylinder) and β the fluid bulk modulus (mineral oils used in hydraulic control systems have a bulk modulus in the region of $1.4 \cdot 10^9$ N/m). This equation can be used if the mechanical structure is perfectly rigid. The pressures in cylinder chambers P_a and P_b can be calculated as (Fig. 3.2) [1]:

$$\frac{dP_a}{dt} = \frac{\beta}{V_a} \cdot \left(Q_a - \frac{dV_a}{dt} \right), \quad \frac{dP_b}{dt} = \frac{\beta}{V_b} \cdot \left(-Q_b - \frac{dV_b}{dt} \right) \tag{5}$$

$$\frac{dV_a}{dt} = A_a \cdot \frac{dx}{dt}, \quad \frac{dV_b}{dt} = -A_b \cdot \frac{dx}{dt} \tag{6}$$

$$V_a = V_{a0} + \Delta V_a = V_{a0} + A_a \cdot x, \quad V_b = V_{b0} - A_b \cdot x \tag{7}$$

where x and v are the position and speed of the piston and A_a, A_b the active area of the piston annulus on "a" and "b" side.

If the seals at the piston are not perfect there is an additional internal leakage oil flow Q_i between the chamber "a" and "b". In the same way an external leakage oil flow Q_e appears when seals between the rod and cylinder are inefficient. C_i and C_e the internal and external cylinder's leakage coefficient can be calculated using R_i and R_e internal and external cylinder's leakage resistance [1]:

$$R_i = \frac{P_a - P_b}{Q_i} = C_i(P_a - P_b), \ \frac{1}{R_i} = C_i, \ Q_i = C_i(P_a - P_b) \tag{8}$$

$$R_e = \frac{P_b}{Q_e}, \ \frac{1}{R_e} = C_e, \ Q_e = C_e \cdot P_b \tag{9}$$

Using equations 2-9 the dynamic pressures in chamber "a" and "b" [1], [4]:

$$P_a = \int \frac{\beta}{V_{a0} + A_a \cdot x} [Q_a - A_a \cdot v - Q_i] dt \tag{10}$$

$$P_b = \int \frac{\beta}{V_{b0} - A_b \cdot x} [-Q_b + A_b \cdot v + Q_i - Q_e] dt \tag{11}$$

The net hydraulic acting force on the piston F_P can be calculated by the pressures and annuluses of the two sides:

$$F_P = A_a \cdot P_a - A_b \cdot P_b \tag{12}$$

The load force F_L can be calculated as:

$$F_L = B_P \cdot v + K_S \cdot x + sign(v) \cdot F_f \tag{13}$$

where K_S is the stiffness coefficient of the spring and there is a shock absorber with damping ratio B_P. F_f frictional force is notoriously difficult to measure and accurate values of these coefficients are unlikely to be known, but order of magnitude estimates can sometimes be made from relatively simple empirical tests. The movement equation of the complete system:

$$F_P - F_L = M \cdot (a + g) \tag{14}$$

where M is the load mass (together with the mass of actuator piston), a acceleration and g gravity.

5 Test Environment and Model Parameters

Plant model runs on a PC using xPC Target real-time kernel and a laptop with MATLAB to generate the code from mathematical model of the electro-hydraulic servo system. The PCI-6251 A/D card voltage input is used to control the servo valve and voltage outputs to examine the time functions of the system variables. The control signal of the servo valve was generated by the F2812 DSP board.

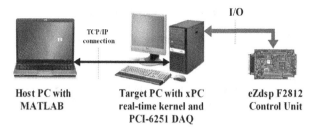

| Host PC with MATLAB | Target PC with xPC real-time kernel and PCI-6251 DAQ | eZdsp F2812 Control Unit |

Fig. 5.1. Test Environment

The parameters of the examined electro-hydraulic servo system:

Table 5.1. Parameters of the Flow Control Servo Valve

Symbol	Description	Value
ζ_s	Damping ratio of the spool	0.9
ω_s	Natural frequency of the spool	200 rad/s
k_t	Proportionality coefficient between the control current and valve spool displacement	0.1 m/A
c_s	Volumetric flow coefficient	6
w	Valve-port width	10^{-2} m

Table 5.2. Parameters of the Actuator

Symbol	Description	Value
V_{a0}	Internal fluid volume when the piston is in middle position	$300 \cdot 10^{-6}$ m^3
A_a	Active area of the piston annulus	$8 \cdot 10^{-4}$ m^2
V_{b0}	Internal fluid volume when the piston is in middle position	$200 \cdot 10^{-6}$ m^3
A_b	Active area of the piston annulus	$3 \cdot 10^{-4}$ m^2
β	Hydraulic fluid bulk modulus	$1.4 \cdot 10^9$ N/m^2
R_i	Internal cylinder's leakage resistance	10^{11} Ns/m^5
R_e	External cylinder's leakage resistance	10^{12} Ns/m^5

Table 5.3. Parameters of the Power Supply

Symbol	Description	Value
P_S	System pressure	$210 \cdot 10^5$ Pa
P_T	Tank pressure	0 Pa
ρ	Volumetric density of the oil	890 kg/m^3

Table 5.4. Parameters of the Load

Symbol	Description	Value
M	Load mass	50 kg
B_P	Damping ratio	10^3 Ns/m
K_S	Spring stiffness coefficient	$5 \cdot 10^4$ N/m
Ff	Frictional force	10 N
g	Gravity	9.81 m/s^2

6 Test Results

The test execution time was 0.5 s and step response was examined (step time was 0.1 s). The observed variables are i servo valve control current, x_s servo valve spool

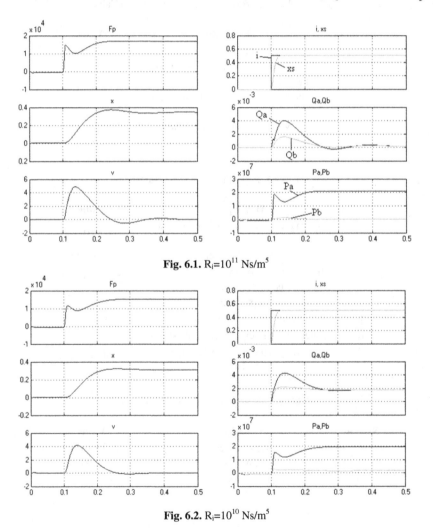

Fig. 6.1. $R_i = 10^{11}$ Ns/m^5

Fig. 6.2. $R_i = 10^{10}$ Ns/m^5

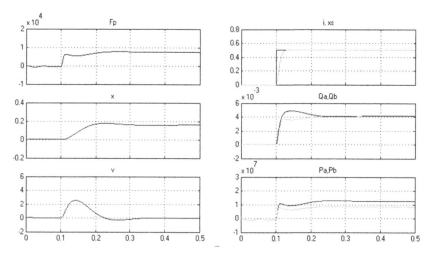

Fig. 6.3. $R_i = 10^9$ Ns/m^5

Table 6.1. Test results at different internal leakage resistance Ri

R_i internal leakage resistance [Ns/m^5]	F_{pmax} maximum of acting force [N]	a_{max} maximum acceleration [m/s^2]	x_∞ position in steady state [m]
10^9	7515	2,55	0,16
10^{10}	15068	4,2	0,311
10^{11}	16776	4,9	0,312

position (these are independent of R_i), Q_a, Q_b load and return flows of the valve, P_1, P_2 load and return pressures, x position of the piston and mass load and F_p acting force of the hydraulic actuator. Time functions of the electro-hydraulic system are shown in the next figures Fig. 6.1-3.

7 Conclusion

It is shown in the step responses (Fig. 6.1-3) that position of servo valve pool x_s rapidly follows the valve control current and opens the flow channel to the hydraulic actuator. While the servo valve opens, the oil flow and pressure increasing immediately (Q_a, P_a). When the servo valve opens, the oil flow is immediately increasing and it causes growing pressure drop in the control valve.

It follows that the increasing of load pressure is cracked (see P_a, t=0,1 s...0,15 s in Figure 6.1). When the quantity of oil flow goes on decreasing, the load pressure starts to grow again and reaches the system pressure. Because of the acting force F_p is proportional to load pressure difference the time function of acting force is similar to this.

When the hydraulic actuator is moving in positive direction and the internal sailing is not perfect, the increased internal leakage oil flow (Q_i) is increasing the necessary load Q_a and return Q_b flow (Fig. 6.1-3). Furthermore these higher oil flows increase the pressure drop on the servo valve (Eq. 2) so the maximum load pressure P_a is less:

$$P_a = P_S - \frac{\rho}{2} \cdot \left(\frac{Q_a}{c_s \cdot w \cdot x_s} \right)^2 \tag{15}$$

and the return pressure P_b is higher:

$$P_b = P_T + \frac{\rho}{2} \cdot \left(\frac{Q_b}{c_s \cdot w \cdot x_s} \right)^2 \tag{16}$$

Therefore the maximum acting force F_p is less:

$$F_P = A_a \cdot P_a - A_b \cdot P_b \tag{17}$$

Additionally it is shown in Eq. 10 and 11 that higher leakage oil flow reduces the dynamic change of load pressure P_a and increases return pressure P_b so the acting force dynamic decreasing also:

$$dP_a / dt = \frac{\beta}{V_{a0} + A_a \cdot x} [Q_a - A_a \cdot v - Q_i] \tag{18}$$

$$dP_b / dt = \frac{\beta}{V_{b0} - A_b \cdot x} [- Q_b + A_b \cdot v + Q_i - Q_e] \tag{19}$$

Table 6.1 consist data about how changes the maximum acting force, acceleration and steady state position while the internal leakage resistance changes. The result goes to show that internal leakage in hydraulic cylinder can extremely change the dynamic and steady state behavior of the electro-hydraulic servo system.

8 Summary

In this paper an advanced mathematical model and real-time HIL test was prepared to examine the behavior of a complex electro-hydraulic servo system. It is shown in the test results that decreasing internal leakage resistance/higher internal leakage oil flow causes decreasing acting force in the cylinder. According to the decreasing acting force the system transients are slower.

The presented HIL plant model is capable to examine the dynamic behavior of the described electro-hydraulic servo system in real-time, and this model can give a basis for further examination of these systems.

References

[1] Kővári, A.: Hardwer-in-the-Loop Testing of an Electro-Hydraulic Servo System. In: 10th International Symposium of Hungarian Researchers on Computational Intelligence and Informatics, Budapest, pp. 631–642 (2009)

[2] Richard Poley, D.S.P.: Control of Electro-Hydraulic Servo Actuators, Texas Instruments Application Report (2005)

[3] Kővári, A.: Programming of TMS320F2812 DSP using MATLAB Simulink, In: Hungarian Science Week at Collage of Dunaújváros, CDROM, Dunaújváros (2007)

[4] Halnay, A., Safta, C.A., Ursu, I., Ursu, F.: Stability of Equilibria in a Four-dimensional Nonlinear Model of a Hydraulic Servomechanism. Journal of Engineering Mathematics 49(4), 391–405 (2004)

Mathematical Model of a Small Turbojet Engine MPM-20

Ladislav Főző[1], Rudolf Andoga[2], and Ladislav Madarász[1]

[1] Department of Cybernetics and Artificial Intelligence FEI TU Košice, Letná 9, Košice, 04001
`ladislav.fozo@tuke.sk`, `ladislav.madarasz@tuke.sk`
[2] Department of Avionics, LF TU Košice, Rampová 7, Košice 04001
`rudolf.andoga@tuke.sk`

Abstract. The growing demands on safety authority and intelligence of control systems requires new approaches in design. One approach that is usable and potentially effective are anytime control algorithms that can be used in the area of modeling and control [6, 11, 14, 15]. Use of such algorithms allows the system to flexibly react on changes of outer environment and be able to survive deficiency of time, information and resources.

Keywords: turbojet engine, small turbojet engine, anytime algorithms, artificial intelligence, analytic modeling.

1 Small Turbojet Engine – MPM 20 as an Object of Modeling

The experimental engine MPM 20 has been derived from the TS – 20 engine, what is a turbo-starter turbo-shaft engine previously used for starting engines AL-7F. The engine has been adapted according to [9]. The engine has been rebuilt to a state, where it represents a single stream engine with radial compressor and single one stage non-cooled turbine and outlet jet. The basic scheme of the engine is shown in Figure 1.

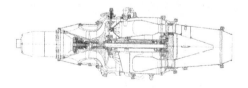

Fig. 1. The scheme of MPM 20 engine [9]

In order to model the engine, it is necessary to measure its characteristics. All sensors, except fuel flow and rotations sensor, are in fact analogue which in and have voltage output. This is then digitalized by a SCXI measurement system and corresponding A/D converter and sent through a bus into computer. Every parameter is measured at the sampling rate of 10 Hz. The data acquisition has been done in LabView environment [1, 4].The digital measurement of parameters of MPM 20 engine

I.J. Rudas et al. (Eds.): Computational Intelligence in Engineering, SCI 313, pp. 313–322.
springerlink.com

in real time is important to create a model and control systems complying with FADEC definition („Full Authority Digital Electronic Engine Control"). Moreover we needed to change the engine from static single regime engine into a dynamic object, what was done by regulation of pressure beyond the compressor according to which the current fuel supply actuator changes actual fuel supply for the engine in real time. The system has been described in [8]. The graph in Figure 2 shows dynamic changes of parameters of the engine to changes of fuel supply input.

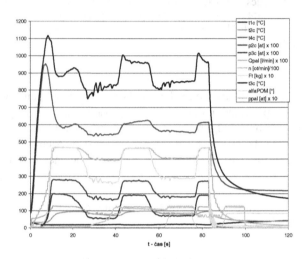

Fig. 2. The measured data by indirect regulation

2 Modeling the MPM 20 Turbojet Engine

Static and dynamic properties of turbojet engines (MPM 20) can also be described by a mathematical model of operation single stream engine under equilibrium or non-equilibrium conditions. This will allow to model thrust, fuel consumption, pressures and temperatures of the engine by different altitudes and velocities in the chosen cuts of the engine.

The steady operation of the engine is such a regime, where in every element of the engine same thermodynamic processes are realized. Operation of an engine in its steady operation can be described by:

1 algebraic equations of balance of mass flow of working materials through nodes of the engine, equations of output balance, equations of regulation rules and equations describing particular oddities of an engine. A system of equations expresses that for given outer conditions of operation of an engine, characteristics of all nodes of an engines and preset values of control parameters (fuel supply, cross section of the output nozzle, angle of compressor blades), operation of the engine will settle itself on one and only one regime [12];

2 graphically,by utilization of knowledge of characteristics of all parts (output, compressor, turbine, etc) of the engine and their preset curves of joint operations

(e.g. lines of stable rations of T_{3c}/T_{1c} in compressor). Designation of all curves of the engine is done in a way that we will try to fulfill continuity conditions for all parts of the engine and characteristics of all these parts are given. These characteristics can be found by direct measurement, computation, etc.

2.1 Mathematical Modeling of Radial Compressor of MPM 20 Turbojet Engine

Steady and transitory operation of the engine can be drawn into the characteristics of compressor, this chapter shows briefly the computation of characteristics of the compressor.

The object of research – MPM-20 engine has single stage single sided compressor of radial construction. The analytic model comes from the theory of similarity of flows in compressors. It is also assumed that compressor that are geometrically similar and flow in individual conditions satisfy the aerodynamic similarity rules have the same characteristics. Even though not all conditions of similarity are satisfied fully, it is possible to find mutual dimensionless parameters characterizing flows in compressors. Behavior of a compressor as a complex can be assessed according to changes of these parameters [10].

The computation is based on characteristics of known measured compressors. The resulting measurements are visualized in subsidiary diagrams that include dependencies of the chosen parameters and on their base characteristics of a compressor is selected. From the phase of design of the compressor, the computational regime is known, which is characterized by pressure ratio, efficiency, speed and mass flow of air. From the knowledge of the computational regime and by use of diagrams of dimensionless parameters characteristics of a compressor can be worked out that is similar to characteristics of compressors used by creation of proportional diagrams.

All the needed mathematical grounds are in the literature [7] and from them the following courses are computed:

- efficiency of the compressor in its working points,
- change of proportional effective work,
- change of mass air flow and pressure ratio of the compressor
- effective work of the compressor on the surge line,

These are needed for creation of mathematic model of the researched engine MPM-20.

Approach to Computation of Characteristics of Centrifugal Compressor

By designation of the characteristics we expected that the computational values are given (in this case obtained from the simulator of the engine (Figure 5) [7]) that means $(\pi KC)vyp$ – pressure ratio (Q) vyp – mass airflow, $(\eta KC)vyp$ – efficiency, that are representative for standard atmospheric conditions of pressure p1C=0.1013 MPa and temperature T1C=288 K. The approach consists then of the following steps:

- Set the surge line for the chosen relative speed $\overline{n}$.
- Compute parameters of the compressor on the line $\overline{n}$=const.

The main computed parameters of the compressor π_{KC}, Q, η_{KC} are then plotted into the characteristics as a dependency of η_{KC}, π_{KC} =f(Q), because the computation is

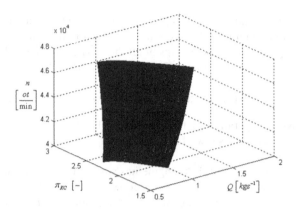

Fig. 3. 3D characteristics of the compressor of MPM-20 engine

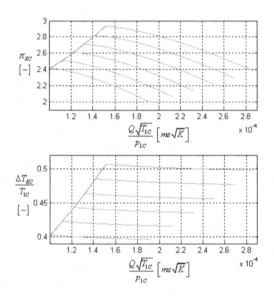

Fig. 4. Characteristics of the compressor of MPM-20 engine

done for the standard conditions where $Q=Q_v$. In some cases the characteristics is depicted in the form of η_{KC}, π_{KC} =f(($Q*sqrt(T_{1C}))/p_{1C}$), for different $\bar{n}$. Figure 3 shows 3D dependency η_{KC}=f(Q) for different speed n and Figure 4 depicts characteristics of the compressor $\Delta T_{KC}/T_{1C}$, π_{KC} =f(($Q*sqrt(T_{1C}))/p_{1C}$) for different $\bar{n}$ (increasing from bottom). These are further used for designation of curves of the mutual work of all parts of the engine MPM-20.

The observed values from the characteristics of the compressor were compared to the measured values on the real object and we obtained the following mean absolute percentage errors (MAPE):

- MAPE=1,328[%] for pressure ratio,
- MAPE=3,6[%] for efficiency and
- MAPE=2,04[%] for increase in temperature.

The mentioned results show that the MAPE of temperature growth on the compressor has overcome the level of 2%, this is caused by accumulation of error by computation through different variables. Virtual real-time simulations in Matlab - Simulink environment have shown that if we obtain better results in one parameter by setting deliberate constants it makes results by the second parameter worse. This result is a compromise between all three observed parameters and the computed characteristics will be used in creation of full analytic model of MPM-20 engine.

2.2 Mathematical Model in Steady and Transitory Operation of the MPM 20 Engine

As was mentioned in Chapter 2, any regime of the turbojet engine has to fulfill the continuity equation which designates dependencies between mass flow of air through the compressor, turbine, combustion chamber and exhaust system:

$$Q_{VS} = Q_k = Q_{SK} = Q_T = Q_{tr} = Q \tag{1}$$

and a condition of no distortion of the shaft

$$n_k = n_T = n \tag{2}$$

where

Q_{VS} – mass flow of air through input system,
Q_k – mass flow of air through the compressor
Q_{SK} – mass flow of air through combustion chamber,
Q_T – mass flow of gases through the turbine,
Q_{tr} – mass flow of gases through exhaust nozzle,
n_k – revolutions of compressor,
n_T – revolutions of turbine.

Another condition for steady operation of the engine has to be fulfilled – the engine doesn't change its revolutions in time.

$$\frac{dn}{dt} = 0 \tag{3}$$

This condition will be fulfilled when output of the turbine will be the same as output taken by the compressor and accessories of the engine

$$W_{KC} = \eta_m W_{TC} \tag{4}$$

where

η_m – mechanical effectiveness of the engine,

W_{KC} – technical work of the compressor,

W_{TC} – technical work of the turbine.

A detailed algorithm of designation of operational points of steady operation of a single stream engine is described in [7].

Non steady operation of an engine is a regime of its operation, where in every element of the engine time changing thermodynamic processes occur. Function of the engine in such non steady regimes can be described by a system of differential and algebraic equations. Such system of equations describes transient processes by change of regime of the engine, when thrust lever is moved or other change of flight regime occurs.

Such non-steady regime occurs when work of the turbine and compressor isn't equal, this means that rotation moments of the turbine M_T and compressor M_K aren't equal. Acceleration of the engine is dependant upon this difference:

$$M_T - M_K - M_{ag} = J \frac{d\omega}{dt}$$
(5)

where

$\dfrac{d\omega}{dt}$ - angular acceleration of the engine,

J - moment of inertia of all rotating masses reduced to the shaft of the engine

M_{ag} - moment needed for actuation of aggregates and overcoming of friction.

As the angular velocity is given by the equation $\omega = \dfrac{\pi n}{30}$ and output is given by equation $P = M\omega$ and incursion of mechanical effectiveness, the basic equation of non-stable operation of the engine is obtained:

$$P_T \eta_m - P_k = J \frac{\pi^2}{900} n \frac{dn}{dt}$$
(6)

Stable operation of the engine is then computed which gives the initial conditions. Differences of revolutions are then computed in a given time space Δt and we repeat this algorithm until the end of the transient process.

Analytic mathematical model of the engine is based on physical rules which characterize properties and operation of different nodes of the engine, thermodynamic and aerodynamic processes obtained by temperature cycle. While we have to take in account range of operation of turbojet engines which give changes of thermodynamic properties of working material.

Contrary to the experimental one, the analytical model of the engine allows us to compute parameters of the engine that cannot be simply simulated by models built upon the experimental data, which use only known parameters. This way we can

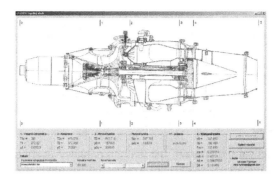

Fig. 5. Temperature circuit calculation worked out in Matlab GUI

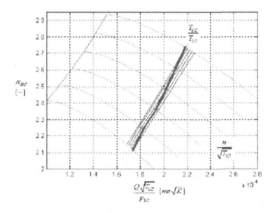

Fig. 6. The curve of steady state of operation of the MPM 20

compute engine surge lines, workloads on shafts, different internal temperatures and also parameters, which are measured and can be used for checking the results of the model. The analytic model allows us to compute parameters of our engine also by different values of outer pressure and temperature of air, different speed of flight and height [12]. Complexity of the model is out of scope of this paper, Figure 6 illustrates computed curve of steady state of operation for the MPM 20 engine. X-axis denotes the air flow through the engine, Y-axis the compression ratio, red line represents surge line, green lines represent different speed (reduced RPM's), and the dark red line represents acceleration of the engine with fast geometry of the exhaust nozzle.

The view of the analytic model in stable regime (and in transitory states) in the form of compressor characteristics (Fig. 6) has the advantage of viewing different important parameters of the engine into one graph. This is advantageous mainly in diagnostics, that means by observation and supervision of important parameters of the engine, if some of the results from sensors doesn't present wrong values, while other parameters of the engine are in normal. This is useful in diagnostics of the engine. Another advantage of the analytic approach is its high precision (we are able to achieve precision within 1% of mean absolute error) [7, 6]. Though there is a

320   L. Főző, R. Andoga, and L. Madarász

deficiency that simulation with analytic model requires high computational power and the needed estimation of certain parameters of the model. We tried to decrease computational demands on simulation of the analytic model by using methods of artificial intelligence to replace complex non-linear equations describing characteristics of the individual parts of the engine.

2.3 Methods of Artificial Intelligence by Analytic Modeling the MPM 20 Turbojet Engine

Resulting from practical expertise of the data and created analytic models we found that adaptive fuzzy inference systems are well suited for replacing the complex equations found in analytic modelling. We used the ANFIS – Adaptive-Network-based Fuzzy Inference System.

This system is based on network architecture just like the neural networks that maps input on the bases of membership fuzzy functions and their parameters to outputs. The network architecture is of feedforward character [15].

To verify the ANFIS method, we are showing a simple physical dependency expressing the pressure ratio of a radial compressor, which is a type of compressor found on the MPM 20 engine.

$$\Pi_{KC} = \left[1 + \frac{u_2^{\,2}(\mu+\alpha)}{c_p T_{1C}} \eta_{KC} \right]^{\frac{\kappa}{\kappa-1}}$$

(7)

The equation can be understood as a static transfer function with two inputs – the temperature T_{1c} and circumferential speed u_2 (speed of the compressor) and one output in the form of pressure ratio. The resulting is shown in Fig. 7. The surface shown in Fig. 7 is equal to numeric computation of the equation 7.

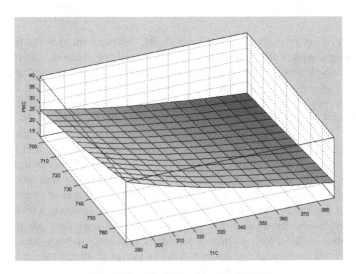

Fig. 7. Equation (7) modeled by ANFIS

The obtained results have confirmed that the chosen method ANFIS is suitable for modeling of any mathematic – physical equation with very low computational demands (trained FIS system is computationaly very simple) by very fine sample period (by very fine interval of values of input parameters). Therefore we will further be oriented on improvement of the complex and highly computationaly demanding analytic model of the MPM 20 engine by use of AI methods, with ANFIS in particular.

3 Conclusions

The developed analytic model and further simulations, drive us to a conclusion that by regulation of the engine by fuel supply, also regulation of exhaust nozzle cross/section plays an important role. This model combined with advanced models can create a backbone for creating control aglorithms with nominal model and for purposes of diagnostics and estimation of parameters of the engine. This can bring increase in efficiency of the engine together with increasing the safety of its operation regardless of operational state of the engine, which represents a key element in the mechatronic complex of an aircraft.

Acknowledgments. The work described in this paper has been supported by VEGA num. 1/0394/08 "Algorithms of situational control and modeling of complex systems".

References

[1] Andoga, R., Főző, L., Madarász, L.: Approaches of Artificial Intelligence in Modeling and Control of a Small Turbojet Engine. In: Proceedings of the microCAD 2007 International Scientific Conference, March 22-23, pp. 1–6. University of Miskolc, Hungary (2007) ISBN 978-963-661-742-4Ö

[2] Andoga, R., Főző, L., Raček, M., Madarász, L.: Inteligentné metódy modelovania a riadenia zložitých systémov. In: Medzinárodná konferencia Kybernetika a informatika, June 28-30, pp. 77–78. Slovenská republika, Michalovce (2006) ISBN 80-227-2431-9

[3] Andoga, R., Madarász, L., Főző, L.: Situational Modeling and Control of a Small Turbojet Engine MPM20. In: IEEE International Conference on Computational Cybernetics, Tallin, Estonia, August 20-22, pp. 81–85 (2006) ISBN 1-4244-0071-6

[4] Andoga, R.: Hybrid Methods of Situational Control of Complex Systems, Dissertation Thesis, Technical University of Košice, p. 120 (2006)

[5] Andoga, R.: Hybridné prístupy v situačnom riadení zložitých systémov., Písomná práca k dizertačnej skúške, FEI TUKE, p. 66 (2005)

[6] Főző, L., Andoga, R., Madarász, L.: Advanced Anytime Control Algorithms and Modeling of Turbojet Engines. In: Computational Intelligence and Informatics: Proceedings of the 10th International Symposium of Hungarian Researchers, November 12-14, 2009, pp. 83–94. Budapest Tech, Budapest (2007) ISBN 978-963-7154-96-6

[7] Főző, L.: Use of Mathematical Model of Steady and Nonsteady Operation of MPM20 Turbojet Engine by Design of Anytime Control Algorithms (in Slovak), Dissertation thesis, Dept. of Cybernetics and AI, Faculty of Electrical Engineerin and Informatics, Technical University of Košice, Slovakia, 144 p (September 2008)

[8] Főző, L., Andoga, R., Madarász, L.: Digital control system for MPM20 engine. In: Proceedings of ICCC 2007, 5th IEEE International Conference on Computational Cybernetics, Gammarth,Tunisia, October 19-20, pp. 281–284 (2007) ISBN 1-4244-1146-7

[9] Hocko, M.: Hodnotenie stavu LTKM na základe zmeny termodynamických parametrov, Dizertačná práca, VLA M.R.Š. Košice (2003)

[10] Lazar, T., et al.: Tendencie vývoja a modelovania avionických systémov, 160 pp, MoSR, Bratislava (2000) ISBN 80-88842-26-3

[11] Madarász, L.: Inteligentné technológie a ich aplikácie v zložitých systémoch. Vydavateľstvo Elfa, s.r.o., TU Košice, p. 349 (2004) ISBN 80–89066–75-5

[12] Považan, J.: Konštrukcia matematických modelov leteckých turbo-kompresorových motorov, VLA M.R.Š. v Košiciach (1999)

[13] Shing, J., Jang, R.: IEEE Transactions on System, Man and Cybernetics 23(3) (May/June 1993)

[14] Takács, O., Várkonyi-Kóczy, A.R.: Szoft számitási eszközök anytime rendszerekben. Híradás-technika 2001/2 május, Volume LVI., pp. 55-59, Takács, O.: Eljárások és hibakorlátok lágy számítási módszerek anytime rendszerekben való alkalmazásához, Doktori (PhD) értekezés, BME Budapest (2004)

[15] Zilberstein, S.: Operational Rationality through Dompilation of Anytime Algorithms. Dissertation for the degree of doctor of philosophy in computer science (1993)

Performance Prediction of Web-Based Software Systems

Ágnes Bogárdi-Mészöly[1], András Rövid[2], and Tihamér Levendovszky[3]

[1] Department of Automation and Applied Informatics, Budapest University of Technology
and Economics, Hungary
agi@aut.bme.hu
[2] Institute of Intelligent Engineering Systems, Óbuda University, Budapest, Hungary
rovid.andras@nik.uni-obuda.hu
[3] Vanderbilt University, USA
tihamer@isis.vanderbilt.edu

Abstract. This paper addresses the issues to establish performance models and evaluation methodologies applying the identified queue limit performance factors. The computational complexity of the novel algorithms is provided. It is demonstrated that the proposed models and algorithms can be used for performance prediction of web-based software systems in ASP.NET environment. The validity of the novel models and algorithms as well as the correctness of the performance prediction is proven with performance measurements. It is shown that the error of the suggested models and algorithms is less than the error of the original model and algorithm. These methods facilitate more efficient performance prediction of web-based software systems.

1 Introduction

The performance of web-based software systems is one of the most important and complicated consideration, because they face a large number of users, and they must provide high-availability services with low response time, while they guarantee a certain throughput level. Performance metrics are influenced by many factors. Several papers have investigated various configurable parameters, and the way in which they affect the performance of web-based software systems [1] [2].

The performance metrics can be predicted at early stages of the development process with the help of a properly designed performance model and an appropriate evaluation algorithm. In the past few years several methods have been proposed to address this issue. Several of them is based on queueing networks or extended versions of queueing networks [3]. Another group is using Petri-nets or generalized stochastic Petri-nets [4]. As the third kind of the approaches, the stochastic extension of process algebras, such as TIPP (Time Processes and Performability Evaluation) [5], EMPA (Extended Markovian Process Algebra) [6], and PEPA (Performance Evaluation Process Algebra) [7] can be mentioned.

I.J. Rudas et al. (Eds.): Computational Intelligence in Engineering, SCI 313, pp. 323–336.
springerlink.com   © Springer-Verlag Berlin Heidelberg 2010

Web-based software systems access some resources while executing the requests of the clients. Typically several requests arrive at the same time, thus, competitive situation is established for the resources. For modeling such situations queueing model-based approaches are widely used. Queueing networks have also been proposed to model web-based software systems [8] [9] [10].

The paper is organized as follows. Section 2 introduces the background and related work. Section 3.1 provides novel models and algorithms to model the queue limit. Section 3.2 shows that the proposed models and algorithms can be applied to performance prediction of web-based software systems in ASP.NET environment. Section 3.3 verifies the correctness of the performance prediction.

2 Background and Related Work

This section is devoted to review the background and research efforts related this work, namely, the concept of thread pools and queued requests as well as the queueing network models for multi-tier software systems, which have led to the extensive research of performance models of web-based software systems.

2.1 Thread Pools and Queued Requests

In case of using a thread pool, when a request arrives, the application adds it to an incoming queue [11]. A group of threads retrieves requests from this queue and processes them. As each thread is freed, another request is executed from the queue.

The architecture of ASP.NET environment can be seen in Fig. 1. If a client is requesting a service from the server, the request goes through several subsystems before it is served.

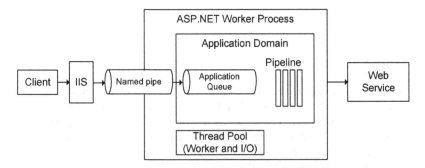

Fig. 1. The architecture of ASP.NET environment

From the IIS, the accepted HTTP connections are placed into a named pipe. This is a global queue between IIS and ASP.NET, where the requests are posted from native code to the managed thread pool. The global queue is managed by the process that runs ASP.NET, its limit is set by the *requestQueueLimit* property. When the Requests Current counter – which includes the requests that are queued, being executed, or waiting to be written to the client – reaches this limit, the requests are rejected [12]. Thus, the Requests Current must be greater than the sum of *maxWorkerThreads* plus *maxIOThreads*. From the named pipe, the requests are placed into an application queue, also known as a virtual directory queue. Each virtual directory has a queue that is used to maintain the availability of worker and I/O threads. The number of requests in these queues increases if the number of available worker and I/O threads falls below the limit specified by *minFreeThreads* property. The application queue limit is configured by the *appRequestQueueLimit* property. When the limit is exceeded, the requests are rejected.

The .NET Framework offers a highly optimized thread pool. This pool is associated with the physical process where the application is running, there is only one pool per process. The *maxWorkerThreads* attribute means the maximum number of worker threads, the *maxIOThreads* parameter is the maximum number of I/O threads in the .NET thread pool. These attributes are automatically multiplied by the number of available CPUs. The *minFreeThreads* attribute limits the number of concurrent requests, because all incoming requests will be queued if the number of available threads in the thread pool falls below the value for this setting. The *minLocalRequestFreeThreads* parameter is similar to *minFreeThreads*, but it is related to the requests from the local host.

In our previous work [2] [13], six influencing parameters of the performance have been found: *maxWorkerThreads*, *maxIOThreads*, *minFreeThreads*, *minLocalRequestFreeThreads*, *requestQueueLimit*, *appRequestQueueLimit*, which is proven by a statistical method, namely, the chi square test of independence. The identified performance factors must be modeled to improve performance models. The behavior of the thread pool has been modeled in our previous work [14]. This paper introduces the modeling of the queue size limit.

2.2 Queueing Models for Multi-tier Software Systems

This section discusses the base queueing network model and the Mean-Value Analysis evaluation algorithm for multi-tier software systems used in this paper.

A product form network should satisfy the conditions of job flow balance, one-step behavior, and device homogeneity [15]. The job flow balance assumption holds only in some observation periods, namely, it is a good approximation for long observation intervals since the ratio of unfinished jobs to completed jobs is small.

Definition 1. The **base queueing model** is defined for multi-tier information systems [9] [16], which are modeled as a network of M queues $Q_1,...,Q_M$ illustrated in Fig. 2. Each queue represents an application tier. S_m denotes the service time of a request at Q_m ($1 \leq m \leq M$). A request can take multiple visits to each queue during its overall execution, thus, there are transitions from each queue to its successor and its predecessor, as well. Namely, a request from queue Q_m either returns to Q_{m-1} with a certain probability p_m, or proceeds to Q_{m+1} with the probability $1-p_m$. There are only two exceptions: the last queue Q_M, where all the requests return to the previous queue ($p_M=1$) and the first queue Q_1, where the transition to the preceding queue denotes the completion of a request. Internet workloads are usually session-based. The model can handle session-based workloads as an infinite server queueing system Q_0 that feeds the network of queues and forms the closed queueing network depicted in Fig. 2. Each active session is in accordance with occupying one server in Q_0. The time spent at Q_0 corresponds to the user think time Z.

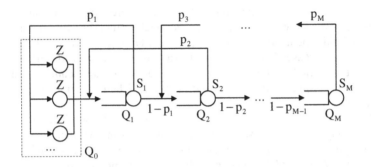

Fig. 2. Modeling a multi-tier information system using a queueing network

Remark 1. There are transitions from each queue only to its successor and its predecessor, because in multi-tier architecture only neighboring tiers communicate each other directly. The transition of Q_1 to the preceding queue denotes the completion of a request, since requests of clients are started from and completed in the presentation tier (in three-tier architecture the presentation tier is represented by Q_1).

Remark 2. In this paper, closed queueing networks are applied, since closed queueing models predict the performance for a fixed number of sessions serviced by the application.

The Mean-Value Analysis (MVA) algorithm for closed queueing networks [3] [17] is applicable only if the network is in product form. In addition, the queues are assumed to be either fixed-capacity service centers or infinite servers, and in both cases, exponentially distributed service times are assumed.

Remark 3. Since the base queueing model satisfies the conditions above, the MVA algorithm can evaluate the base queueing model.

Definition 2. The *Mean-Value Analysis* for the base queueing model is defined by Algorithm 1, and the associated notations are in Table 1.

Algorithm 1. Pseudo code of the MVA algorithm

1: **for all** $m = 1$ to M **do**
2:     $L_m = 0$
3: **for all** $n = 1$ to N **do**
4:     **for all** $m = 1$ to M **do**
5:         $R_m = V_m \cdot S_m \cdot (1 + L_m)$
6:     $R = \sum_{m=1}^{M} R_m$
7:     $\tau = n / (Z + R)$
8:     **for all** $m = 1$ to M **do**
9:         $L_m = \tau \cdot R_m$

Table 1. Notations of Mean-Value Analysis evaluation algorithm

Notation	Meaning	I/O	Comment
L_m	queue length of Q_m $(1 \leq m \leq M)$	O	
M	number of tiers	I	
N	number of customers	I	number of sessions in the base queueing model
R	response time	O	
R_m	response time for Q_m $(1 \leq m \leq M)$	O	
S_m	service time for Q_m $(1 \leq m \leq M)$	I	
τ	throughput	O	
V_m	visit number for Q_m $(1 \leq m \leq M)$	I	derived from transition probabilities p_m of the base queueing model
Z	user think time	I	

The model can be evaluated for a given number of concurrent sessions, note that a session in the model corresponds to a customer in the evaluation algorithm. The algorithm uses visit numbers instead of transition probabilities. The visit number is the number of times when a tier is invoked during the lifetime of a request, and visit numbers can be easily derived from transition probabilities.

Remark 4. The visit number concept of the MVA algorithm is more general than the transition probabilities from a queue to its successor and its predecessor in case of the base queueing model. Thus, the model could be more general, but in practice only neighboring tiers communicate each other directly.

3 Modeling the Queue Limit

The queue size must be limited to prevent requests from consuming all the memory for the server, for an application queue. By taking the queue limit (Section 2) into consideration, the base queueing model (Definition 1) and the MVA evaluation algorithm (Definition 2) can be effectively enhanced. The enhancement of the baseline model handles such concurrency limits, when each tier has an individual concurrency limit [9]. This approach manages the case in which all tiers have a common concurrency limit.

This section discusses modeling the queue limit [18]. Firstly, this chapter proposes novel models and algorithms to model the global and the application queue limit performance factors. Then, it validates the proposed models and algorithms, and verifies the correctness of performance prediction with the novel models and algorithms.

3.1 Novel Models and Algorithms

Definition 3. The ***extended queueing model with global queue limit*** (QM-GQL) is defined by Fig. 3, where the Q_{drop} is an infinite server queueing system, the Z_{drop} is the time spent at Q_{drop}, the *GQL* is the global queue limit, which corresponds to the *requestQueueLimit* parameter in ASP.NET environment. If the *GQL* is less than the queued requests sum, the next requests proceed to Q_{drop}. Requests from Q_{drop} proceed back to Q_0, namely, these requests are reissued.

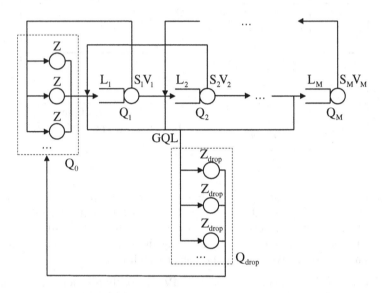

Fig. 3. Extended queueing model with global queue limit

Definition 4. The ***extended MVA with global queue limit*** (MVA-GQL) is defined by Algorithm 2, where the Z_{drop} is the time spent at Q_{drop}, the *GQL* is the global queue limit, which corresponds to the *requestQueueLimit* parameter in ASP.NET environment.

Algorithm 2. Pseudo code of the MVA-GQL

1: **for all** $m = 1$ to M **do**
2:     $L_m = 0$
3: $nql = 1$
4: **for all** $n = 1$ to N **do**
5:     **for all** $m = 1$ to M **do**
6:         $R_m = V_m \cdot S_m \cdot (1 + L_m)$
7:     $R = \sum_{m=1}^{M} R_m$
8:     $\tau = nql / (Z + Z_{drop} + R)$

9:   **for all** $m = 1$ to M **do**

10:     $L_m = \tau \cdot R_m$

11:   **if** $\displaystyle\sum_{m=1}^{M} L_m > GQL$ **then**

12:     **for all** $m = 1$ to M **do**

13:       $L_m = oldL_m$

14:   **else**

15:     $nql = nql + 1$

16:   **for all** $m = 1$ to M **do**

17:     $oldL_m = L_m$

Considering the concept of the global queue, if the current requests – queued plus executing requests (by ASP.NET) – exceed the global queue limit (GQL), the next incoming requests will be rejected. In these cases, the queue length does not have to be updated (see Steps 10 and 13 of Algorithm 2). The queued requests sum of the model and algorithm contains not only the queued requests by ASP.NET but the working threads of ASP.NET, as well.

Proposition 1. The novel MVA-GQL (Definition 4) can be applied as an approximation method to the proposed QM-GQL (Definition 3).

Proof. The QM-GQL model does not satisfy the condition of job flow balance (see in Section 2.2). Thus, the MVA-GQL evaluation algorithm can be applied as an approximation method to the QM-GQL model. In Step 8 of Algorithm 2, when it computes the throughput, the Z_{drop} of the model is taken into consideration similarly to Z. In Steps 11 and 13 of the algorithm, if the GQL is less than the queued requests sum, the next requests proceed to Q_{drop} in the model, the queue length does not have to be updated in the algorithm.

Proposition 2. The computational complexity of the novel MVA-GQL (Definition 4) is $\theta(N \cdot M)$, where N is the number of customers and M is the number of tiers.

Proof. Assume that each execution of the ith line takes time c_i, where c_i is a constant. The total running time is the sum of running times for each statement executed. A statement that takes c_i time to execute and is executed n times contributes $c_i \cdot n$ to the total running time.

The worst-case running time of this novel algorithm can be seen in Equation 1. If the number of customers N and the number of tiers M is finite, the computational time is finite, the algorithm is terminating.

Consider only the leading term of the formula, since the lower-order terms are relatively insignificant for large N and M. The constant coefficient of the leading term can be ignored, since constant factors are less significant than the order of growth in determining computational efficiency for large inputs.

Since the order of growth of the best-case and worst-case running times is the same, the asymptotic lower and upper bounds are the same, thus, the computational complexity is $\theta(N \cdot M)$.

$$(c_5 + c_6 + c_9 + c_{10} + c_{12} + c_{13} + c_{16} + c_{17}) \cdot N \cdot M +$$
$$(c_4 + c_5 + c_7 + c_8 + c_9 + c_{11} + c_{12} + c_{16}) \cdot N + \qquad (1)$$
$$(c_1 + c_2) \cdot M + (c_1 + c_3 + c_4)$$

Definition 5. The *extended queueing model with application queue limit* (QM-AQL) is defined by Fig. 4, where the Q_{drop} is an infinite server queueing system, the Z_{drop} is the time spent at Q_{drop}, the *AQL* is the application queue limit, which corresponds to the *appRequestQueueLimit* parameter in ASP.NET environment, in addition, the *WT* is the maximum number of working threads, which equals *maxWorkerThreads+maxIOThreads-minFreeThreads* in ASP.NET environment. If the *AQL+WT* is less than the queued requests sum, the next requests proceed to Q_{drop}. Requests from Q_{drop} proceed back to Q_0, namely, these requests are reissued.

Definition 6. The *extended MVA with application queue limit* (MVA-AQL) is defined by Algorithm 3, where the Z_{drop} is the time spent at Q_{drop}, the *AQL* is the application queue limit, which corresponds to the *appRequestQueueLimit* parameter in ASP.NET environment, in addition, the *WT* is the maximum number of working threads, which equals *maxWorkerThreads+maxIOThreads-minFreeThreads* in ASP.NET environment.

Considering the concept of the application queue, if the number of queued requests (by ASP.NET) exceeds the application queue limit (*AQL*), the next incoming requests will be rejected. In these cases, the queue length has not to be updated (see Steps 10 and 13 of Algorithm 3). Since the queued requests sum of the model and the algorithm contains not only the queued requests by ASP.NET but the working threads of ASP.NET, as well. Thus, *WT* has to be subtracted from the number of queued requests of the model and algorithm to obtain the queued requests by ASP.NET.

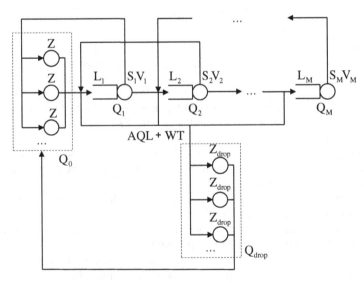

Fig. 4. Extended queueing model with application queue limit

Algorithm 3. Pseudo code of the MVA-AQL

```
1: for all m = 1 to M do
2:     Lₘ = 0
3: nql = 1
4: for all n = 1 to N do
5:     for all m = 1 to M do
6:         Rₘ = Vₘ · Sₘ · (1 + Lₘ)
```

$$7: \quad R = \sum_{m=1}^{M} R_m$$

```
8:     τ = nql / (Z + Z_drop + R)
9:     for all m = 1 to M do
10:        Lₘ = τ · Rₘ
```

$$11: \quad \text{if } \sum_{m=1}^{M} L_m - WT > AQL \text{ then}$$

```
12:        for all m = 1 to M do
13:            Lₘ = oldLₘ
14:    else
15:        nql = nql + 1
16:    for all m = 1 to M do
17:        oldLₘ = Lₘ
```

Proposition 3. The novel MVA-AQL (Definition 6) can be applied as an approximation method to the proposed QM-AQL (Definition 5).
Proof See the proof of Proposition 1.

Proposition 4. The computational complexity of the novel MVA-AQL (Definition 6) is $\theta(N \cdot M)$, where N is the number of customers and M is the number of tiers.
Proof See the proof of Proposition 2.

These extensions do not increase the complexity of the evaluation algorithm, because the computational complexity of the original algorithm is the same.

3.2 Performance Prediction and Validation

In this section, it is shown that the proposed models and algorithms can be used for performance prediction of web-based software systems. Firstly, the proposed models and algorithms have been implemented with the help of MATLAB. Secondly, the input values have been measured or estimated. For model parameter estimation and model evaluation, only one measurement or estimation in case of one customer is required in ASP.NET environment. Then, the proposed models have been evaluated by the novel algorithms to predict the performance metrics.

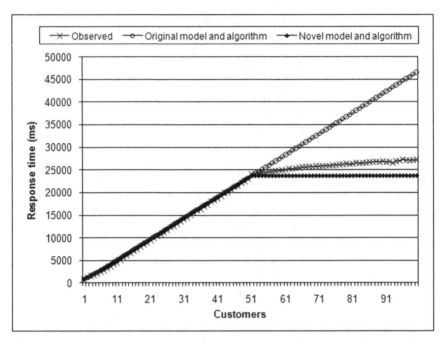

Fig. 5. Observed response time, predicted with original MVA, and predicted with novel MVA-GQL

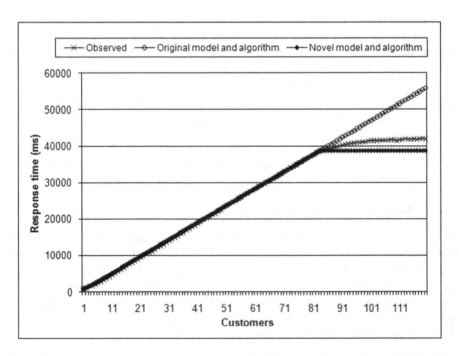

Fig. 6. Observed response time, predicted with original MVA, and predicted with novel MVA-AQL

Proposition 5. The novel QM-GQL and MVA-GQL as well as QM-AQL and MVA-AQL (Definitions 3 and 4, 5 and 6) can be applied to performance prediction of web-based software systems in ASP.NET environment, these proposed models and algorithms predict the performance metrics more accurate than the original base queueing model and MVA (Definitions 1 and 2).

Proof. The novel models and algorithms have been validated and the correctness of the performance prediction with the proposed models and algorithms has been verified with performance measurements in ASP.NET environment. These validation and verification have been performed by comparing the observed values provided by performance measurements in ASP.NET environment, the predicted values with the original model and algorithm, and the predicted values with the proposed models and algorithms. The results depicted in Figs. 5 and 6 have shown that the outputs of the proposed models and algorithms approximate the measured values much better than the outputs of the original model and algorithm considering the shape of the curve and the values, as well. Thus, the proposed models and algorithms predict the response time and throughput performance metrics much more accurate than the original model and algorithm in ASP.NET environment.

3.3 Error Analysis

Finally, the error has been analyzed to verify the correctness of the performance prediction with the proposed models and algorithms. Two methods have been applied: the average absolute error function and the error histogram.

Definition 7. The *average absolute error function* is defined by Equation 2, where the index *alg* corresponds to the values provided by the algorithm, the index *obs* is related to the observed values, R is the response time, and N is the number of measurements.

$$error_{alg-obs} = (\sum_{i=1}^{N} |R_{alg_i} - R_{obs_i}|)/N \qquad (2)$$

Proposition 6. The average absolute error of the response time performance metric provided by the novel QM-GQL and MVA-GQL as well as QM-AQL and MVA-AQL (Definitions 3 and 4, 5 and 6) is less than the average absolute error of the response time computed by the original base queueing model and MVA (Definitions 1 and 2).

Proof The results of the average absolute error function are presented in Table 2. The results have been shown that the error of the novel models and algorithms is substantially less than the error of original model and algorithm in ASP.NET environment.

Table 2. The results of the average absolute error function

	Global queue	Application queue
Original	4774.7	2139.3
Novel	1284.3	888.4417

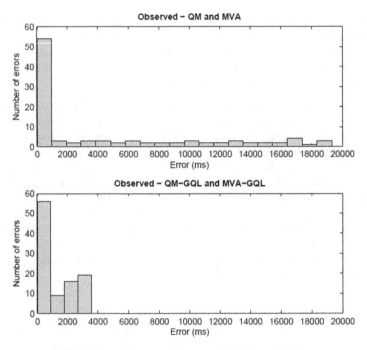

Fig. 7. Error histogram in case of global queue limit

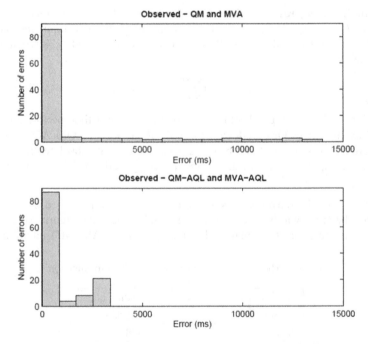

Fig. 8. Error histogram in case of application queue limit

The error histograms of the original and the proposed models and algorithms are depicted in Figs. 7 and 8.

The values of the error (x-axis) with the proposed models and algorithms are substantially less than with the original model and algorithm. The number of errors in the first bin (y-axis) of the original and the novel models and algorithms are approximately the same, because the response time and the throughput computed by the proposed models and algorithms are the same as the performance metrics computed by the original model and algorithm, until all the requests have successfully been served. With the novel models and algorithms, the number of errors in the first bins (y-axis) is greater, since all of the errors are only in these first bins. In case of the original model and algorithm, the errors increase at much greater rate.

4 Conclusions

The queue limit parameters are dominant factors considering the response time and throughput performance metrics. These identified performance factors must be modeled to improve performance models. In this paper, novel models and algorithms have been provided to model the global and application queue limit. It has been proven that the novel algorithms can be applied as an approximation method to the proposed models. The computational complexity of the proposed algorithms has been proven. It has been shown that the proposed models and algorithms can be applied to performance prediction of web-based software systems in ASP.NET environment, the novel models and algorithms predict the performance metrics more accurate than the original model and algorithm. The proposed models and algorithms have been validated and the correctness of performance prediction with novel models and algorithms has been verified with performance measurements in ASP.NET environment. Error has been analyzed to verify the correctness of the performance prediction. The results have shown that the proposed models and algorithms predict the performance metrics much more accurate than the original model and algorithm.

With the novel models and algorithms of this paper, the global and application queue limit performance factors can be modeled, in addition, performance prediction can be performed much more accurate. The extended MVA with thread pool [14] and the extended MVA with queue limit are independent and additive, thus, they can be applied together or separately, as well.

Acknowledgments. This research was supported by the János Bolyai Research Scholarship of the Hungarian Academy of Sciences.

References

[1] Sopitkamol, M., Menascé, D.A.: A Method for Evaluating the Impact of Software Configuration Parameters on E-commerce Sites. In: ACM 5th International Workshop on Software and Performance, pp. 53–64 (2005)
[2] Bogárdi-Mészöly, Á., Szitás, Z., Levendovszky, T., Charaf, H.: Investigating Factors Influencing the Response Time in ASP. In: Bozanis, P., Houstis, E.N. (eds.) PCI 2005. LNCS, vol. 3746, pp. 223–233. Springer, Heidelberg (2005)

[3] Jain, R.: The Art of Computer Systems Performance Analysis. John Wiley and Sons, Chichester (1991)

[4] Bernardi, S., Donatelli, S., Merseguer, J.: From UML Sequence Diagrams and Statecharts to Analysable Petri Net Models. In: ACM International Workshop Software and Performance, pp. 35–45 (2002)

[5] Herzog, U., Klehmet, U., Mertsiotakis, V., Siegle, M.: Compositional Performance Modelling with the TIPPtool. Performance Evaluation 39, 5–35 (2000)

[6] Bernardo, M., Gorrieri, R.: A Tutorial on EMPA: A Theory of Concurrent Processes with Nondeterminism, Priorities, Probabilities and Time. Theoretical Computer Science 202, 11–54 (1998)

[7] Gilmore, S., Hillston, J.: The PEPA Workbench: A Tool to Support a Process Algebra-Based Approach to Performance Modelling. In: International Conference Modelling Techniques and Tools for Performance Evaluation, pp. 353–368 (1994)

[8] Menascé, D.A., Almeida, V.: Capacity Planning for Web Services: Metrics, Models, and Methods. Prentice Hall PTR, Englewood Cliffs (2001)

[9] Urgaonkar, B.: Dynamic Resource Management in Internet Hosting Platforms, Dissertation, Massachusetts (2005)

[10] Smith, C.U., Williams, L.G.: Building Responsive and Scalable Web Applications. In: Computer Measurement Group Conference, pp. 127–138 (2000)

[11] Carmona, D.: Programming the Thread Pool in the.NET Framework,NET Development (General) Technical Articles (2002)

[12] Marquardt, T.: ASP.NET Performance Monitoring, and When to Alert Administrators, ASP.NET Technical Articles (2003)

[13] Bogárdi-Mészöly, Á., Levendovszky, T., Charaf, H.: Performance Factors in ASP.NET Web Applications with Limited Queue Models. In: 10th IEEE Int. Conference on Intelligent Engineering Systems, pp. 253–257 (2006)

[14] Bogárdi-Mészöly, Á., Hashimoto, T., Levendovszky, T., Charaf, H.: Thread Pool-based Im-provement of the Mean-Value Analysis Algorithm. Lecture Notes in Electrical Engineering, vol. 28(2), pp. 347–359 (2009)

[15] Denning, P., Buzen, J.: The Operational Analysis of Queueing Network Models. Computing Surveys 10, 225–261 (1978)

[16] Urgaonkar, B., Pacifici, G., Shenoy, P., Spreitzer, M., Tantawi, A.: An Analytical Model for Multi-tier Internet Services and its Applications. ACM SIGMETRICS Performance Evaluvation Review 33(1), 291–302 (2005)

[17] Reiser, M., Lavenberg, S.S.: Mean-Value Analysis of Closed Multichain Queuing Networks. Association for Computing Machinery 27, 313–322 (1980)

[18] Bogárdi-Mészöly, Á., Rövid, A., Levendovszky, T.: Novel Models and Algorithms for Queue Limit Modeling. In: 10th International Symposium of Hungarian Researchers on Computational Intelligence and Informatics, pp. 579–590 (2009)

Optimization in Fuzzy Flip-Flop Neural Networks

Rita Lovassy[1,2], László T. Kóczy[1,3], and László Gál[1,4]

[1] Inst. of Informatics, Electrical and Mechanical Engineering, Faculty of Engineering Sciences,
Széchenyi István University Győr, Hungary
[2] Inst. of Microelectronics and Technology, Kandó Kálmán Faculty of Electrical Engineering,
Óbuda University, Budapest, Hungary
[3] Dept. of Telecommunication and Media Informatics, Budapest University of Technology and
Economics, Hungary
[4] Dept. of Technology, Informatics and Economy University of West Hungary
lovassy.rita@kvk.uni-obuda.hu, koczy@sze.hu,
gallaci@ttmk.nyme.hu

Abstract. The fuzzy J-K and D flip-flops present s-shape transfer characteristics in same particular cases. We propose the fuzzy flip-flop neurons; single input-single output units derived from fuzzy flip-flops as sigmoid function generators. The fuzzy neurons-based neural networks, Fuzzy Flip-Flop Neural Networks (FNN) parameters are quasi optimized using a second-order gradient algorithm, the Levenberg-Marquardt method (LM) and an evolutionary algorithm, the Bacterial Memetic Algorithm with Modified Operator Execution Order (BMAM). The quasi optimized FNN's performance based on Dombi and Yager fuzzy operations has been examined with a series of test functions.

1 Introduction

Artificial neural networks and fuzzy logic models can approximate any multivariate nonlinear function [7]. The aim of this paper is essentially to show how fuzzy models, neural networks, and bacterial algorithms can be usefully merged and deployed to solve parameter optimization and function approximation problems. The combination of these three techniques tries to minimize their weakness and to profit their advantages. The main idea is using the property of learning from the data of the network using a parallel structure which can be implemented in hardware.

In analogy with the binary flip-flops which are the basic units in every synchronous sequential digital circuit, the fuzzy flip-flops can be reckoned as the basic functional blocks for developing dynamical fuzzy systems. The fuzzy flip-flops used as neurons are interlinking to form a complex fuzzy neural network. How these neurons are organized is itself a highly complex problem. This study has been focused more or less on the neuro-computations direction. A Fuzzy Flip-Flop Neural Network (FNN) as a novel implementation possibility of multilayer perceptron neural networks is investigated and its optimization and learning algorithm is proposed.

I.J. Rudas et al. (Eds.): Computational Intelligence in Engineering, SCI 313, pp. 337–348.
springerlink.com

The paper is organized as follows. After the Introduction, in Section 2 we present the fuzzy J-K and D flip-flops in general. The fuzzy neurons derived from fuzzy J-K flip-flop with feedback and fuzzy D flip-flop are given. In Section 3 the concept of the fuzzy flip-flop (F^3) based multilayer perceptron is introduced. The neuron element of FNN may be any F^3 with more or less sigmoidal transfer characteristics. The FNN can approximate any multidimensional input functions. The Levenberg-Marquardt method (LM) [11] and a special kind of bacterial memetic algorithm [14], the Bacterial Memetic Algorithm with Modified Operator Execution Order (BMAM) [3] is proposed for FNN parameter optimization and network training. The near optimal values of Q are given which are quasi independent from the input function complexity and FNN neuron number for a fix fuzzy operation parameter value. Finally, in Section 4 the quasi optimized FNN's performance based on various fuzzy flip-flop types has been examined with a series of multidimensional input functions.

2   Fuzzy J-K and D Flip-Flops

The concept of fuzzy flip-flop was introduced in the middle of 1980's by Hirota (with his students) [4]. The Hirota Lab recognized the essential importance of the concept of a fuzzy extension of a sequential circuit and the notion of fuzzy memory. From this point of view they proposed alternatives for "fuzzifying" digital flip-flops. The starting elementary digital units were the binary J-K flip-flops. Their definitive equation was used both in the minimal disjunctive and conjunctive forms. As fuzzy connectives do not satisfy all Boolean axioms, the fuzzy equivalents of these equations resulted in two non-equivalent definitions, "reset and set type" fuzzy flip-flops, using the concepts of fuzzy negation, t-norm and t-conorm operations. In [5] Hirota et al. recognized that the reset and set equations cannot be easily used as elements of memory module, because of their asymmetrical nature. In their paper [13] Ozawa, Hirota and Kóczy proposed a unified form of the fuzzy J-K flip-flop characteristic equation, involving the reset and set characteristics, based on min-max and algebraic norms. A few years later, the hardware implementation of these fuzzy flip-flop circuits in discrete and continuous mode was presented in [14]. The fuzzy flip-flop was proposed as the basic unit in fuzzy register circuits. The unified formula of the fuzzy J-K flip-flop was expressed as follows:

$$Q_{out} = (J \vee \overline{K}) \wedge (J \vee Q) \wedge (\overline{K} \vee \overline{Q}) \tag{1}$$

Where $\wedge$, $\vee$ and $^-$ denote fuzzy operations, in particular fuzzy conjunction, disjunction and negation respectively (e.g. $\overline{K} = 1 - K$). As a matter of course, it is possible to substitute the standard operations by any other reasonable fuzzy operation triplet (e.g. De-Morgan triplet), thus obtaining a multitude of various fuzzy flip-flop pairs.

2.1   Fuzzy Flip-Flop Neuron

A single input-single output unit derived from fuzzy J-K flip-flop where $\overline{Q}$ is fed back to K ($K = 1 - Q$) and (old) Q is fixed is proposed (Fig. 1).

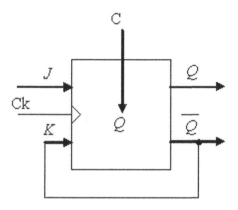

Fig. 1. Fuzzy J-K flip-flop neuron

The clocked fuzzy J-K flip-flop neuron circuit can be built up using hardware blocks (denoted by i, u and n symbols) to realize various t-norms, t-conorms and fuzzy negations [16]. Since t-norms and t-conorms are functions from the unit square into the unit interval, the fuzzy J-K flip-flop block diagram differs from the binary J-K flip-flop structure. The input J is driven by a synchronized clock pulse in the sample-and- hold (S/H) circuit (Fig 2).

The output of fuzzy J-K flip-flop neuron depends from the value of Q_{fix} and input values of J:

$$Q_{out} = (J \ u \ Q_{fix}) \ i \ (J \ u \ Q_{fix}) \ i \ \left(Q_{fix} \ u \ (1 - Q_{fix})\right) \tag{2}$$

The concept of a novel fuzzy D flip-flop type was introduced in [8]. Connecting the inputs of the fuzzy J-K flip-flop in a particular way, namely, by applying an inverter in the connection of the input J to K, case of $K = 1 - J$, a fuzzy D flip-flop is obtained (Fig. 3).

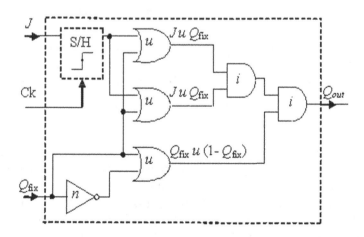

Fig. 2. Fuzzy J-K flip-flop neuron block diagram

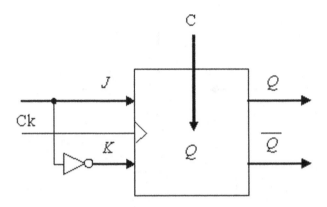

Fig. 3. Fuzzy D flip-flop neuron

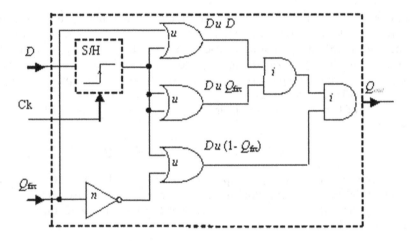

Fig. 4. Fuzzy D flip-flop neuron block diagram

Interconnecting the blocks of fuzzy operations in a different way a fuzzy D flip-flop neuron is obtained.

Substitute $\overline{K} = J$ in equation (1) and let $D = J$, the characteristic equation of fuzzy D flip-flop neuron is

$$Q_{out} = (D\ u\ D)\ i\ (D\ u\ Q_{fix})\ i\ \left(D\ u\ \left(1 - Q_{fix}\right)\right) \tag{3}$$

In our previous papers [8], [10] the unified approach based on various norms combined with the standard negation, was analyzed in order to investigate, whether and to what degree they present more or less sigmoid (s-shaped) $J \rightarrow Q(t+1)$ characteristics in particular cases, when $K = 1 - Q$ (unified fuzzy J-K flip-flop with feedback), $K = 1 - J$ (new fuzzy D flip-flop derived from the unified fuzzy J-K one) with fixed value of Q. We have conducted extensive investigations by simulations and found that the $J \rightarrow Q(t+1)$ transfer characteristics of fuzzy J-K flip-flops with feedback based on Łukasiewicz, Yager, Dombi and Hamacher norms, further the $D \rightarrow Q(t+1)$ characteristics of (new) fuzzy D flip-flops of Łukasiewicz, Yager and Hamacher operations show quasi sigmoid curvature. The fuzzy J-K and D F^3s based on algebraic norms, furthermore the fuzzy D F^3s based on Dombi norms have non-sigmoidal behavior.

3 Multilayer Perceptrons Constructed of Fuzzy Flip-Flop Neurons (Fuzzy Neural Network, FNN)

In general, two trainable layer networks with sigmoid transfer functions in the hidden layers and linear transfer function in the output layer are universal approximators [6]. The neuro-fuzzy system proposed here is based on two hidden layers constituted from fuzzy flip-flop neurons. The FNN is a supervised feedforward network, applied in order to approximate a real life application, two dimensional trigonometric functions and a six dimensional benchmark problem.

3.1 Parameter Optimization in the Fuzzy Neural Network

A change of t-norms, Q and fuzzy operations parameter value pairs in the fuzzy flip-flops characteristic equations leads to the modification of the slope of the transfer function, which will affect the learning rate in the implementation of neural networks. In the next it will be show that the near-optimal values of Q are independent from the input function type and FNN neuron number for a fix fuzzy operation parameter value. For simplicity the value of the parameters were fixed (equal to 2 and respectively to 5 in case of Yager and Dombi fuzzy operations). It is known that simple parametrical t-norms, furthermore simple fuzzy flip-flop characteristic equation are uncomplicated for tuning and hardware realization. Rudas et al. [15] proposed the hardware implementation of a generation of parametric families of fuzzy connectives together with min-max, Łukasiewicz and drastic t-norms and t-conorms. In [16] Zavala et al. used FPGA technology to implement the above mentioned fuzzy operations into an 8 bit single circuit that allows operation selection.

In this section optimal FNN parameter searching methods are presented. In particular for predefined network architecture a gradient descent optimization method and an evolutionary algorithm tune the values of the parameter Q for fixed fuzzy operation parameter values. The application of several model identification algorithms in order to achieve as good as possible approximation features of the FNN have been proposed. The Levenberg-Marquardt algorithm is applied to determine the optimal Q intervals in subsection 3.1.1. In subsection 3.1.2 the BMAM method is proposed to identify in the established interval the problem dependent quasi optimal values.

3.1.1 Levenberg-Marquardt Algorithm (LM)

A second-order gradient algorithm, the Levenberg-Marquardt method is applied to find the optimal Q intervals for every combination of fuzzy J-K and D flip-flops with Yager and Dombi fuzzy operations.

To train this kind of network with the usual Levenberg-Marquard technique the derivatives of the transfer functions have to be calculated. Then the FNN can be used and trained in the usual way. During network training, the weights and thresholds are first initialized to small, random values and the network was trained in order to minimize the network performance function according with Levenberg-Marquardt algorithm with 100 maximum numbers of epochs as more or less sufficient. This can be used for training any network as long as its inputs, weights, and transfer functions can be differentiated. During the simulations the input data patterns were used like this: the first 60 percent for training and the remaining 20-20 percents for validation and test results. In order to check the goodness of the training in the test phase, the Mean Squared Error (MSE) as a measure of the error made by the FNN was used.

The errors for all input patterns were propagated backwards, from the output layer towards the input layer. The corrections to the weights were selected to minimize the residual error between actual and desired outputs. The chosen target activation function is the *tansig* (hyperbolic tangent sigmoid transfer function).

The lowest MSE value indicated the optimal Q value which might lead to good learning and approximation properties. The result obtained for the training and test sets give the 300 runs average approximation goodness value. During simulations the median MSE values have been compared, considering them as the most important indicators of trainability. The median is a robust estimate of the center of a data sample, since outliers have little effect on it.

3.1.2   Bacterial Memetic Algorithm with Modified Operator Execution Order Algorithm (BMAM)

In the Bacterial Memetic Algorithm with Modified Operator Execution Order Algorithm, the learning of a neural network is formulated as a weight optimization problem, usually using the mean square error as fitness evaluation scheme. It has been shown that evolutionary algorithms work efficient for solving nonlinear and constrained optimization problems. These methods do not use derivatives of the functions, such as the gradient-based training algorithms. Similar to biological recombination, these methods are based on the search for a large number (population) of solutions. The algorithm starts with several alternative solutions to the optimization problem, which are the population individuals. The solutions are coded as binary strings. The basic steps followed by the algorithm embrace the bacterial mutation operation and the LM method.

The FNN weights, biases and Q values have been encoded in a chromosome and participated in the bacterial mutation cycle. In this application, according to the network size, a population with different parameters was initialized. Therefore a procedure is working on changing the variables, testing the model obtained in this way and selecting the best models. During simulations 30 generations of 5 individuals with 5 clones were chosen to obtain the best fitting variable values, with the lowest performance. Then the same part or parts of the chromosome is choose and mutate randomly, except one single clone that remains unchanged during this mutation cycle. The LM method nested into evolutionary algorithm is applied for 5 times for each individual. The selection of the best clone is made and transfers its parts to the other clones. The part choosing-mutation-LM method-selection-transfer cycle is repeated until all the parts are mutated, improved and tested. The best individual is remaining in the population, and all other clones are deleted. This process is repeated until all the individuals have gone through the modified bacterial mutation. The Levenberg-Marquardt method is applied 3 times for each individual, executing several LM cycles during the bacterial mutation after each mutation step.

In this way the local search is done for every global search cycle. The quasi optimal values can be identified at the end of the BMAM training algorithm.

Using the BMAM algorithm, executing several LM cycles after each bacterial mutation step, there is no need to run a few hundred training cycles to find the lowest MSE value. After just one run the optimal parameter values could be identify. The optimal Q intervals obtained with the LM algorithm have been reduced in this case to a single quasi optimal value identified at the end of the training process. In this approach, with only one training cycle an acceptable model whose error does not exceed an acceptable level has been obtained.

4   Numerical Simulation and Results

The FNN architecture is predefined, depending on the input function complexity. The proposed network topologies approximate with sufficient accuracy the test functions. The FNN training algorithm applied to different architectures will have different performance. In this approach the choice of an optimal network design for a given problem is a guessing process. With a sufficient neuron number the network can be trained with a sufficient accuracy in sufficient training time. In particular, the application of LM and BMAM algorithms are proposed for training various FNNs with different structures in order to approximate a real-life application, two dimensional trigonometric functions and a benchmark problem which dates were selected from the input/output test points of a six dimensional non-polynomial function. The quasi optimization of FNN is demonstrated in the following case studies as in [9]. The test functions are arranged in the order of complexity.

A Simple Real - Life Application: Approximation of a Nickel-Metal Hydride Battery Cell Charging Characteristics

In this particular case, the FNN approximates a Nickel-Metal Hydride (NiMH) Battery Cell charging characteristics [2], a one-input real-life application.

The nickel-metal hydride batteries can be repeatedly charged and discharged for about 500 cycles. The charging process duration can be different, from 15 minutes to 20 hours. In our experiment was more than 1 hour. The charge characteristics are affected by current, time and temperature. The test function is a characteristic between the battery capacity input and the cell voltage. The battery type was GP 3.6 V, 300 mAH, 3x1.2V NiMH, charged for 1.5 hours with 300 mA and 25°C.

Two - Input Trigonometric Functions

We used the next two two dimensional polynomial input functions as test functions

$$y_3 = \left(\sin\left(c_1 \cdot x_1\right)^5 \cdot \cos\left(c_2 \cdot x_2\right)^3 \right) / 2 + 0.5 \qquad (4)$$

$$y_4 = e^{-\frac{r^2}{100}} \cdot \cos\left(\frac{r}{2}\right) \text{ where } r = \sqrt{x_1^2 + x_2^2}, ; \; x_1, \, x_2 \in [-20, 20] \qquad (5)$$

Six Dimensional Non-Polynomial Function

This widely used target function originates from paper [1]. Its expression is:

$$y_5 = x_1 + x_2^{0.5} + x_3 x_4 + 2e^{2(x_5 - x_6)} \qquad (6)$$

where $x_1, \, x_2 \in [1,5]$, $x_3 \in [0,4]$, $x_4 \in [0,0.6]$, $x_5 \in [0,1]$, $x_6 \in [0,1.2]$.

4.1   Case Study 1: Optimization of Q with LM Algorithm

Figure 1 shows the fluctuation of the Q values for fuzzy J-K and D flip-flops based on Yager norms. The real-life application (denoted with 1D) is approximated with a 1-2-2-1 FNN size, described by a set of 543 input/output data sets selected equidistantly from a set of 2715 test points. 1-20-20-1 feedforward neural network structures based on F³s was proposed to approximate the two two input trigonometric function, equations (4) and (5), (labeled as 2D trig and 2D), represented by 1600 input/output data sets. To approximate the six dimensional benchmark problem we studied a 1-10-5-1 FNN (6D-1) furthermore a 1-10-10-1 FNN (6D-2) sizes given by 200 dates.

The number of neurons was chosen after experimenting with different size hidden layers. Smaller neuron numbers in the hidden layer result in worse approximation properties, while increasing the neuron number results in better performance, but longer simulation time.

Evaluating the simulation results, low MSE values appears in the same domains which fact leads to assess optimal Q *intervals*, depending less from the input function type and FNN neuron number for a fix fuzzy operation parameter value. In case

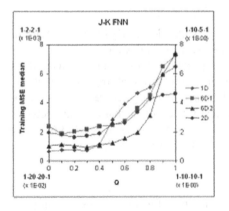
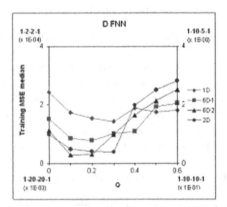

Fig. 5. J-K FNN and D FNN based on Yager norms

of J-K FNN based on Yager norms the Q optimum interval is $[0; 0.3]$, in case of D type FNN based on the same norms $Q \in [0.1; 0.3]$ (see Fig. 5). The LM training algorithm is very sensitive to the parameter's initial position in the search space. An inconvenient generated random parameter set leads in a hardly trainable neural network with bad performance, because the LM method is a local searcher.

4.2 Case Study 2: Optimization of Q with BMAM Algorithm

Using the BMAM algorithm, executing several LM cycles after each bacterial mutation step, there is no need to run 300 training cycles to find the lowest MSE value. After just one run the optimal parameter values could be identified. The quasi optimal Q values for various FNNs identified by the BMAM algorithm are listed in Table 1. The optimal Q intervals obtained with the LM algorithm have been reduced in this case to a single quasi optimal value identified at the end of the training process. In this approach, with only one training cycle an acceptable model whose error does not exceed an acceptable level has been obtained.

4.3 Comparison of Different FNNs with Respect to the Training Algorithm

The FNN architecture is predefined, depending on the input function complexity. The FNN training algorithm applied to different architectures will have different performance. In this approach the choice of an optimal network design for a given problem is a guessing process. In particular, the application of a recently improved BMAM algorithm is applied for training various FNNs with different structures. This new, complex software is able to train all the FNN parameters which have been encoded in a bacterium (chromosome) with the BMAM, eliminating completely the imprecision accused by training them with the LM algorithm. The simulation results published in our previous papers obtained under the same conditions could turn out to be different because the LM method is very sensitive to the initial values of the search space.

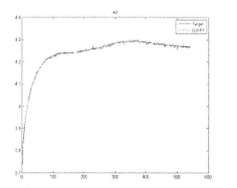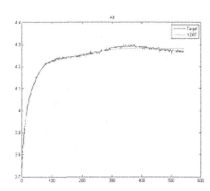

Fig. 6. Simulation results of J-K FNN based on Dombi norms and D FNN based on Yager norms

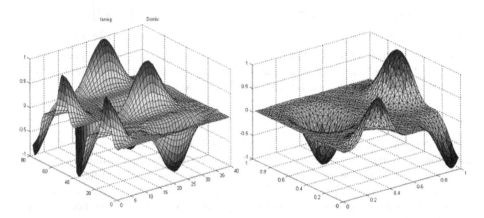

Fig. 7. Simulation results of traditional NN and J-K FNN based on Dombi norms (2D trig)

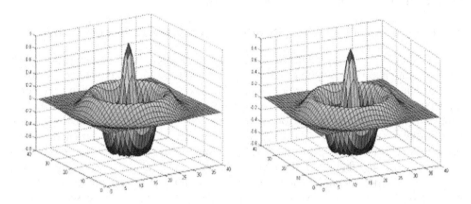

Fig. 8. Simulation results of traditional NN and J-K FNN based on Dombi norms (2D)

Fig. 6 (real-life application) presents the graphs of the simulations in case of fuzzy J-K and D flip-flop neurons trained with BMAM algorithms based on Yager and Dombi norms, using quasi optimized values.

Figs. 7-8 compare the approximation goodness of the traditional neural network and J-K FNN based on Dombi norms in order to approximate the two dimensional test functions.

Table 1 summarizes the near optimal Q values for fixed fuzzy operation parameter values furthermore the average approximation goodness made by BMAM algorithm, indicating the median MSEs.

The BMAM is efficient in searching the global minimum, but this method still search in a subset of all possible parameters. The number of generations and the population size are fixed, chosen corresponding to the difficulty of the problem to be solved. There is no guarantee that a FNN will converge to the local optimum after those generations. The mentioned training algorithms produce different training results.

Table 1. Training MSE median and quasi optimal Q values

Test Func.	*tansig* NN	J-K FNN				D FNN			
		Yager type		Dombi type		Yager type		Dombi type	
		Q	MSE	Q	MSE	Q	MSE	Q	MSE
1D 1-2-2-1	2.11 $x10^{-5}$	0.06	9.33 $x10^{-5}$	0.27	**4.83 $x10^{-5}$**	0.26	1.17 $x10^{-4}$	0.21	4.78 $x10^{-4}$
2D trig 1-20-20-1	9.13 $x10^{-7}$	0.10	1.53 $x10^{-2}$	0.24	**8.75 $x10^{-6}$**	0.24	8.21 $x10^{-3}$	0.22	2.93 $x10^{-2}$
2D 1-20-20-1	1.36 $x10^{-6}$	0.13	7.53 $x10^{-3}$	0.21	**8.12 $x10^{-5}$**	0.24	9.65 $x10^{-3}$	0.22	3.22 $x10^{-2}$
6D 1-10-10-1	1.12 $x10^{-4}$	0.25	5.92 $x10^{-1}$	0.25	**2.98 $x10^{-1}$**	0.28	1.45 $x10^{-1}$	0.19	1.58 $x10^{-0}$

Several test function have been used to test that the values of the parameter Q are less dependent (with values in a narrow interval) from the input functions and network size for a fixed fuzzy operation parameter value.

5 Conclusion

The benefits arising from the combination between artificial neural networks, fuzzy systems and bacterial memetic algorithms have been investigated in this paper. A hybrid algorithm is proposed combining the mentioned methods for FNN parameter optimization and training. The BMAM method achieves better function approximation results than the non-evolutionary LM algorithm. The best training algorithm is always problem dependent. Future scope for research lies in the investigation of a simultaneous optimization of the FNN learning parameters and network architecture.

Acknowledgments. This paper was supported by the Széchenyi István University Main Research Direction Grant, National Scientific Research Fund Grant OTKA K75711, further by Óbudai Egyetem Grants.

References

[1] Botzheim, J., Hámori, B., Kóczy, L.T., Ruano, A.E.: Bacterial Algorithm Applied for Fuzzy Rule Extraction. In: IPMU 2002, Annecy, France, pp. 1021–1026 (2002)

[2] Fan, R.: NiMH Battery Charger Reference Design, Designer Refernce Manual (2003)

[3] Gál, L., Botzheim, J., Kóczy, L.T.: Improvements to the Bacterial Memetic Algorithm used for Fuzzy Rule Base Extraction. In: Proc. of CIMSA 2008, Computational Intelligence for Measurement Systems and Applications, Istanbul, Turkey, pp. 38–43 (2008)

[4] Hirota, K., Ozawa, K.: Concept of Fuzzy Flip-flop, Preprints of 2nd IFSA Congress, Tokyo, pp. 556–559 (1987)

[5] Hirota, K., Ozawa, K.: Fuzzy Flip-Flop as a Basis of Fuzzy Memory Modules. In: Gupta, M.M., et al. (eds.) Fuzzy Computing. Theory, Hardware and Applications, pp. 173–183. North Holland, Amsterdam (1988)

[6] Hornik, K.M., Stinchcombe, M., White, H.: Multilayer Feedfordward Networks are Universal Approximators. Neural Networks 2(5), 359–366 (1989)

[7] Kecman, V.: Learning and Soft Computing: Support Vector Machines, Neural Networks, and Fuzzy Logic Models. The MIT Press, Cambridge (2001)

[8] Lovassy, R., Kóczy, L.T., Gál, L.: Applicability of Fuzzy Flip-Flops in the Implementation of Neural Networks. In: Proc. of CINTI 2008, 9th International Symposium of Hungarian Researchers on Computational Intelligence, Budapest, Hungary, pp. 333–344 (2008)

[9] Lovassy, R., Kóczy, L.T., Gál, L.: Quasi Optimization of Fuzzy Neural Networks. In: Proc. of CINTI 2009, 10th International Symposium of Hungarian Researchers on Computational Intelligence, Budapest, Hungary, pp. 303–314 (2009)

[10] Lovassy, R., Kóczy, L.T., Gál, L.: Function Approximation Capability of a Novel Fuzzy Flip-Flop Based Neural Network. In: Proc. of IJCNN 2009, International Joint Conference on Neural Networks, Atlanta, USA, pp. 1900–1907 (2009)

[11] Marquardt, D.: An Algorithm for Least-Squares Estimation of Nonlinear Parameters. SIAM J. Appl. Math. 11, 431–441 (1963)

[12] Nawa, N.E., Furuhashi, T.: Fuzzy Systems Parameters Discovery by Bacterial Evolutionary Algorithms. IEEE Tr. Fuzzy Systems 7, 608–616 (1999)

[13] Ozawa, K., Hirota, K., Kóczy, L.T., Omori, K.: Algebraic Fuzzy Flip-Flop Circuits, Fuzzy Sets and Systems, vol. 39(2), pp. 215–226. North Holland, Amsterdam (1991)

[14] Ozawa, K., Hirota, K., Kóczy, L.T.: Fuzzy Flip-Flop. In: Patyra, M.J., Mlynek, D.M. (eds.) Fuzzy Logic. Implementation and Applications, pp. 197–236. Wiley, Chichester (1996)

[15] Rudas, I.J., Batyrshin, I.Z., Zavala, A.H., Nieto, O.C., Horváth, L., Vargas, L.V.: Generators of Fuzzy Operations for Hardware Implementation of Fuzzy Systems. In: Gelbukh, A., Morales, E.F. (eds.) MICAI 2008. LNCS (LNAI), vol. 5317, pp. 710–719. Springer, Heidelberg (2008)

[16] Zavala, A.H., Nieto, O.C., Batyrshin, I., Vargas, L.V.: VLSI Implementation of a Module for Realization of Basic t-norms on Fuzzy Hardware. In: Proc. of FUZZ-IEEE 2009, IEEE Conference on Fuzzy Systems, Jeju Island, Korea, August 20-24, pp. 655–659 (2009)

Author Index